METHODS IN MOLECULAR BIOLOGY

For further volumes:
http://www.springer.com/series/7651

For over 35 years, biological scientists have come to rely on the research protocols and methodologies in the critically acclaimed *Methods in Molecular Biology* series. The series was the first to introduce the step-by-step protocols approach that has become the standard in all biomedical protocol publishing. Each protocol is provided in readily-reproducible step-by-step fashion, opening with an introductory overview, a list of the materials and reagents needed to complete the experiment, and followed by a detailed procedure that is supported with a helpful notes section offering tips and tricks of the trade as well as troubleshooting advice. These hallmark features were introduced by series editor Dr. John Walker and constitute the key ingredient in each and every volume of the Methods in Molecular Biology series. Tested and trusted, comprehensive and reliable, all protocols from the series are indexed in PubMed.

CRISPR Guide RNA Design

Methods and Protocols

Edited by

Tudor A. Fulga, David J. H. F. Knapp, and Quentin R. V. Ferry

Radcliffe Department of Medicine, Weatherall Institute of Molecular Medicine, University of Oxford, Oxford, UK

Humana Press

Editors
Tudor A. Fulga
Radcliffe Department of Medicine
Weatherall Institute of Molecular Medicine
University of Oxford
Oxford, UK

David J. H. F. Knapp
Radcliffe Department of Medicine
Weatherall Institute of Molecular Medicine
University of Oxford
Oxford, UK

Quentin R. V. Ferry
Radcliffe Department of Medicine
Weatherall Institute of Molecular Medicine
University of Oxford
Oxford, UK

ISSN 1064-3745 ISSN 1940-6029 (electronic)
Methods in Molecular Biology
ISBN 978-1-0716-0686-5 ISBN 978-1-0716-0687-2 (eBook)
https://doi.org/10.1007/978-1-0716-0687-2

This Humana imprint is published by the registered company Springer Science+Business Media, LLC part of Springer Nature.
The registered company address is: 1 New York Plaza, New York, NY 10004, U.S.A.

Preface

The ability to target and modify DNA sequences is a critical component of natural regulatory processes. Such systems can be co-opted to delete, insert, or modify DNA at a specific site in a user-controlled manner. Hitherto, DNA targeting has been the purview of proteins alone. Early attempts to develop such technologies relied on protein evolution for changing target specificity. Even second-generation tools such as zinc finger nucleases and transcription activator-like effector nucleases (TALENs), for which DNA targeting rules could be defined, required arduous molecular biology to alter target specificity. The repurposing of the clustered regularly interspaced short palindromic repeats (CRISPR) active immune system from bacteria has alleviated much of these issues. Such CRISPR-based systems have two components: a CRISPR-associated protein (Cas) and a short guide RNA that directs the protein to a genomic region following simple base-paring interactions. This system, exemplified by CRISPR/Cas9, allows targeting of almost any specific DNA sequence with near single-nucleotide resolution by simply reprogramming the short (~20 base pair) spacer region of the guide RNA. Consequently, DNA modifications can be produced and validated with unprecedented ease. This has resulted in CRISPR/Cas rapidly becoming a staple technique in most biological laboratories. The massive surge in the scope of this technology has also facilitated a rapid diversification of applications and CRISPR systems. Cas9 variants with altered sequence specificities, target fidelities, and enzymatic activities have been developed, which enabled not only gene editing but also controlling transcriptional activity, modifying epigenetic states, RNA localization, and visualization of genomic loci. The backbone of all of these techniques is one simple molecule: the guide RNA.

In this volume, we focus on the CRISPR-associated guide RNA and how it can be designed, modified, and validated for a broad repertoire of purposes. The chapters fall into several sections. We begin with methods for the computational design of target-specific guide RNAs. We next discuss chemical modifications which can be used to improve RNA stability, specificity, and efficiency. From there we cover additional modifications which can be used to create inducible guide RNAs, append additional functional domains, and express guide RNAs in a conditional manner. Finally, we cover methods for measuring off-target guide RNA activity. Overall, these chapters provide a comprehensive pipeline for guide RNA design, which we hope will become an invaluable resource in applying this powerful technology to basic research and therapeutic applications.

We would like to thank all contributors to this volume, as well as the series editor, Prof. John M. Walker, for their work on this valuable volume. Finally, we thank you the readers and wish you the best of luck with your CRISPR adventures.

Oxford, UK

Tudor A. Fulga
David J. H. F. Knapp
Quentin R. V. Ferry

Contents

Part IV Characterization of CRISPR Efficacy and Specificity

Contributors

DANIEL ALLEN • *The Mina and Everard Goodman Faculty of Life Sciences, Advanced Materials and Nanotechnology Institute, Bar-Ilan University, Ramat-Gan, Israel*

SANGSU BAE • *Department of Chemistry, Hanyang University, Seoul, South Korea; Research Institute for Convergence of Basic Sciences, Hanyang University, Seoul, South Korea*

ROBERTO BALLARINO • *Science for Life Laboratory (SciLifeLab), Research Division of Genome Biology, Department of Medical Biochemistry and Biophysics, Karolinska Institutet, Stockholm, Sweden*

MICHAEL BOUTROS • *Division Signaling and Functional Genomics, German Cancer Research Center (DKFZ), Heidelberg, Germany; Department of Cell and Molecular Biology, Medical Faculty Mannheim, Heidelberg University, Heidelberg, Germany*

BRITTA A. M. BOUWMAN • *Science for Life Laboratory (SciLifeLab), Research Division of Genome Biology, Department of Medical Biochemistry and Biophysics, Karolinska Institutet, Stockholm, Sweden*

TOM BROWN • *Department of Chemistry, Chemistry Research Laboratory, University of Oxford, Oxford, UK*

CHRISTOPHER R. CROMWELL • *Department of Pharmacology, University of Alberta, Edmonton, AB, Canada*

NICOLA CROSETTO • *Science for Life Laboratory (SciLifeLab), Research Division of Genome Biology, Department of Medical Biochemistry and Biophysics, Karolinska Institutet, Stockholm, Sweden*

AFAF H. EL-SAGHEER • *Department of Chemistry, Chemistry Research Laboratory, University of Oxford, Oxford, UK; Chemistry Branch, Department of Science and Mathematics, Faculty of Petroleum and Mining Engineering, Suez University, Suez, Egypt*

QUENTIN R. V. FERRY • *Radcliffe Department of Medicine, Weatherall Institute of Molecular Medicine, University of Oxford, Oxford, UK*

TUDOR A. FULGA • *Radcliffe Department of Medicine, Weatherall Institute of Molecular Medicine, University of Oxford, Oxford, UK*

FLORIAN HEIGWER • *Division Signaling and Functional Genomics, German Cancer Research Center (DKFZ), Heidelberg, Germany; Department of Cell and Molecular Biology, Medical Faculty Mannheim, Heidelberg University, Heidelberg, Germany*

AYAL HENDEL • *The Mina and Everard Goodman Faculty of Life Sciences, Advanced Materials and Nanotechnology Institute, Bar-Ilan University, Ramat-Gan, Israel*

BASIL P. HUBBARD • *Department of Pharmacology, University of Alberta, Edmonton, AB, Canada*

GUE-HO HWANG • *Department of Chemistry, Hanyang University, Seoul, South Korea*

ORTAL IANCU • *The Mina and Everard Goodman Faculty of Life Sciences, Advanced Materials and Nanotechnology Institute, Bar-Ilan University, Ramat-Gan, Israel*

JUAN JOVEL • *The Applied Genomics Core, Office of Research, University of Alberta, Edmonton, AB, Canada*

DAESIK KIM • *Center for Genome Engineering, Institute for Basic Science (IBS), Seoul, Republic of Korea; Genome Editing Research Center, KRIBB, Daejeon, Republic of Korea*

JIN-SOO KIM • *Center for Genome Engineering, Institute for Basic Science (IBS), Seoul, South Korea; Department of Chemistry, Seoul National University, Seoul, South Korea*

YONGSUB KIM • *Department of Biomedical Sciences, Asan Medical Institute of Convergence Science and Technology, Asan Medical Center, University of Ulsan College of Medicine, Seoul, Republic of Korea; Stem Cell Immunomodulation Research Center, Asan Medical Center, University of Ulsan College of Medicine, Seoul, Republic of Korea*

DAVID J. H. F. KNAPP • *Radcliffe Department of Medicine, Weatherall Institute of Molecular Medicine, University of Oxford, Oxford, UK*

JIYEON KWEON • *Department of Biomedical Sciences, Asan Medical Institute of Convergence Science and Technology, Asan Medical Center, University of Ulsan College of Medicine, Seoul, Republic of Korea; Stem Cell Immunomodulation Research Center, Asan Medical Center, University of Ulsan College of Medicine, Seoul, Republic of Korea*

SHUHAN LU • *Department of Pathology, Medical Research Institute, Zhongnan Hospital of Wuhan University, Wuhan, Hubei, China; Department of Urology, Medical Research Institute, Zhongnan Hospital of Wuhan University, Wuhan, Hubei, China*

MOIRA A. MCMAHON • *Ionis Pharmaceuticals Inc., Carlsbad, CA, USA*

MEGHDAD RAHDAR • *Ionis Pharmaceuticals Inc., Carlsbad, CA, USA*

JENNY SHAPIRO • *The Mina and Everard Goodman Faculty of Life Sciences, Advanced Materials and Nanotechnology Institute, Bar-Ilan University, Ramat-Gan, Israel*

DAVID M. SHECHNER • *Department of Pharmacology, University of Washington, Seattle, WA, USA*

HA RIM SHIN • *Department of Biomedical Sciences, Asan Medical Institute of Convergence Science and Technology, Asan Medical Center, University of Ulsan College of Medicine, Seoul, Republic of Korea; Stem Cell Immunomodulation Research Center, Asan Medical Center, University of Ulsan College of Medicine, Seoul, Republic of Korea*

ARUN SHIVALINGAM • *Department of Chemistry, Chemistry Research Laboratory, University of Oxford, Oxford, UK*

LAPATRADA TAEMAITREE • *Department of Chemistry, Chemistry Research Laboratory, University of Oxford, Oxford, UK*

ADI TOVIN • *The Mina and Everard Goodman Faculty of Life Sciences, Advanced Materials and Nanotechnology Institute, Bar-Ilan University, Ramat-Gan, Israel*

XIAOLING WANG • *Guangzhou Medical University, Guangzhou, Guangdong, China*

YOUJUN WU • *Department of Translational Research & Cellular Therapeutics, Beckman Research Institute, City of Hope National Medical Center, Duarte, CA, USA*

JIING-KUAN YEE • *Department of Translational Research & Cellular Therapeutics, Beckman Research Institute, City of Hope National Medical Center, Duarte, CA, USA*

HAO YIN • *Department of Pathology, Medical Research Institute, Zhongnan Hospital of Wuhan University, Wuhan, Hubei, China; Department of Urology, Medical Research Institute, Zhongnan Hospital of Wuhan University, Wuhan, Hubei, China*

YING ZHANG • *Medical Research Institute, Renmin Hospital of Wuhan University, Wuhan, China*

Part I

In Silico Design and Optimization of Guide RNA Sequences

Chapter 1

Cloud-Based Design of Short Guide RNA (sgRNA) Libraries for CRISPR Experiments

Florian Heigwer and Michael Boutros

Abstract

CRISPR/Cas-based genome editing in any biological application requires the evaluation of suitable genomic target sites to design efficient reagents. Considerations for the design of short guide (sg) RNAs include the assessment of possible off-target activities, the prediction of on-target efficacies and mutational outcome. Manual design of sgRNAs taking into account these parameters, however, remains a difficult task. Thus, computational tools to design sgRNA reagents from small scale to genome-wide libraries have been developed that assist during all steps of the design process. Here, we will describe practical guidance for the sgRNA design process using the web-based tool E-CRISP used in the design of individual sgRNAs. E-CRISP (www.e-crisp.org) has been the first web-based sgRNA design tool and uniquely features simple, yet efficient, scoring schemes in combination with fast evaluation and simple usage. We will also discuss the installation of a dockerized version of CRISPR Library Designer (CLD) that can be deployed locally or in the cloud to support the end-to-end design of sgRNA libraries for more than 50 different organisms. CLD was built upon E-CRISP to further increase the scope of sgRNA design to more experimental modalities (CRISPRa/i, Cas12a, all possible protospacer adjacency motifs) offering the same flexibility as E-CRISP, plus the scalability through local and cloud installation. Together, these tools facilities the design of small and large-scale CRISPR/Cas experiments.

Key words CRISPR/Cas, sgRNA design, Computational methods, Web services, Command-line tools, Multispecies

1 Introduction

CRISPR/Cas has been developed into one of the most versatile toolboxes for editing large genomes with broad applications in the life sciences, biotechnology, and medicine. Based on an adaptive immune response system of bacteria, CRISPR/Cas applications comprise two essential parts: the chimeric single guide (sg)RNA, a fusion of the guide- and tracrRNA, and the CRISPR associated endonuclease that is guided to its sequence homologous target by the sgRNA. Most commonly used in research applications, a codon-optimized Cas9 endonuclease variant derived from the bacterium *S. pyogenes* is used in conjunction with an sgRNA construct

Tudor A. Fulga et al. (eds.), *CRISPR Guide RNA Design: Methods and Protocols*, Methods in Molecular Biology, vol. 2162, https://doi.org/10.1007/978-1-0716-0687-2_1,

[1, 2]. The first 20 nucleotides of each sgRNA confer sequence homology–dependent target specificity and can thus be used to "program" target binding of the Cas9–sgRNA complex and subsequent introduction of double-strand breaks [3].

Applications of CRISPR/Cas9 range from functional annotation of genes and disease variants in a variety of organisms to gene therapy and the development of new pharmaceuticals [1, 4–6]. Given the broad applicability of the CRISPR/Cas system, rational design of targeting agents is important to avoid unspecific side effects (off-targets) and maximize targeting efficacy (on-target). While this might seem a trivial task for single genes, applications in high-throughput experimentation or for in vivo studies pose significant challenges on the rational search for best ranking targeting agents [7, 8]. The challenge design algorithms face, is to balance specificity and efficacy of gene perturbation reagents by identifying optimal target sites through assessment of their sequence and location. Design software can be employed to aid researchers in these tasks. Today, many different design tools have been developed that cover various design and analysis aspects, organisms, and experimental approaches, such as E-CRISP, CLD, CHOPCHOP, CCTop, CRISPOR, or GUIDEscan as also reviewed in Graham et al. or Cui et al. [9–16]. And, future developments also explore unsupervised deep learning of sequence features for guide RNA design optimization, though still with limited accessibility in terms of different applications and organisms [17, 18]. Most computational tools perform the following steps: (1) target site scanning for protospacer adjacency motifs (PAM, needed for enzymatic double strand cutting) within the input sequences, (2) annotation of putative target sites with efficiency scores based on target sequence and location features, and (3) specificity assessment using realignment of the target sites to the respective target genome.

In this chapter we describe how to use the CRISPR Library designer (CLD) and E-CRISP, which guide their users through these steps with a special emphasis on multispecies (also nonmodel organisms) applicability, scalability, reproducibility, and usability throughout the design process [15, 16]. E-CRISP was developed as a web-based tool to assist in identifying sgRNAs and presents the user with many options to flexibly adjust design criteria. CLD, in contrast, was originally developed as a command-line tool to enable, based on a similar workflow, the design of small to genome-scale sgRNA libraries. Both tools assess the following key considerations for each target site: binding efficacy as determined by nucleic acid content and sequence features [19–21], genomic features that determine functional penetrance of the genetic perturbation [22–25] and finally targeting specificity as determined by sequence homology and the fault tolerance of sgRNA/Cas9-DNA recognition [13, 17, 26–32]. We will also give practical advice on how to solve relevant practical tasks such as large scale sgRNA design using the new docker-based version of CLD, single gene-based design using E-CRISP or evaluating efficacy and specificity of given sgRNAs using E-CRISP reevaluation. We also discuss comparability with other tools and give practical advice on the fine-tuning of parameters when using our design software.

2 Materials

2.1 Materials to Use CLD

- Multiprocessor computer with ≥16 GB RAM and up to date operating system.
- Docker v ≥ 2.0: follow providers instructions on how to install docker.
  - On Windows or Mac: https://www.docker.com/products/docker-desktop
  - On Linux: https://docs.docker.com/install/linux/docker-ce/centos/
- Tested on Windows 10, MacOS-Mojave, CentOS v7.5, Ubuntu 16.04,
- Enabled internet connection such that "docker run hello-world" can run.
  - If your computer operates behind a proxy server follow the instructions given at https://docs.docker.com/config/daemon/systemd/
- To be able to use the graphical user interface (GUI).
  - MacOS:

    Install xquartz: https://www.xquartz.org/

    In the terminal type: *open -a XQuartz.*

    go to XQuartz → Preferences → Security → tick "*Allow connections from network clients*",

    In the terminal type: *IP = $(ifconfig en0 | grep inet | awk '$1=="inet" {print $2}').*

    In the terminal type: *xhost + $IP.*

    adapt docker-compose.yaml change to folder and In the terminal type: *docker-compose up* or *docker run -e DISPLAY = $IP:0 -v ~/Desktop:/data boutroslab/cld_docker cld_gui*,
  - Windows:

    Install a command line package manager for windows https://chocolatey.org/

Follow this guide to install the graphical interface manager for windows: https://dev.to/darksmile92/run-gui-app-in-linux-docker-container-on-windows-host-4kde

In the terminal type: *docker run -e DISLPAY= $IP:0.0 -v ~/Desktop:/data boutroslab/cld_docker cld_gui.*

– GUIed linux:

In the terminal type: *docker run -e DISLPAY= $IP:0 -v ~/Desktop:/data boutroslab/cld_docker cld_gui.*

- on any operating system for using the simple command-line interface:
  - open a terminal, bash-shell, CMD, Cygwin or similar,
  - In the terminal type: *docker run -v ~/Desktop/cld_demo:/data boutroslab/cld_docker cld.*
  - needed for library design are,

    preformatted libraries prepared using either –*task=make_database* or downloaded from http://www.dkfz.de/signaling/crispr-downloads/DATABASES

    a parameter file that comes with the cld_docker from git hub (https://github.com/boutroslab/cld_docker/blob/master/params.txt),

    a gene list or coordinate file (e.g., https://github.com/boutroslab/cld_docker/blob/master/gene_list_examples.txt),

    an in-/output folder mountable to docker (e.g., a folder on your Desktop).

2.2 Materials to Use for E-CRISP

- Any computer or similar device with a standard internet connection.
- Tested on common web browsers Firefox (v 64), Google-chrome (v 71), Safari (v 12).
- Text-editing tools to open tabular output files.

3 Methods

In this section we will provide a detailed stepwise explanation of the commands and steps that need to be performed when (1) designing an sgRNA library against a targeted set of genes for use in a CRISPR screen, (2) choosing sgRNA targets in the human gene BAX for creating a null mutation, and (3) evaluating a set of sgRNA for their potential targets and off-targets. These are simple examples that should represent most of the day-to-day tasks that can be solved with CLD and E-CRISP (*see* **Notes 1–4**). All file or directory names, code to execute in a terminal, or interface button names are in *italic*.

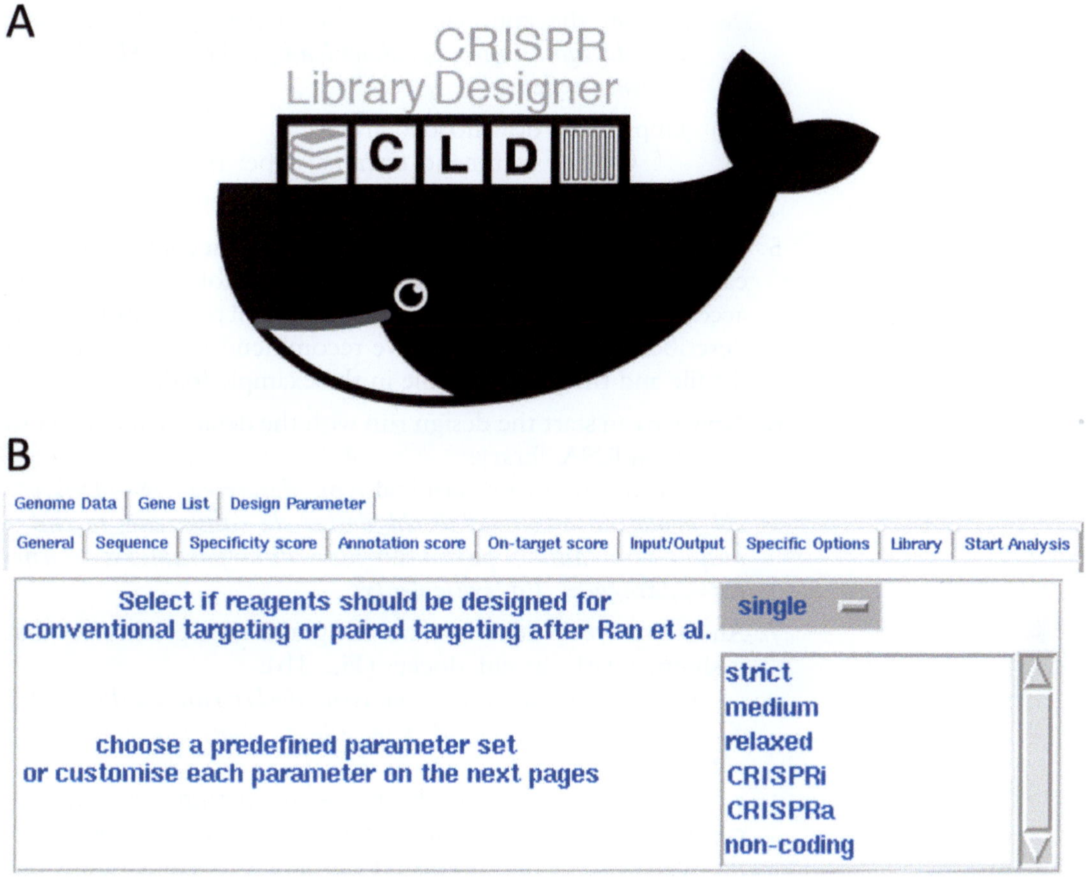

Fig. 1 Overview about CLD. (**a**) CRISPR Library Designer (CLD) is now shipped with all its dependencies and capacities using docker container technology. The downloadable version (https:/hub.docker.com/r/boutroslab/cld_docker) contains all necessities to run and use CLD in the command-line or using its X11 based graphical user interface. (**b**) CLD's graphical user interface guides users through all sgRNA design parameters and allows to directly reiterate over the design process

3.1 Design of a Pooled sgRNA Library Covering a Focused Set of Genes

1. Download the suggested file templates from GitHub by checking out https://github.com/boutroslab/cld_docker (Fig. 1a).
2. Create an Input/Output directory in the path of choice. In this example we chose *~/Desktop/cld_demo (~ = your home directory, usually /Users/myname/...)*.
3. Copy *params.txt* and *sgRNA_target_examples.txt* over to *~/Desktop/cld_demo*.
4. Download an unzip the *Homo sapiens* genome from https://www.dkfz.de/signaling/crispr-downloads/DATABASES/.
   (a) Download the human precompiled database.

On the command terminal type: *wget https://www.dkfz.de/signaling/crispr-downloads/DATABASES/homo_sapiens.tar.gz*

(b) Unpack the downloaded archive

On the command terminal type: *tar -xzvf homo_sapiens.tar.gz.*

5. Copy the resulting *crispr_databases* folder that contains a folder called *homo_sapiens* over to the *cld_demo* folder so it can be accessed from within Docker and prepare all prerequisite files as described in Subheading 3. We recommend to place the gene list file and the parameter file in the example folder.
6. Now we can start the design run with the default parameters to create a sgRNA library:

   On the command terminal type: *docker run -v ~/Desktop/cld_demo:/data boutroslab/cld_docker cld --task=end_to_end --output-dir=/data --parameter-file=/data/params.txt --gene-list=/data/gene_list_examples.txt*
7. An alternative solution is to use the graphical user interface that is shipped with the cld_docker (Fig. 1b):

   On the command terminal type: *docker run -e DISPLAY=$IP:0 -v ~/Desktop/cld_demo:/data boutroslab/cld_docker cld_gui*

   There the user can do the same designs as with the command-line version but all parameters that have been formerly defined in the params.txt can now be changed individually guided by the graphical user interface.
8. As a result one will receive summary statistics depending of the filter parameters that were defined in the parameter file such as the total number of possible sgRNA targets, the targets excluded by exon filtering or the ones filtered out because of low specificity (*see* **Notes 6–11**).
9. In addition one does receive a folder named after the time the analysis was triggered. This way individual design runs can be simultaneous but still remain uniquely traceable. This time-stamped folder eventually contains all results of the design process in unfiltered data called "*all_results_together.tab*" and "*all_results_together.gff*". The tab file contains all information needed for filtering and formatting a sgRNA library. The GFF format file contains the same target sites but can be visualized in a genome browser such as ENSEMBL (*see* **Note 8**).
10. When cld is run in *end_to_end* task mode it will continue to filter and assemble an sgRNA library from the *all_results_together.tab* file. The resulting text files will be named prepending the prefix that can be defined within the command to run cld (*--library_name=<string>*). Using those parameters one can

also define the sgRNA coverage per gene and the total number of sgRNAs in the library. CLD will then provide a formatted sgRNA library in ready-to-order FASTA format that depends on the adapters specified in the design command (*--5-prime=<string>, --3-prime=<string>*). In addition, CLD provides a variety of information on each sgRNA target site in the *.tab* and *.gff* files (*see* **Note 10** for detailed explanations of the different parameters and output information).

11. One can now alter parameters iteratively until a library is found that suits the needs of your experiments (*see* **Note 5**).

3.2 Design sgRNAs to Knockout BAX

1. Go to http://e-crisp.org and, if not selected automatically, click on *Design* in the blue main navigation bar (Fig. 2a).
2. Go to *1. Select Organism* and select the Homo sapiens GRCH38 genome as design target. It will determine where the gene sequence is searched and what the genetic background for off-target search is. (*See* **Note 5** the Drosophila genome is selected by default.)
3. Got to step *2. Select Target region* and enter the example gene symbol *BAX* (*see* **Note 10**).
4. You can now click *Start sgRNA search.*
   (a) Alternatively you can change the preselected default options for restricting or relaxing the target filter criteria. Medium criteria (default) should be sufficient to deliver a set of specific and active sgRNAs for perturbing protein coding genes.
   (b) You might also want to click on *show advanced options* and change many different parameters that restrict or relax filtering criteria for selecting optimal target ranges (*see* **Note 2**).
5. You will automatically be taken to a new page that is structured into four basic compartments (Fig. 2b):
   (a) The header where you can download the results files (*see* **Note 10**).
   (b) A detailed explanation of reasons why the target sites were filtered (this gives hints as to why there might only be a few or none sgRNAs, *see* **Note 2**).
   (c) A table presenting sgRNA targets in a ranked list which includes the efficacy, specificity and annotation scoring as well as the alignment to the genome in a pop-up window (Matchstring).
   (d) A genome browser-like image showing the sgRNA targets in their genomic annotation context.
6. sgRNA target site sequences are now evaluated given the information presented in either the html table or the downloadable

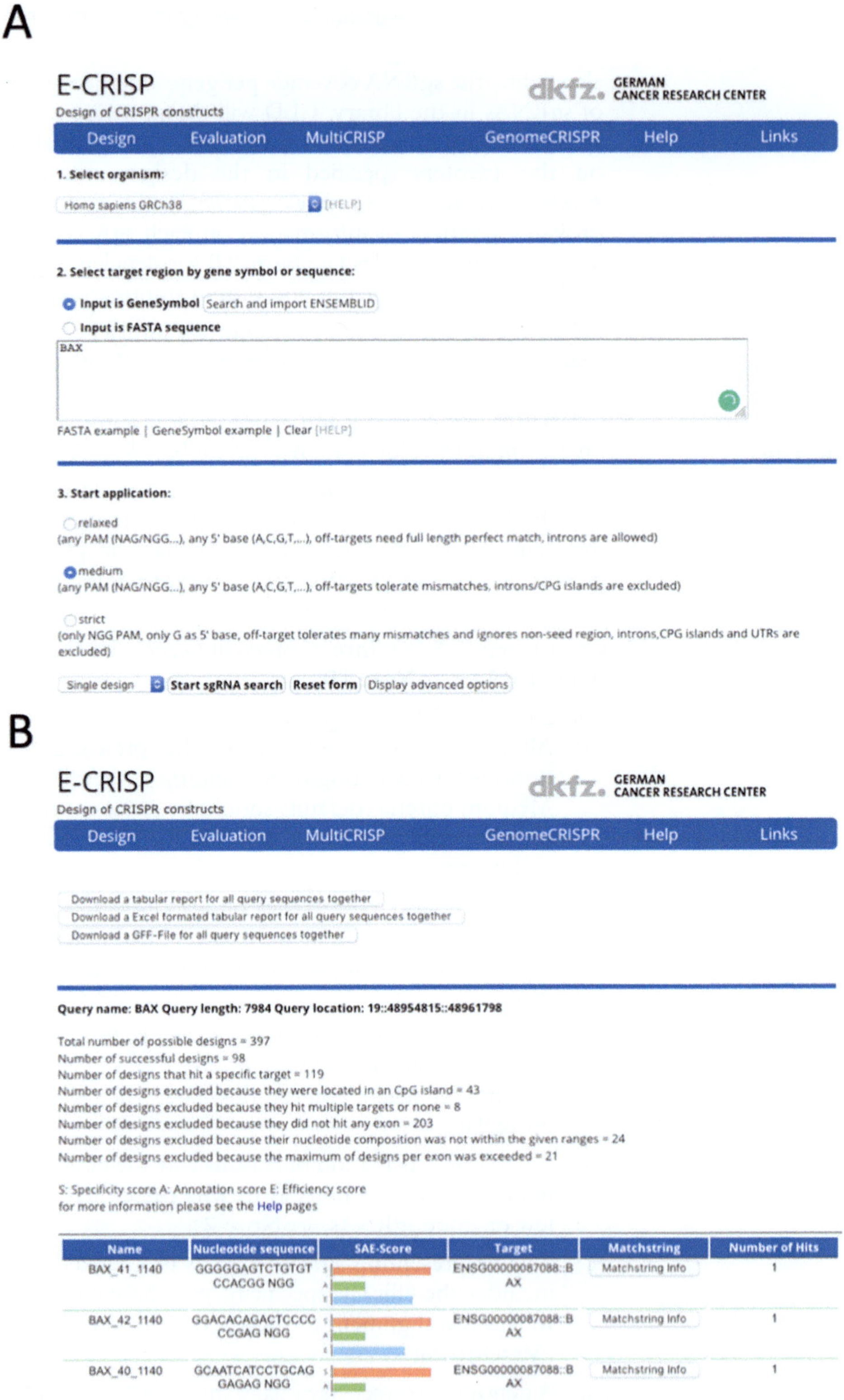

Fig. 2 Overview on E-CRISP. (**a**) E-CRISP.org allows for sgRNA design on a gene-by-gene basis. It requires no programming or computing infrastructure knowledge and allows for fast and uncomplicated iteration through sgRNA design and parameter optimizations. (**b**) Results of sgRNA design using E-CRISP are available for download in flat file or spreadsheet format. There is also detailed information, why sgRNA have failed the filter criteria, plus a rank ordered list of sgRNAs below

results table (XLSX-Spread sheet). All lists are sorted hierarchically by specificity (highest to lowest), annotation (e.g., coding exons/transcripts targeted, best to worst for functional perturbation) and last by efficacy (based on sequence features). Thus, the best target sites can be chosen from the top of the lists. Experience and user feedback recommend testing up to three sgRNAs to identify at least one functional sgRNA.

7. Assumed you chose the first listed sgRNA target, it has the 23 nt sequence GGGGGAGTCTGTGTCCACGG NGG. This includes the protospacer adjacency motif (PAM, here NGG) at its 3′ end. This motif shall only be present in the target but not the sgRNA. Thus, only 20 nt GGGGGAGTCTGTGTCCACGG are constituting the 5′ end of the final sgRNA when adding the improved sgRNA scaffold introduced by Dang et al. (GTTTAAGAGCTATGCTGGAAACAGCATAGCAAGTTT AAATAAGGCTAGTCCGTTATCAACTTGAAAAAGTGGC ACCGAGTCGGTGCTTTTTT, 3). The resulting sgRNA would then have the sequence GGGGGAGTCTGTGTCCAC GG GTTTAAGAGCTATGCTGGAAACAGCATAGCAAG TTTAAATAAGGCTAGTCCGTTATCAACTTGAAAAAGT GGCACCGAGTCGGTGCTTTTTT.
8. In order to clone this sequence into a functional DNA delivery and RNA expression construct one can, for example, follow the lentiCRISPR.v2 protocol [33].
9. Repeat these steps for more sgRNAs targeting BAX or start over design against another gene.

3.3 Evaluate 5 sgRNAs for Their Potential Targets

This evaluation protocol assumes that target recognition is evaluated for Cas9 from *Streptococcus pyogenes* (spCas9, PAM: NGG).

1. Go to http://e-crisp.org/E-CRISP/reannotate_crispr.html and find sgRNA targets to be evaluated. Example sgRNA targets can be found at https://github.com/boutroslab/cld/blob/master/sgRNA_target_examples.txt
2. Every sgRNA target must be entered in FASTA format of the following format:
   >sgRNA_target_1
   GCTTGTAGATGGCCATGGCG
3. Under *1. Select organism* the organism to find sgRNAs in and evaluate off-target matches needs to be selected. For this example, we select: *Homo sapiens GRCh38.*
4. Further, one may define the *number of 5′ mismatches tolerated by the program.* Reports have shown that mismatches in the PAM-distal region of the sgRNA target site are often well tolerated [26]. Thus, this can be considered when evaluating off-targets by tolerating mismatches within this region. By

default, this behavior is, however, switched off, and mismatches are equally considered at the entire target site.

5. One may also alter the tolerated edit distance that determines when a target match is considered valid. The edit distance includes mismatches, insertions and deletions. A tolerated edit distance of 2 and a number of ignored 5′ base pairs of 3 can be considered very relaxed criteria. This way, also target sites with rather unlikely cutting efficacy will be matched by the algorithm.
6. Press *Start* and your input target sites will be evaluated. You can also upload files listing sgRNA targets in FASTA format and therewith reannotated entire libraries of sgRNA designs.
7. The results are presented similarly as in the sgRNA design step. You will be given a list ordered by sgRNA target and target/off-target site similarity. This way you are also enabled to cross-compare sgRNA targets that were designed by different means. At the top of the page you will also find a button which lets you download a tab-delimited text file that contains more in-depth information on each sgRNA target and off-target. These include Name, Target-chrome(s), Start, End, Gene targets, Spec-Score, Anno-Score, Eff-Score, Matchstring, Sequence, Direction, CDS_score, exon_score, seed_GC, doench_score, xu_score, doench_30_mer. Detailed explanations of those scores may be obtained under http://www.e-crisp.org/E-CRISP/aboutpage.html.

4 Notes

1. CLD and E-CRISP offer extensive help pages.

   In cases of questions that this protocol did not fully or just briefly addressed see the extensive help material, glossary and manual pages provided with E-CRISP (http://www.e-crisp.org/E-CRISP/aboutpage.html) or CLD (https://github.com/boutroslab/cld/blob/master/MANUAL.md).
2. CLD and E-CRISP often provide fewer designs than other tools.

   It has been reported that E-CRISP and CLD, in many cases, offer fewer designs than other comparable tools. This is caused by the rather restrictive filtering that is used by default when using both software and is thought to provide only high-quality sgRNA designs to their users. To not show or filter out designs of inferior quality or excess designs against specific genes or exons was a purposeful design decision. This way we can provide users always with top ranked sgRNA and do not distract by providing possible target sites that we discourage researchers from trying to target. However, in some situations it might be useful to be informed about these target sites and be less sensitive while filtering. To this end we recommend that users gradually and carefully lower the sgRNA filter restrictions and inspect unfiltered sgRNA lists for their properties. Users may then also use their own filter criteria based on the information E-CRISP and CLD provide.

3. For designing sgRNA for more than five genes, CLD should be used.

   E-CRISP is provided using a web server hosted within the German Cancer Research Center. Due to limited resources we cannot back this web service with unlimited capacities. Thus, we recommend that all users interested in designing sgRNA libraries exceeding five genes use our dockized version of CLD locally on their own hardware that now also works an all available modern operating systems (https://github.com/boutroslab/cld_docker).

4. Off-target prediction is based on sequence homology-based heuristics.

   E-CRISP and CLD assess off-target site binding by the Cas–sgRNA complex solely based on sequence homology with the targeting part of the sgRNA. As has been shown, some parts of the sgRNA are more tolerant to mismatches then others [26]. This has been respected in the software and can be tuned using available parameters (*see* **Note 5**). However, all PAM or sgRNA homology-independent binding of the Cas–sgRNA complex cannot be excluded this way [27, 28]. In addition, when comparing different sgRNA designers one needs to consider that E-CRISP and CLD rely on bowtie and bowtie2 for off-target prediction. While these alignment software can be tuned to be as sensitive as possible, their heuristic nature of seed alignment will never make them fully comprehensive tools to find every (distantly homologous) possible target site in any genome [34, 35]. Still, this is sufficient to rank-order sgRNAs by their specificity [14].

5. Default presets should be questioned.

   E-CRISP and CLD offer a number of diverse input parameters that can be optimized during the design process. Table 1 summarizes them (reproduced in brief form from Heigwer et al. [16]). Both sgRNA design software have a number of preset parameter configurations, for example purposeful design. This considers specific requirements that depend on the experimental procedure that the sgRNA are being used for. This purpose exclusivity is a parameter set that overwrites all other filter criteria and is rather restrictive and should be considered as a restrictive starting point from which to relax single parameters. In addition, E-CRISP allows the

Table 1
CLD/E-CRISP parameters explained

Parameter	Definition/effect	Input-type
purpose_exclusive	Defines if the purpose exclusive choice should affect and overwrite the design filtering criteria	Boolean (1 or 0)
purpose	Defines following purposes: knockdown/out which requires designs to hit in coding sequences near the start codon of a gene, N-terminal tagging requires the start codon to be targeted and C-terminal tagging requires the stop codon to be targeted by sgRNAs	Knockout, n-tagging, c-tagging, noncoding, CRISPRa or CRISPRi
PAM_location	Defines if the PAM motif is 3′ or 5′ located with respect to the protospacer	3_prime or 5_prime
PAM	Defines the PAM sequence, which can be NAG, NGG, or any (allowing both)	IUPAC coded nucleotides
min_length	Defines the minimum length of the protospacer (5′ sequence before the PAM)	Numeric
max_length	Defines the maximum length of the protospacer (5′ sequence before the PAM)	Numeric
min_G	Defines minimum total G content	Numeric
max_G	Defines maximum total G content	Numeric
min_A	Defines minimum total A content	Numeric
max_A	Defines maximum total A content	Numeric
min_C	Defines minimum total C content	Numeric
max_C	Defines maximum total C content	Numeric
min_T	Defines minimum total T content	Numeric
max_T	Defines maximum total T content	Numeric
right_homology	If homology arms are chosen, this number defines the length of the arm 5′ of the target site	Numeric
left_homology	If homology arms are chosen, this number defines the length of the arm 3′ of the target site	Numeric
downstream_window	Defines the 5′ nucleotide distance window of a design to the start or stop codon for tagging analysis	Numeric
upstream_window	Defines the 3′ nucleotide distance window of a design to the start or stop codon for tagging analysis	Numeric

(continued)

Table 1 (continued)

Parameter	Definition/effect	Input-type
number_of_CDS	Defines the allowed nucleotide number within CDS downstream of the start codon, if knockout is chosen as the purpose criterion	Numeric
minspacerlength	Defines the minimum spacer length, if paired design is chosen as the purpose criterion (as for the double nickase or FokI-Cas9 approach)	Numeric
maxspacerlength	Defines the maximum spacer length, if paired design is chosen as the purpose criterion (as for the double nickase or FokI-Cas9 approach)	Numeric
preceding	defines if the protospacer should begin with a specific base (e.g., U6 promoter would favor G at this position)	IUPAC coded Nucleotide
gene_exclusive	Defines if the sgRNA needs to target a region within the targeted gene (for CRISPRa/i 500 before and after the gene are parsed too)	Boolean (1 or 0)
exon_exclusive	Defines if the sgRNA needs to target a region within an exon	Boolean (1 or 0)
CDS_only	Defines if the sgRNA needs to target a region within a coding region	Boolean (1 or 0)
CpG_exclusive	Defines if the sgRNA is allowed to target a region within a CpG island	Boolean (1 or 0)
specific_exon	Defines if a specific exon number is to be targeted	Numeric
retrieve_recomb_matrix	Defines if the sequences for homology arms should be computed and reported	Boolean (1 or 0)
bowtie_version	Defines which version of bowtie or blast should be used for off-target analysis. Bowtie is more sensitive to mismatches of single designs, and bowtie2 is optimized for paired alignments of sequences. Here Blast tends to be the most sensitive toward less homologous sequences. For all mapping algorithms, only full-length alignments are counted.	bowtie, bowtie2 or blast
offtargetdb	Defines if off-targets should be searched in genomic sequences, sequences of annotated genes or exons of protein coding sequences	genomeDNA, gDNA or cDNA
off-targets-allowed	Defines how many off-targets per design are tolerated before it is excluded from the report	Numeric

(continued)

Table 1
(continued)

Parameter	Definition/effect	Input-type
unspecific_leading_bases	Defines the number of 5′ base pairs of the target site to be ignored for the off-target mapping	Numeric
edit_distance_allowed	Defines the edit distance (sum of all mismatch or INDEL positions) allowed during alignment to be still counted as off-target	Numeric
bowtie_mode	Defines the bowtie mode as referenced in the bowtie2 manual	Sensitive, very sensitive, fast, very-fast
ignore_intergenic	Defines if off-targets which are not in any gene should be ignored	True or false
sec_off_target	Defines if sgRNA targets sites should be checked for binding sites in sequences that are not in the genome of interest (for example: GFP etc.). Those sequences need to be provided in an extra fasta formatted file in the database path and named **'secondary_off_targets.fasta'**	Boolean (1 or 0)
max_per_exon	Defines the maximum number of sgRNA allowed to be reported per exon	numeric
out_gff	Defines if a gff should be generated	Boolean (1 or 0)
specific_transcript	Defines if only a specific transcript (provided as an ENSEMBL TR ID) should be targeted. This is not applicable if more than one gene is searched.	ENSEMBL transcript ID or any
match_info	Defines if a detailed alignment information should be printed on any sgRNA mapping	Boolean (1 or 0)
draw_html_report	Defines if an html report should be created	Boolean (1 or 0)
working_path	Defines which unix path to the results should be used. Results are created in the current working directory (.) by default.	e.g., /data/workdir/
databasepath	Defines if a unix path to the folder containing CLD formatted databases should be generated	e.g., /data/databases/
ref_organism	Defines the reference organism as given in the name of the database and the subdirectories, e.g., if the organism is homo_sapiens, the database needs to have the prefix homo_sapiens	e.g., homo_sapiens, dmel

(continued)

Table 1
(continued)

Parameter	Definition/effect	Input-type
data_type	Defines if the input file contains official gene symbols, ENSEMBL IDs, or genomic coordinates. Coordinates need to be given as ID, chromosome (Ensembl_type), start, end. Coordinate data need to be tab separated and different entries need to be newline separated	Should be either ensemble_acc, gene_symbol or coordinates
ignore_missing_id	Defines if the program should die if IDs are faced, that cannot be found in the currently used database	Boolean (1 or 0)
kind	Defines if sgRNA target sites should be found in a single or paired mode (suitable for the paired nickase or FokI paired nuclease approach)	Single or double
exclude_overlapping_genes	Defines if sgRNA designs targeting multiple overlapping genes/antisense transcripts should be excluded	Boolean (1 or 0)
sort_by_rank	Defines if sgRNAs should be ranked additionally by an on-target score	Boolean (1 or 0)
scores	Defines the on-target score to be used. The preset scores are derived from the algorithms proposed by Doench et al. [8]. However, they are only defined for a 20 nt protospacer adjacent to a NGG PAM.	xu_score, doench_old or custom
custom_score	Defines a custom scoring function in perl code. The function needs to be unnamed and dependent on sequence information of the 30mer described in Doench et al. [8]. Results of the function need to be numeric.	String in perl language defining an anonymous function, which acts on the Doench 30mer
cover_many_transcripts	Defines if priority in sgRNA choice for the final library should be given on maximum coverage of all transcripts of a gene. All other scores will be ignored but shown in the resulting tables.	Boolean (1 or 0)

user to apply parameter presets that set parameters such that resulting sgRNAs are more or less specific and restricted to certain target sites. These are separated in default, medium and relaxed and should be considered as rough standard guidelines from which to start to explore the parameter space.

6. Target site identification using E-CRISP takes 30 s on average. A genome wide library design using CLD can take up to 12 h using CLD.

   Users need to plan the time they'll spend on designing sgRNAs depending on the size of the library and the complexity of the task. While single gene design in E-CRISP is relatively fast and iterative (30–60 s per run), a whole genome library design run using CLD resulting in about 180,000 sgRNAs will take about 12–16 h depending on the computing infrastructure. For larger Jobs or analyzing many different hyperparameters in parallel we recommend using CLD with its command-line interface on a high-performance cluster.

7. Using alternative design tools as orthogonal measure.
   Sensitivity and Precision of sgRNA design tools can sometimes be hard to judge upon. Thus, many researchers employ a number of different sgRNA designers in order to gain an overview about the average genomic areas, that the designer find suitable for targeting. We recommend GuideScan (large libraries for application in human and mouse), CRISPOR (web-based design against single genes for applications in a vast number of organisms), Protospacer-Workbench (local app for CRISPR target site analysis in also less annotated genomes), and CHOPCHOP (web-based service for single gene design of sgRNA and TALEN constructs) for these purposes [11, 13, 14, 36].

8. Visualizing sgRNA in genome browser context using GFF file outputs.

   E-CRISP and CLD both provide their output in a GFF file format. This enables it to be compatible with most modern online genome-browsers such that sgRNA designs can be visualized next to various genome annotation tracks offered by the genome data bases. The following steps allow visualization of sgRNA targets as ENSEMBL genome track:

   (a) Locate the *.gff file that you downloaded from E-CRISP or that has been created as a result of CLD.

       - In E-CRISP the GFF file output needs to be activated in the advanced options menu section 8 at the bottom of the page.

   (b) Make sure your computer is properly connected to the Internet.

   (c) Go to http://www.ensembl.org/index.html in your web browser.

   (d) Select Human as the organism off choice in this example.

   (e) Go to the upper-left panel and select "Display your own data with ENSEMBL."

(f) Give your data track a name (e.g., "mysgRNAtrack").

(g) Click upload file and select your GFF file.

(h) Next click upload and your track will be added to the tracks displayed in the genome browser. When working fine there will appear a message where you can click "*go to closest region with data.*"

(i) You are now able to inspect all sgRNAs in comparison to all genomic annotation available in ENSEMBL (e.g., coding sequences, promoters, regulatory element, transcription start sites).

9. E-CRISP allows input as FASTA sequence, gene symbol and allows fuzzy search for the correct (ENSEMBL) gene symbol.

   For all design tasks a target must be given. For sgRNA library or single gene design this has been implemented differently. For library design, official gene symbols as used in ENSEMBL, ENSEMBL gene IDs or genomic coordinates in the form of (1 100 3000; chromosome start end) can be the input. Therefore, ENSEMBL IDs or gene symbols must be exactly matched to the names within the database. By default, CLD will result in an error message if an ID could not be matched. This behavior can be set to "*ignore_missing_ids=0*" such that unmatched IDs will be collected in the missing genes file. E-CRISP, on the other hand, cannot ignore missing IDs. Instead users are given a tool with which they might search the exact ID or symbol of their gene using a fuzzy search (similar to Google autocomplete).

   (a) Go to http://www.e-crisp.org/E-CRISP/.

   (b) In section 2, click "*Search and import ENSEMBL IDs.*"

   (c) In the section that appeared below, you can now search for suitable IDs that fit your gene of interest (e.g., p53).

   (d) All similar ids are now shown in a scrollable list to the right.

   (e) Click on genes in that list to select them for design. They will appear in the lower list and in the text input window.

   (f) They can be removed from the selection by clicking them in the lower right scrollable list.

10. Interpreting E-CRISP/CLD result files

    CLD will produce a number of output files that provide the user with useful information about the final designed library and the design process as such. Consider Table 2 for details (reproduced from Heigwer et al. [16]). All outputs can be handled using text editors. Use the "open with" method of your computer and select some favorite text editor to open those files. We recommend *TextWrangler* when working with

Table 2
CLD output file formats

File ending	Content
∗.coverage.tab	A tab-delimited table with two columns: first ENSEMBL IDs in the library, second how many designs are in the library targeting this gene.
∗.fasta	A fasta file containing the ready to order oligonucleotides with the respective target site and the vector parts.
∗.gff	"General Feature Format" file containing all designs that made it into the library for more information see. http://www.sanger.ac.uk/resources/software/gff/spec.html. For view in the ENSEMBL genome browser see http://www.ensembl.org/info/website/upload/index.html.
∗.tab	Tab-delimited CRISPR result table with the following columns: Name, Length, Start, End, Strand, Nucleotide sequence, Gene Name, Transcripts, Transcript:: Exon, Number of CpG Islands hit, Sequence around the cut side, %A %C %T %G, S-Score, A-Score, E-Score, per cent of total transcripts, hit Target, Match-start, Match-end, Match string, Edit-distance, Number of Hits, Direction. This table contains one row for every design that made it into the library. The table is sorted by gene, annotation score, specificity score, efficiency score. The table is tab-delimited.
∗.large.gff	"General Feature Format" file containing all designs that could have made it into the library. For more information see http://www.sanger.ac.uk/resources/software/gff/spec.html. For view in the ENSEMBL genome browser see http://www.ensembl.org/info/website/upload/index.html.
∗.large.tab	Tab-delimited CRISPR result table with the following columns: Name, Length, Start, End, Strand, Nucleotide sequence, Gene Name, Transcripts, Transcript:: Exon, Number of CpG Islands hit, Sequence around the cut side, %A %C %T %G, S-Score, A-Score, E-Score, per cent of total transcripts, hit Target, Match-start, Match-end, Match string, Edit-distance, Number of Hits, Direction.
∗.missing.tab	Table like the coverage table for every gene from the initial library that has not made into the subset library.

MacOS and *TextPad* when working on windows machines. The following files are the result of the design process (∗ represents library name, brief reproduction from Heigwer et al. [16]):

(a) ∗.large.gff & ∗.large.tab: contain all suitable target sites for each gene before library filtering. Contains separate entries for each off-target site any sgRNA might have. The output file is in GFF file format are compliant with the custom track function of ENSEMBL (*see* **Note 8**)

(b) ∗.gff & ∗.tab: files filtered to neither contain off-target sites nor excess sgRNA targets per gene.

(c) ∗.missing.tab: lists all gene identifiers which are not covered in final library either because inferior coverage or the maximum size of the complete library has been exceeded.

(d) ∗.coverage.tab: lists genes with their final coverage.

(e) ∗.fasta: contains the ready-to-order sequences of constructs for on-chip synthesis, one entry for each construct. External negative or positive or negative controls need to be added directly to this file. Remapping of sequencing reads after a screening experiment is also performed using an index constructed on this file.

11. CLD can handle multiple cores.
To enable multicore processing adapt the ***docker run*** command by adding the *--cpus 6* argument.

References

1. Doench J, Am G (2018) I ready for CRISPR? A user's guide to genetic screens. Nat Rev Genet 19:67–80
2. Zhan T, Rindtorff N, Betge J, Ebert MP, Boutros M (2018) CRISPR/Cas9 for cancer research and therapy. Semin Cancer Biol. https://doi.org/10.1016/j.semcancer.2018.04.001
3. Cong L et al (2013) Multiplex genome engineering using CRISPR/Cas systems. Science 339:823–826
4. Wang T et al (2017) Gene essentiality profiling reveals gene networks and synthetic lethal interactions with oncogenic Ras. Cell 168:890–903.e15
5. Shalem O et al (2014) Genome-scale CRISPR-Cas9 knockout screening in human cells. Science 343:84–87
6. LaFountaine JS, Fathe K, Smyth HDC (2015) Delivery and therapeutic applications of gene editing technologies ZFNs, TALENs, and CRISPR/Cas9. Int J Pharm 494:180–194
7. Chuai G-H, Wang Q-L, Liu Q (2017) In silico meets in vivo: towards computational CRISPR-Based sgRNA design. Trends Biotechnol 35:12–21
8. Doench JG et al (2014) Rational design of highly active sgRNAs for CRISPR-Cas9-mediated gene inactivation. Nat Biotechnol 32:1262–1267
9. Graham DB, Root DE (2015) Resources for the design of CRISPR gene editing experiments. Genome Biol 16:260
10. Cui Y, Xu J, Cheng M, Liao X, Peng S (2018) Review of CRISPR/Cas9 sgRNA design tools. Interdiscip Sci 10:455–465
11. Montague TG, Cruz JM, Gagnon JA, Church GM, Valen E (2014) CHOPCHOP: a CRISPR/Cas9 and TALEN web tool for genome editing. Nucleic Acids Res 42: W401–W407
12. Stemmer M, Thumberger T, Del Sol Keyer M, Wittbrodt J, Mateo JL (2015) CCTop: an intuitive, flexible and reliable CRISPR/Cas9 target prediction tool. PLoS One 10:e0124633
13. Haeussler M et al (2016) Evaluation of off-target and on-target scoring algorithms and integration into the guide RNA selection tool CRISPOR. Genome Biol 17
14. Perez AR et al (2017) GuideScan software for improved single and paired CRISPR guide RNA design. Nat Biotechnol 35:347–349
15. Heigwer F, Kerr G, Boutros M (2014) E-CRISP: fast CRISPR target site identification. Nat Methods 11:122–123
16. Heigwer F et al (2016) CRISPR library designer (CLD): software for multispecies design of single guide RNA libraries. Genome Biol 17:55
17. Doench JG et al (2016) Optimized sgRNA design to maximize activity and minimize off-target effects of CRISPR-Cas9. Nat Biotechnol 34:184–191
18. Chuai G et al (2018) DeepCRISPR: optimized CRISPR guide RNA design by deep learning. Genome Biol 19:80
19. Dang Y et al (2015) Optimizing sgRNA structure to improve CRISPR-Cas9 knockout efficiency. Genome Biol 16:280
20. Singh D, Sternberg SH, Fei J, Doudna JA, Ha T (2016) Real-time observation of DNA recognition and rejection by the RNA-guided endonuclease Cas9. Nat Commun 7:12778
21. Hilton IB, Gersbach CA (2015) Enabling functional genomics with genome engineering. Genome Res 25:1442–1455
22. Horlbeck MA et al (2016) Nucleosomes impede Cas9 access to DNA in vivo and in vitro. elife 5:e12677

23. Shi J et al (2015) Discovery of cancer drug targets by CRISPR-Cas9 screening of protein domains. Nat Biotechnol 33:661–667
24. Uusi-Mäkelä MIE et al (2018) Chromatin accessibility is associated with CRISPR-Cas9 efficiency in the zebrafish (*Danio rerio*). PLoS One 13:e0196238
25. Bae S, Kweon J, Kim HS, Kim J-S (2014) Microhomology-based choice of Cas9 nuclease target sites. Nat Methods 11:705–706
26. Tsai SQ et al (2015) GUIDE-seq enables genome-wide profiling of off-target cleavage by CRISPR-Cas nucleases. Nat Biotechnol 33:1–12. https://doi.org/10.1038/nbt.3117
27. Yang H et al (2018) Base editing generates substantial off-target single nucleotide variants. bioRxiv 480145. https://doi.org/10.1101/480145
28. Kuscu C, Arslan S, Singh R, Thorpe J, Adli M (2014) Genome-wide analysis reveals characteristics of off-target sites bound by the Cas9 endonuclease. Nat Biotechnol 32:677–683
29. Pattanayak V et al (2013) High-throughput profiling of off-target DNA cleavage reveals RNA-programmed Cas9 nuclease specificity. Nat Biotechnol 31:839–843
30. Kleinstiver BP et al (2016) High-fidelity CRISPR–Cas9 nucleases with no detectable genome-wide off-target effects. Nature. https://doi.org/10.1038/nature16526
31. Wu X, Kriz AJ, Sharp PA (2014) Target specificity of the CRISPR-Cas9 system. Quant Biol 2:59–70
32. Bolukbasi MF, Gupta A, Wolfe SA (2016) Creating and evaluating accurate CRISPR-Cas9 scalpels for genomic surgery. Nat Methods 13:41–50
33. Weiner A (2018) Cloning guides to lentiCRISPR v2. https://doi.org/10.17504/protocols.io.qx3dxqn
34. Langmead B, Trapnell C, Pop M, Salzberg SL (2009) Ultrafast and memory-efficient alignment of short DNA sequences to the human genome. Genome Biol 10:R25
35. Langmead B, Salzberg SL (2012) Fast gapped-read alignment with Bowtie 2. Nat Methods 9:357–359
36. MacPherson CR, Scherf A (2015) Flexible guide-RNA design for CRISPR applications using Protospacer Workbench. Nat Biotechnol 33:805–806

Chapter 2

Web-Based CRISPR Toolkits: Cas-OFFinder, Cas-Designer, and Cas-Analyzer

Gue-Ho Hwang, Jin-Soo Kim, and Sangsu Bae

Abstract

The CRISPR-Cas system facilitates highly efficient genome editing; thus, it has been applied in many research fields such as biological science, medicine, and gene therapy. However, CRISPR nucleases can cleave off-target sites as well as on-target sites, causing unwanted mutations. Furthermore, after CRISPR treatments are delivered into cells or organisms, it is important to estimate the resulting mutation rates and to determine the patterns of mutations, but these tasks can be difficult. To address these issues, we have developed a tool for identifying potential off-target sites (Cas-OFFinder), a tool for designing CRISPR targets (Cas-Designer), and an assessment tool (Cas-Analyzer). These programs are all implemented on our website so that researchers can easily design CRISPR guide RNAs and assess the resulting mutations by simply clicking on the appropriate buttons; no login process is required.

Key words CRISPR-Cas system, Off-target, Web-based tool, Cas-OFFinder, Cas-Designer, Cas-Analyzer

1 Introduction

RNA-guided endonucleases (RGENs), derived from clustered regularly interspaced short palindromic repeats (CRISPR)—CRISPR associated (Cas) prokaryotic adaptive immune systems, are widely used in biological research, medicine, biotechnology, and gene therapy [1–6]. CRISPR-Cas effectors recognize protospacer-adjacent motif (PAM) sequences in target DNA or RNA with the help of guide RNAs (gRNAs) and induce double strand breaks (DSBs) or cleavages in the target DNA or RNA, respectively [7–9]. In cells or organisms, DNA DSBs are repaired by endogenous DNA repair systems: nonhomologous end joining (NHEJ), microhomology-mediated end joining (MMEJ), or homology directed repair (HDR) [10–12]. NHEJ frequently causes small insertions or deletions (indels) at target sites and MMEJ induces DNA sequence-based deletions. As a result, in coding sequences both systems often cause frameshift mutations, generating gene knockouts. In

Tudor A. Fulga et al. (eds.), *CRISPR Guide RNA Design: Methods and Protocols*, Methods in Molecular Biology, vol. 2162, https://doi.org/10.1007/978-1-0716-0687-2_2, © Springer Science+Business Media, LLC, part of Springer Nature 2021

contrast, HDR repairs cleaved DNA based on donor DNA templates, facilitating exact gene corrections or knockins. As a result, researchers can edit DNA target sites as desired by using RNA-guided CRISPR nucleases.

However, CRISPR nucleases can cleave not only on-target sites but also off-target sites that differ by up to several nucleotides from the on-target sites, causing unwanted off-target mutations [13, 14]. To address this issue, we implemented Cas-OFFinder, a tool that searches for potential off-target sites [15]. Cas-OFFinder first reads the whole genome sequencing data from the organism of interest and finds all targets that have PAM sequences for the specified CRISPR nuclease. Cas-OFFinder then compares all of the identified targets with a reference sequence (or on-target sequence) and counts the number of mismatched bases, after which it generates output files. The main Cas-OFFinder program is written in OpenCL (open computing language), enabling operation in diverse platforms such as central processing units (CPUs) and graphics processing units (GPUs). Cas-OFFinder is available as a command-line program or accessible via our website, *CRISPR RGEN Tools* (http://www.rgenome.net/).

To aid researchers in choosing appropriate target sites in a gene, we next developed the user-friendly Cas-Designer program [16]. When users input DNA sequences of interest, the desired CRISPR type, and a target organism, Cas-Designer provides a list of all possible gRNA candidates, along with useful information. This information includes genome-wide potential off-target sites, including those sites involving bulges in the target site or gRNA, an out-of-frame score to help users choose appropriate sites for gene knockout, and GC content. Cas-Designer displays the results in an interactive table and provides filter functions for users' convenience. Cas-Designer is also available on our website.

After CRISPR treatment, not all cells contain mutated DNA and the patterns of mutations in the cells that do are diverse. To estimate mutation rates in cells, several methods, such as the surveyor nuclease assay, the T7 endonuclease I (T7E1) assay, polyacrylamide gel electrophoresis, droplet digital PCR, and next generation sequencing (NGS), have been used. Among these methods, targeted deep sequencing using a NGS machine results in the highest sensitivity and precision, but analysis of NGS data is not easy. To address this issue, we constructed a web-based tool, named Cas-Analyzer, for assessing genome editing results using NGS data. Cas-Analyzer was mainly implemented by a JavaScript language so that it runs completely at client-side web browsers; uploading a large NGS dataset to a server or local tool installation is not necessary [17].

All three programs are available on our website, and users can freely access them without a login process. We continuously maintain and update these tools. Currently, our website supports

17 different PAMs and 339 different organisms, including a variety of vertebrates, insects, plants, and bacteria.

2 Materials

2.1 Software

All programs have been constructed to run on the *CRISPR RGEN Tools* website (Fig. 1), but we also provide a command line version of Cas-OFFinder. The off-line version of Cas-OFFinder can be run in Linux (with a proprietary driver installed), Max OS X (Snow leopard or higher), or Windows 7 or higher (XP or below is not supported). For running Cas-OFFinder with OpenCL language, an OpenCL-enabled device should be prepared and the corresponding runtime driver should be preinstalled. Links for downloading the OpenCL driver for each device are shown in Table 1.

2.2 Hardware

The off-line version of Cas-OFFinder requires an OpenCL-enabled device; devices can be checked at the Khronos group website (https://www.khronos.org/conformance/adopters/conformant-products/#opencl).

2.3 Data

The off-line version of Cas-OFFinder requires access to whole genome sequence information from the organism of interest, which can be found at the NCBI (https://www.ncbi.nlm.nih.gov/), Ensembl (https://www.ensembl.org/), and UCSC (http://hgdownload.soe.ucsc.edu/downloads.html) websites.

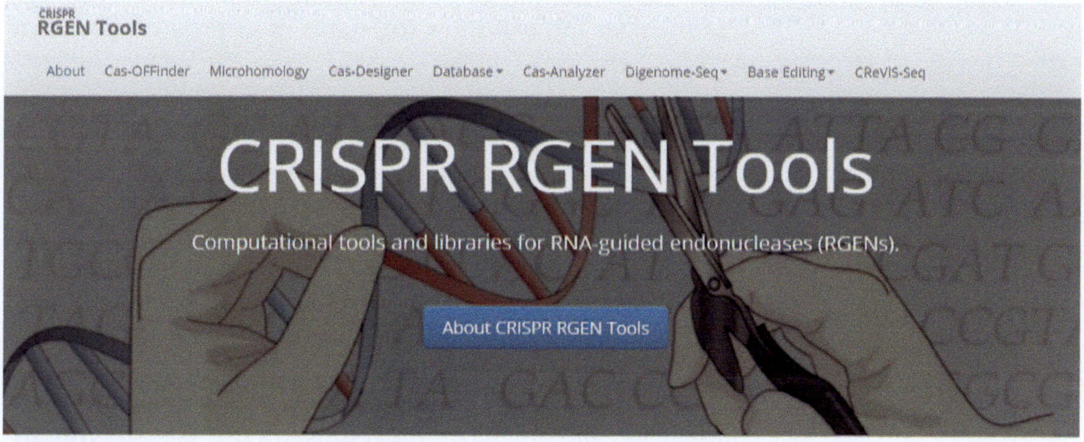

Fig. 1 The *CRISPR RGEN Tools* main page

Table 1
List of links for downloading OpenCL drivers

Device	Download link
Intel	http://software.intel.com/en-us/vcsource/tools/opencl-sdk
Nvidia	http://www.nvidia.com/Download/index.aspx
AMD	http://support.amd.com/en-us/download

3 Methods

3.1 Cas-OFFinder

As mentioned above, Cas-OFFinder can be used both online and off-line. This section introduces the usage of the online web tool and describes how to install and perform analysis with the off-line version.

3.1.1 Use of Cas-OFFinder via the Online Web Tool

1. Access Cas-OFFinder (http://www.rgenome.net/cas-offinder/) from the *CRISPR RGEN Tools* website (*see* **Note 1**).
2. Indicate the type of PAM recognized by the selected CRISPR endonuclease, as well as the reference organism, in the appropriate boxes (Fig. 1a; *see* **Note 2**).
3. Paste or type the target sequences (not including the PAM sequence) in the query sequence box. Enter the maximum allowed number of mismatches and number of nucleotides in potential DNA and RNA bulges. Target sequence lengths can range from 15 to 25 nucleotides and all target sequences must be of the same length (Fig. 1; *see* **Note 3**).
4. Click the submit button. Cas-OFFinder will then start to analyze the data and will show the status of its progress (*see* **Note 4**).
5. When the analysis is finished, Cas-OFFinder will display two tables (the Summary and Details tables), as shown in Fig. 2. The Summary table shows, for each target sequence, the numbers of predicted off-target sites for each condition: bulge type, bulge size, and number of mismatches. The Details table shows each off-target sequence, together with its position, direction, number of mismatches, bulge size, and bulge type. Filter and download buttons appear between the two tables. These buttons allow users to filter results according to the bulge type (whether the bulge is in the DNA or RNA) and number of mismatches and download the results in a text file (*see* **Note 5**).

3.1.2 Use of the Off-line Version of Cas-OFFinder

1. Check whether your device is OpenCL-enabled and install the latest suitable driver (*see* **Note 6**).

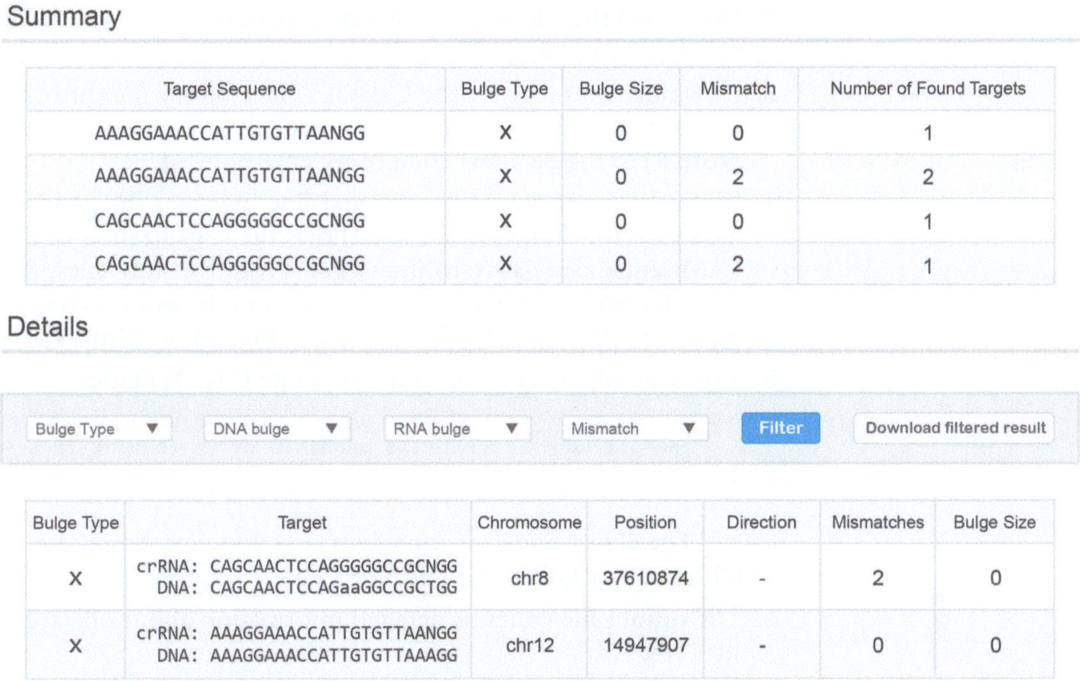
Summary

Target Sequence	Bulge Type	Bulge Size	Mismatch	Number of Found Targets
AAAGGAAACCATTGTGTTAANGG	X	0	0	1
AAAGGAAACCATTGTGTTAANGG	X	0	2	2
CAGCAACTCCAGGGGGCCGCNGG	X	0	0	1
CAGCAACTCCAGGGGGCCGCNGG	X	0	2	1

Details

Bulge Type	Target	Chromosome	Position	Direction	Mismatches	Bulge Size
X	crRNA: CAGCAACTCCAGGGGGCCGCNGG DNA: CAGCAACTCCAGaaGGCCGCTGG	chr8	37610874	-	2	0
X	crRNA: AAAGGAAACCATTGTGTTAANGG DNA: AAAGGAAACCATTGTGTTAAAGG	chr12	14947907	-	0	0

Fig. 2 A sample Cas-OFFinder results page showing summarized and detailed information as tables. Users can filter and download the desired information as a text file

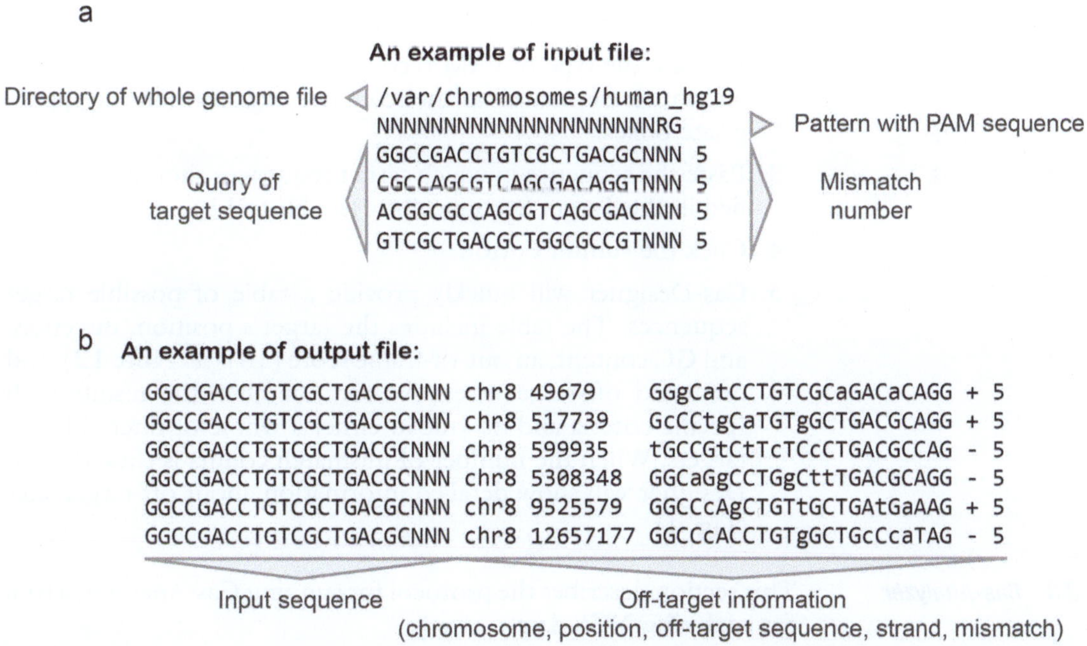

Fig. 3 File formats for the off-line version of Cas-OFFinder. (**a**) Format of the input file. (**b**) Format of the output file

2. Download the whole genome sequence of the target organism in FASTA format (*see* **Note 7**).
3. Download and register the Cas-OFFinder binary file (http://sourceforge.net/projects/cas-offinder/files/Binaries/; *see* **Note 8**) in the personal computing environment.
4. Write the input file. Here, the first line indicates the genome file's directory. The second line shows the organization of the PAM sequence relative to the target. From the next lines, the input file receives the target sequences and the associated mismatch counts, separated by one space (Fig. 3a; *see* **Note 9**).
5. Type the following command to operate Cas-OFFinder.

```
cas-offinder {input_file_path} {G or C} {output_file_-
path}
```

   The G or C indicates whether Cas-OFFinder should use all CPUs (C) or GPUs (G) (*see* **Note 10**).

6. The output file contains detailed information about off-target sites (Fig. 3b).

3.2 Cas-Designer

This section introduces the protocol for running Cas-Designer, which identifies CRISPR target sequences.

1. Access Cas-Designer (http://www.rgenome.net/cas-designer/) from the *CRISPR RGEN Tools.*
2. Indicate the type of PAM recognized by the selected CRISPR endonuclease, as well as the reference organism, in the appropriate boxes.
3. Paste the sequence in which target sequences should be identified in the Target Sequence box (*see* **Note 11**).
4. Click the submit button.
5. Cas-Designer will quickly provide a table of possible target sequences. The table includes the target's position, direction, and GC content, an out-of-frame score [18] (*see* **Note 12**), and predicted off-target sites. A user can filter the results with specific criteria and download either a whole or filtered list of targets. When the number of mismatch counts is clicked, Cas-Designer will show detailed information about off-target sites (Fig. 4).

3.3 Cas-Analyzer

This section describes the protocol for running Cas-Analyzer, a tool for analyzing NGS data.

1. Access Cas-Analyzer (http://www.rgenome.net/cas-analyzer/) from the *CRISPR RGEN Tools* site.

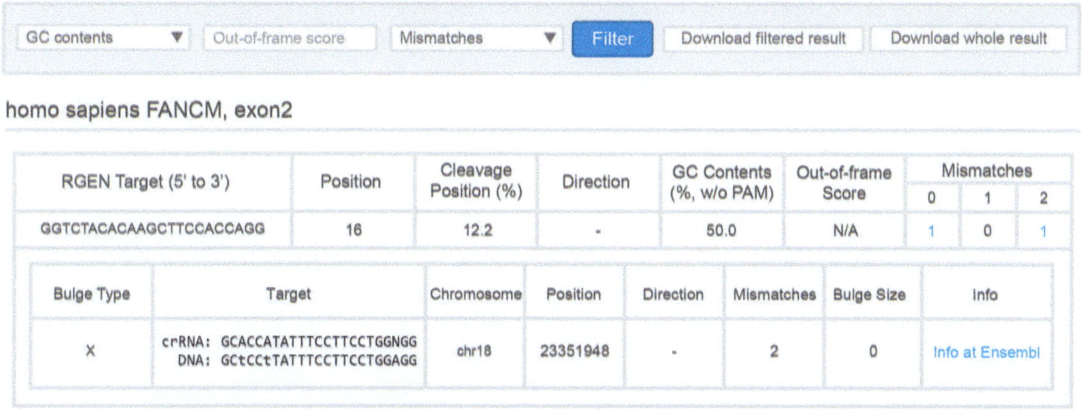

Fig. 4 A sample Cas-Designer results page showing all possible target sites and associated useful information

2. Upload single-end reads, paired-end reads, or merged NGS data after selecting the file type.
3. Input experimental information, including the reference sequence, nuclease type, specific nuclease, and target DNA sequence, as well as analysis parameters (*see* **Note 13**).
4. Click the Submit button. A bar will indicate Cas-Analyzer's progress.
5. When the analysis is finished, Cas-Analyzer will display a results page that shows the count of each mutation pattern detected and statistical graphs (Fig. 5). Aligned sequences will also be displayed at the bottom of this page and can be sorted by clicking the following tabs: All, WT and Substitutions, Insertions and Deletions. Additionally, users can see which of the aligned sequences contain a donor sequence. The data can be downloaded by clicking the appropriate button.

4 Notes

1. The *CRISPR RGEN Tools* site (http://www.rgenome.net) offers web tools to identify targets for and analyze NGS data resulting from CRISPR experiments. In addition to Cas-OFFinder, Cas-Designer, and Cas-Analyzer, there are additional useful tools: an analysis tool for microhomology calculations (Microhomology-Predictor) [18], the Digenome-sequencing analysis tool [19, 20] for profiling CRISPR-Cas9 specificity, gRNA library design tools for Cas9 (Cas-Database) [21] and Cpf1 (Cpf1-Database) [22], and design and analysis toolkits for CRISPR base editors (BE-designer and BE-Analyzer) [23].
2. Currently, Cas-OFFinder provides 17 choices of PAM sequences and 339 choices of organisms. If the desired

Fig. 5 A sample Cas-Analyzer result page showing (**a**) summarized results as a table, (**b**) statistical graphs, and (**c**) query sequences aligned to the reference sequence with filter function

organism is not available, click the sentence "Send request for a new organism," which will open an email message addressed to the *CRISPR RGEN Tools* administrator.

3. Cas-OFFinder supports mixed bases in target sequences.
4. The progress page gives information about the status of analysis. In the *CRISPR RGEN Tools* server, Cas-OFFinder uses two sections for analysis. Thus, Cas-OFFinder usually starts analysis and shows "Running . . ." in the status column, but if Cas-OFFinder is undertaking many tasks, it analyzes query sequences in the order in which they were received, showing a "Waiting for available slot. . ." message in the status column to waiting users.

5. All result files will be stored for three days; the user can see these results by going to the URL at the top of the results page.
6. Cas-OFFinder is written in C++ and OpenCL, which helps multiprocessing by GPUs.
7. The off-line version of Cas-OFFinder receives the directory for a whole genome sequence file. If users want to increase the speed of Cas-OFFinder, we recommend converting the FASTA file format to a 2-bit format to reduce the time required for reading the whole genome sequence file.
8. The Cas-OFFinder source code has been uploaded to GitHub (https://github.com/snugel/cas-offinder). If a user wants to use Cas-OFFinder directly on command, the user should set the Cas-OFFinder binary file at the personal computing environment.
9. In the off-line version, there are no limitations for PAM sequences or number of mismatches.
10. If Cas-OFFinder is not registered in the personal computing environment, the user must input the Cas-OFFinder binary directory with the file name instead of "cas-offinder."

```
        /directory/of/cas-offinder {input_file_path} {G or C}
{output_file_path}
```

11. Sequences do not need to be written in the Target Sequence box. Cas-Designer can also receive FASTA files that contain the target sequence. Additionally, users can find target sequences in exon regions by writing "Exon" in the Target searching range box.
12. The out-of-frame score indicates the predicted ratio (or probability) of frameshift mutations in the target sequence.
13. If a DNA template inserts into the target sequence, the user can detect this event by inputting the template sequence in the Donor DNA sequence box. Cas-Analyzer will then find the template sequence and mark the sequences that contain the template sequence. The user can filter the aligned results to show sequences that contains the template sequence.

Acknowledgments

This work was supported by the Institute for Basic Science grant No. IBS-R021-D1 to J.-S.K. and by National Research Foundation of Korea (NRF) Grants (No. 2016R1D1A1B03934811 and 2018M3A9H3022412) to S.B.

References

1. Horvath P, Barrangou R (2010) CRISPR/Cas, the immune system of Bacteria and Archaea. Science 327:167–170. https://doi.org/10.1126/science.1179555
2. Garneau JE, Dupuis M-È, Villion M, Romero DA, Barrangou R, Boyaval P, Fremaux C, Horvath P, Magadán AH, Moineau S (2010) The CRISPR/Cas bacterial immune system cleaves bacteriophage and plasmid DNA. Nature 468:67–71. https://doi.org/10.1038/nature09523
3. Kim H, Kim J-S (2014) A guide to genome engineering with programmable nucleases. Nat Rev Genet 15:321–334. https://doi.org/10.1038/nrg3686
4. Baek K, Kim DH, Jeong J, Sim SJ, Melis A, Kim J-S, Jin E, Bae S (2016) DNA-free two--gene knockout in *Chlamydomonas reinhardtii* via CRISPR-Cas9 ribonucleoproteins. Sci Rep 6:30620. https://doi.org/10.1038/srep30620
5. Lee SH, Yu J, Hwang G-H, Kim S, Kim HS, Ye S, Kim K, Park J, Park DY, Cho Y-K, Kim J-S, Bae S (2017) CUT-PCR: CRISPR-mediated, ultrasensitive detection of target DNA using PCR. Oncogene 36:6823–6829. https://doi.org/10.1038/onc.2017.281
6. Rubin AJ, Parker KR, Satpathy AT, Qi Y, Wu B, Ong AJ, Mumbach MR, Ji AL, Kim DS, Cho SW, Zarnegar BJ, Greenleaf WJ, Chang HY, Khavari PA (2019) Coupled single-cell CRISPR screening and epigenomic profiling reveals causal gene regulatory networks. Cell 176:361–376. https://doi.org/10.1016/j.cell.2018.11.022
7. Nishimasu H, Ran FA, Hsu PD, Konermann S, Shehata SI, Dohmae N, Ishitani R, Zhang F, Nureki O (2014) Crystal structure of Cas9 in complex with guide RNA and target DNA. Cell 156:935–949. https://doi.org/10.1016/j.cell.2014.02.001
8. Cong L, Ran FA, Cox D, Lin S, Barretto R, Habib N, Hsu PD, Wu X, Jiang W, Marraffini LA, Zhang F (2013) Multiplex genome engineering using CRISPR/Cas systems. Science 339:819–823. https://doi.org/10.1126/science.1231143
9. Jinek M, Chylinski K, Fonfara I, Hauer M, Doudna JA, Charpentier E (2012) A programmable dual-RNA-guided DNA endonuclease in adaptive bacterial immunity. Science 337:816–821. https://doi.org/10.1126/science.1225829
10. Lieber MR, Ma Y, Pannicke U, Schwarz K (2003) Mechanism and regulation of human non-homologous DNA end-joining. Nat Rev Mol Cell Biol 4:712–720. https://doi.org/10.1038/nrm1202
11. Rothkamm K, Krüger I, Thompson LH, Löbrich M (2003) Pathways of DNA double-strand break repair during the mammalian cell cycle. Mol Cell Biol 23:5706–5715. https://doi.org/10.1128/MCB.23.16.5706-5715.2003
12. Mao Z, Bozzella M, Seluanov A, Gorbunova V (2008) DNA repair by nonhomologous end joining and homologous recombination during cell cycle in human cells. Cell Cycle 7:2902–2906. https://doi.org/10.4161/cc.7.18.6679
13. Fu Y, Foden JA, Khayter C, Maeder ML, Reyon D, Joung JK, Sander JD (2013) High-frequency off-target mutagenesis induced by CRISPR-Cas nucleases in human cells. Nat Biotechnol 31:822–826. https://doi.org/10.1038/nbt.2623
14. Cho SW, Kim S, Kim Y, Kweon J, Kim HS, Bae S, Kim J-S (2014) Analysis of off-target effects of CRISPR/Cas-derived RNA-guided endonucleases and nickases. Genome Res 24:132–141. https://doi.org/10.1101/gr.162339.113
15. Bae S, Park J, Kim J-S (2014) Cas-OFFinder: a fast and versatile algorithm that searches for potential off-target sites of Cas9 RNA-guided endonucleases. Bioinformatics 30:1473–1475. https://doi.org/10.1093/bioinformatics/btu048
16. Park J, Bae S, Kim J-S (2015) Cas-Designer: a web-based tool for choice of CRISPR-Cas9 target sites. Bioinformatics 31:4014–4016. https://doi.org/10.1093/bioinformatics/btv537
17. Park J, Lim K, Kim J-S, Bae S (2017) Cas-analyzer: an online tool for assessing genome editing results using NGS data. Bioinformatics 33:286–288. https://doi.org/10.1093/bioinformatics/btw561
18. Bae S, Kweon J, Kim HS, Kim J-S (2014) Microhomology-based choice of Cas9 nuclease target sites. Nat Methods 11:705–706. https://doi.org/10.1038/nmeth.3015
19. Kim D, Bae S, Park J, Kim E, Kim S, Yu HR, Hwang J, Kim J-I, Kim J-S (2015) Digenome-seq: genome-wide profiling of CRISPR-Cas9 off-target effects in human cells. Nat Methods 12:237–243. https://doi.org/10.1038/nmeth.3284
20. Park J, Childs L, Kim D, Hwang G-H, Kim S, Kim S-T, Kim J-S, Bae S (2017) Digenome-seq

web tool for profiling CRISPR specificity. Nat Methods 14:548–549. https://doi.org/10.1038/nmeth.4262

21. Park J, Kim J-S, Bae S (2016) Cas-Database: web-based genome-wide guide RNA library design for gene knockout screens using CRISPR-Cas9. Bioinformatics 32:2017–2023. https://doi.org/10.1093/bioinformatics/btw103
22. Park J, Bae S (2018) Cpf1-Database: web--based genome-wide guide RNA library design for gene knockout screens using CRISPR-Cpf1. Bioinformatics 34:1077–1079. https://doi.org/10.1093/bioinformatics/btx695
23. Hwang G-H, Park J, Lim K, Kim S, Yu J, Yu E, Kim S-T, Eils R, Kim J-S, Bae S (2018) Web-based design and analysis tools for CRISPR base editing. BMC Bioinformatics 19:542. https://doi.org/10.1186/s12859-018-2585-4

Part II

Chemically-Modified Guide RNAs

Chapter 3

Chemical Modification of Guide RNAs for Improved CRISPR Activity in CD34+ Human Hematopoietic Stem and Progenitor Cells

Jenny Shapiro, Adi Tovin, Ortal Iancu, Daniel Allen, and Ayal Hendel

Abstract

Human $CD34^+$ hematopoietic stem and progenitor cells (HSPCs) have the unique ability to repopulate the entire hematopoietic system and thus are at the center of diverse, therapeutically relevant studies. The recent development of the CRISPR/Cas9 tool made the powerful research technique of genome editing highly accessible. Our previous studies demonstrated that high editing efficiency is reached when the CRISPR/Cas9 is introduced to CD34+ HSPCs as a ribonucleoprotein (RNP) complex with chemically modified guide RNAs (gRNAs). The current protocol details a quick 4-day procedure for ex vivo genome editing in CD34+ HSPCs by RNP complexes that are targeted to a specific locus by either a single gRNA (sgRNA) or a 2-part gRNA. The delivery of RNP complexes is performed by electroporation in the presence of a nonspecific, ssDNA electroporation enhancer, which highly improves editing efficiency under the described conditions. This approach is simple and effective with the potential to accelerate many biotechnological and therapeutic applications of the CRISPR/Cas9 technology.

Key words CRISPR, Cas9, gRNA, CD34, HSPCs, Electroporation, Genome editing, Modifications

1 Introduction

Human hematopoietic stem and progenitor cells (HSPCs) are a widely heterogeneous population, which includes a small group of long-term stem cells, able to generate the entire hematopoietic system, de novo [1, 2]. Human HSPCs are commonly identified by the $CD34^+$ cell surface protein that is used as a marker for their enrichment. Due to their place at the origin of the hematopoietic system, CD34+ HSPCs are a target for basic, preclinical, and clinical research. Genome editing in CD34+ HSPCs is a powerful tool, which can unravel developmental mechanisms, establish disease models, and test the feasibility of various therapeutic approaches. Induction of a locus-specific double-strand break (DSB) commences the process of genome editing [3]. This event is followed by two possible DNA repair mechanisms, executed by the cells. The

Tudor A. Fulga et al. (eds.), *CRISPR Guide RNA Design: Methods and Protocols*, Methods in Molecular Biology, vol. 2162, https://doi.org/10.1007/978-1-0716-0687-2_3,

broken ends can be rejoined by nonhomologous end joining (NHEJ), with the occasional insertions and deletions (indels) of nucleotides at the cut site. Indels may cause gene inactivation, and their quantification by chromatogram decomposition algorithms [4] or by next-generation sequencing (NGS) can be used for measuring the efficiency of DSB introduction. The second repair mechanism, the homology-directed repair (HDR), can be harnessed for gene correction or the introduction of new gene variants [5].

Due to its simplicity and affordability, the clustered regularly interspaced short palindromic repeats (CRISPR) and the CRISPR-associated protein 9 (Cas9), known as CRISPR/Cas9 technology, is currently the most widespread system that uses programmable nucleases for the induction of a site-specific DSBs. The Cas9 nuclease is guided by a bound gRNA complementary to a specific 20 bp-long target site within the genome. While high editing efficiency can be achieved in cell lines with a plasmid-based CRISPR/Cas9 system, primary cells demand a different strategy. We have previously shown that the protection of sgRNA termini with chemical modifications can vastly increase editing efficiency in CD34+ HSPCs, probably due to reduced RNase-induced degradation [6]. In addition, the editing efficiency and specificity in CD34+ HSPCs are also elevated when sgRNAs are delivered as part of a ribonucleoprotein (RNP) complex [7]. Therefore, for best results, it is desirable to deliver CD34+ HSPCs with chemically modified, synthetic gRNAs as part of an RNP complex.

Lastly, the synthetic gRNA component for a specific genomic locus can appear in one of two available formulations [8]. The 2-part system is comprised from a CRISPR RNA (crRNA), which carries the 20 nt-long, specific target sequence on its 5′ end, and an additional sequence on its 3′ end, allowing for the annealing to a nontarget specific trans-activating CRISPR RNA molecule (tracrRNA). Alternatively, the sgRNA is synthesized as a ~100 nt-long molecule in which the crRNA and the tracrRNA are covalently bound together by a stem-loop structure. The current protocol details the process of either sgRNA or 2-part synthetic gRNAs complexation with Cas9, and the subsequent delivery of the RNP complex into $CD34^+$ HSPCs by electroporation. It also recommends performing the electroporation in the presence of a nonspecific ssDNA oligonucleotide, termed the electroporation enhancer [9], which can highly increase the editing efficiency of CD34+ HSPCs under the described conditions. Furthermore, cells can be either freshly isolated or thawed, and the procedure can be practiced with a wide cell number range. In conclusion, this approach is simple and effective with the potential to improve a wide array of CRISPR/Cas9 biotechnological and therapeutic applications.

2 Materials

2.1 CD34⁺ HSPC Culture

1. $CD34^+$ HSPC culture medium: various media compositions can be used. The main basic media components are listed below [6, 7, 10]:
   (a) Basic medium for culture and expansion of hematopoietic cells, usually serum-free (optional: StemSpan SFEMII (Stemcell Technologies, USA). Supplemented media may be stored at 4 °C for up to 1 week after preparation).
   (b) A combination of cytokines for culture and expansion of CD34+ HSPCs (optional combination: SCF (100 ng/ml), TPO (100 ng/ml), Flt3-Ligand (100 ng/ml), IL-6 (100 ng/ml)).
   (c) Antiseptic reagents: antibiotics (e.g., 0.1 mg/ml streptomycin and 100 units/ml penicillin) and antifungal (optional).
   (d) Small molecules intended for increased proliferation and decreased differentiation of CD34+ HSPCs (optional: StemRegenin 1 (SR1) (0.75 mM) and UM171 (35 nM, Stemcell Technologies, USA)).
2. Tissue culture plates: choose a size suitable for culturing at a concentration of 0.25×10^6 cells/ml.
3. Low-oxygen tissue culture incubator adjusted to 37 °C, 5% CO_2, and 5% O_2.
4. Phosphate buffered saline (PBS) 1×, sterile.

2.2 CRISPR RNP Complex Preparation

1. Cas9 nuclease: Cas9 recombinant protein with nuclear localization signal (optional: Alt-R S.p. Cas9 Nuclease 3NLS (# 1074181, 1074182, Integrated DNA Technologies (IDT), USA) [6, 7, 10]. Keep in −20 °C freezer.
2. Synthetic gRNA with end modifications improves editing in CD34+ HSPCs [6]. Choose between the two alternative gRNA formulations, targeting the same genomic locus:
   (a) sgRNA (optional: Alt-R™ CRISPR/Cas9 sgRNA, predesigned by IDT).
   (b) 2-part system: including a target-specific crRNA and a general tracrRNA (optional: Alt-R™ CRISPR/Cas9 crRNA, predesigned, and Alt-R™ CRISPR/Cas9 tracrRNA, (# 1072532, IDT)).
3. Include a positive control gRNA: An example of a positive control is the *EMX1* gene. *EMX1* editing levels in $CD34^+$ HSPCs can be seen in Fig. 1 (*see* **Note 1** for details regarding *EMX1* gRNA and PCR primers).
4. A solution for gRNA reconstitution (optional): if gRNAs are delivered lyophilized, follow the manufacturer's instructions regarding the best molecular-grade solution for their

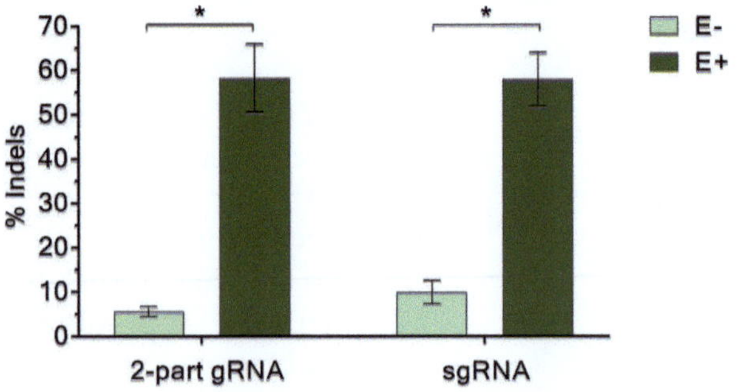

Fig. 1 EMX1 editing in CD34$^+$ HSPCs is increased by the electroporation enhancer. Mobilized CD34$^+$ cells from three independent donors were electroporated (4D-Nucleofector system, Lonza) using 4 μM of CRISPR/Cas9, complexed with an EMX1-targeting gRNA, either in the 2-part gRNA or the sgRNA formulation. The electroporation was performed with or without an electroporation enhancer (E$^-$ and E$^+$, respectively). After 48 h, genomic DNA (gDNA) was extracted from the cells and primers flanking the expected cut site were used for PCR amplification of the EMX1 gene (for primer *see* **Note 1**). Indel frequencies were calculated for each treatment by TIDE software. Error bars represent the standard deviation of the mean. $*p < 0.0001$ (Sidak's multiple comparisons test for 2-way ANOVA)

reconstitution. Unless specified otherwise, reconstitute with TE solution (1×, pH 7.5). Reconstituted gRNA should be kept in −20 °C freezer.

5. PBS 1×, sterile.
6. PCR thermal cycler (optional): can be used for 2-part gRNA annealing and for maintaining a stable room temperature during RNP complex preparation.
7. Plastics: sterile PCR tubes, sterile nuclease-free tips.

2.3 Electroporation and Subsequent Steps

1. Electroporation solution: use either purchased (e.g., P3 Primary Cell 4D-Nucleofector® X Kit S (# V4XP-3032, Lonza, Switzerland), or homemade (e.g., 1 M [10]) electroporation solution suitable for primary cells.
2. Electroporation enhancer (optional): Alt-R™ Cas9 Electroporation enhancer is an auxiliary component for electroporation offered by IDT (# 1075915). It is a 100-nt, single-stranded, carrier DNA that is nonhomologous to the human, rat, and mouse genomes (*see* **Note 2** for sequence details). It was shown to increase editing in cell lines [9], and we have found that editing efficiency is vastly improved by its presence during CD34$^+$ HSPC electroporation (Fig. 1). The enhancer should be reconstituted to a concentration of 100 μM in TE solution (1×, pH 7.5).

3. Electroporator and suitable electroporation cuvettes: consult the electroporator's manufacturer regarding the recommended electroporation program for CD34+ HSPCs (*see* **Note 3**). Select electroporation cuvettes, suitable for the electroporator in use and for the number of cells intended for electroporation in a given experiment.
4. Plastics for electroporation:
   (a) Thin, sterile, disposable, plastic pipettes.
   (b) Multichannel pipette (optional if working with a strip cuvette).
   (c) Sterile reservoirs (optional for recovery medium if working with a strip cuvette).
5. Crude genomic DNA (gDNA) extraction reagents (no need for phase or column separations).
6. PCR primers flanking the expected cut site of interest, a PCR mix, and a PCR thermal cycler.

3 Methods

Cell handling is conducted in a biosafety cabinet, while RNP complex preparation is performed on the bench with sterile and nuclease-free plastics. Note that it is recommended to proceed with electroporation immediately after the complexation of Cas9 with gRNA. In addition, it is advisable to limit the time that the cells spend outside of their culture medium and optimal environment (i.e., while resuspended in the electroporation solution and out of the incubator). Thus, it is best to coordinate steps in Subheadings 3.2 and 3.3, as described in the workflow. A summary of the suggested workflow can be found in Fig. 2. In addition, our recommendation for optimized composition of electroporation reaction is summarized in Table 1.

3.1 Culturing CD34$^+$ HSPCs—Day 0

Isolate or thaw CD34$^+$ HSPCs 48 h before electroporation (*see* **Note 4**). Culture cells in HSPC complete culture medium (see Materials section for details) at a density of 0.25×10^6 cells/ml in a low-oxygen incubator (37 °C, 5% CO_2, and 5% O_2).

3.2 Preparation of CD34$^+$ HSPCs for Electroporation—Day 2

1. Ensure the healthy appearance of the cells by microscopy on the day of electroporation. Next, gently pipet the cells inside the culture medium and collect them into a sterile tube. Rinse the culture dish surface with a prewarmed complete culture medium to collect the remaining cells.
2. Make sure to reach a homogeneous cell suspension by gently pipetting prior to cell counting.

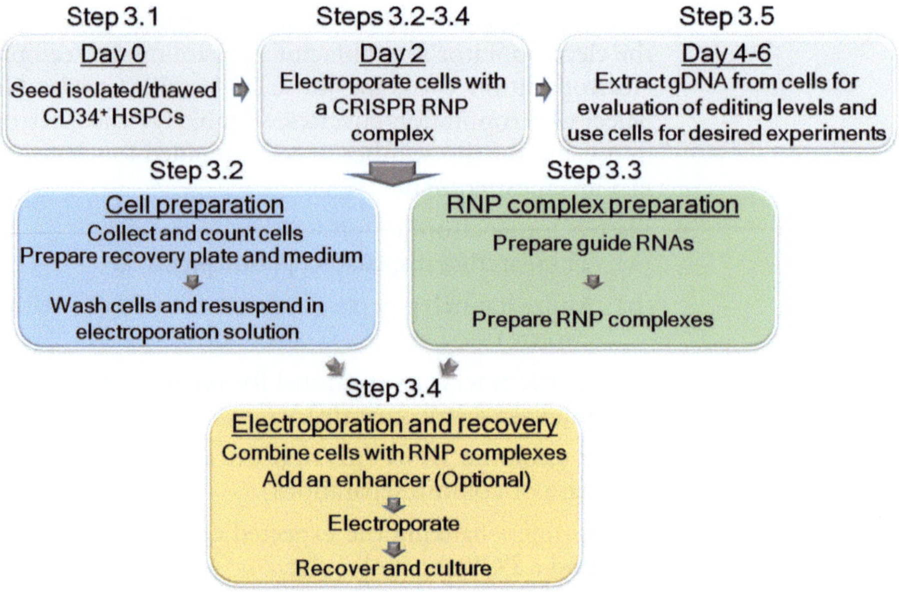

Fig. 2 A schematic workflow of the entire protocol

Table 1
The recommended setup of an electroporation reaction

		Volume fractionation for one final electroporation reaction	
Electroporation reaction component	**Recommended concentration in the final electroporation reaction**	**e.g., One well of the 4D Lonza 20 μl Nucleocuvette™ Strip**	**e.g., 4D Lonza 100 μl Single Nucleocuvette™**
Cells in electroporation solution	5×10^4–1×10^6 cells per each 20 μl of electroporation solution	20 μl	100 μl
RNP complex	4 μM	5 μl	25 μl
Electroporation enhancer (optional)	3.85 μM	1 μl	5 μl
Total volume of electroporation reaction		26 μl	130 μl

3. Calculate the number of cells intended for each electroporation in the range mentioned by the manufacturer for the chosen cuvette (consult Table 1 and **Note 5**). Remember to designate cells for a mock control. These cells will undergo electroporation in the absence of the CRISPR system and will serve as an unedited reference.

4. Transfer the exact volume containing the total number of cells needed for the planned experiment into a new sterile tube, and place it inside a low-oxygen incubator. This should allow the coordination of cell preparation with the simultaneous RNP complexing (Subheading 3.3).
5. Allow the electroporation solution to reach room temperature.
6. Follow the manufacturer's instructions regarding the necessary recovery medium volume for the specific chosen cuvette type. Preheat enough media at 37 °C for all of the cuvettes of a given experiment.

   Prepare a recovery tissue culture plate. Based on the size of the plate, calculate the necessary amount of recovery medium to be added to each well so that upon addition of the postelectroporation cell suspension (after the introduction of the recovery medium to the cuvette), a concentration of 0.25×10^6 cell/ml will be achieved. Until needed (Subheading 3.4), store the plate in a low-oxygen incubator (37 °C).
7. *Coordination tip*: to effectively prepare the cells and the RNP complex simultaneously, skip to Subheading 3.3. Use RNP incubation time for cell centrifuge cycles.

   Remove cells from the incubator, and centrifuge at $300 \times g$ for 5 min at room temperature.
8. Discard medium and rinse the original tube with 650 μl of PBS. Transfer the cells to a sterile microcentrifuge tube (*see* **Note 6**). Collect the remaining cells by rinsing the original tube with an additional 650 μl of PBS and add the volume to the microcentrifuge tube. Centrifuge again at $300 \times g$ for 5 min at room temperature.
9. Discard as much supernatant as possible without disturbing the pellet. Resuspend the cells in an appropriate volume of electroporation solution, calculated for all of the cuvettes in a given experiment (including a mock control). As can be seen in Table 1, it is recommended to add 20 μl of electroporation solution per every 5×10^4–1×10^6 cells. Pipet gently for homogeneous suspension and transfer the cells for each electroporation into a separate, sterile, PCR tube.
10. Let the cells sit at room temperature. Proceed as soon as possible to Subheading 3.4.

3.3 Preparation of CRISPR RNP Complex for Electroporation—Day 2

1. Resuspend all gRNAs in TE buffer (unless otherwise indicated by the manufacturer) and keep them on ice. The sgRNA concentration should be 100 μM, while the crRNA and tracrRNA should reach the concentration of 200 μM. Once resuspended, the sgRNA is ready for use.
2. The 2-part system oligos should be annealed to form a single conjoined gRNA. Mix the crRNA and the tracrRNA in

Step 3.3 - example
2-part gRNA annealing

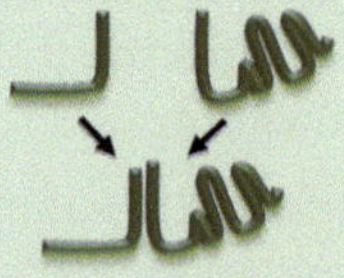

Annealing tubes (Gene X, EMX1)

		Gene of interest	Positive control
Total annealed 2-part gRNA needed for RNP complex assembly	pmol	120	120
crRNA (200 μM)	type	*Gene X*	*EMX1*
	pmol	300	300
	volume (μl)	1.5	1.5
tracrRNA (200 μM)	pmol	300	300
	volume (μl)	1.5	1.5
Annealed 2-part gRNA (100 μM)	pmol	300	300
	volume (μl)	3	3

Incubate: 95°C, 5 minutes.
Cool down gradually to RT (~10 minutes)

Fig. 3 A specific example for the annealing setup of the 2-part system gRNAs

Step 3.3 - example
RNP complex preparation

RNP complex tubes (Gene X, EMX1)

		Gene of interest	Positive control
Number of electroporation reactions		1	1
Total volume of RNP complex (μl) *		5	5
PBSX1 volume (=total volume of RNP complex - Cas9 nuclease - annealed gRNA) (μl)		5-Y-1.2	5-Y-1.2
Cas9 nuclease (Concentration depends on the supplier)	RNP concentration in electroporation reaction (μM)	4	4
	Amount (pmol)/ electroporation reaction	104	104
	volume (μl)	Y	Y
Annealed 2-part gRNA (100 μM)	type	*Gene X*	*EMX1*
	Cas9:gRNA molar ratio	1:1.15	1:1.15
	pmol/electroporation reaction	120	120
	volume (μl)	1.2	1.2

* Use extra reactions to compensate for pipetting inaccuracies.

Incubate: room temperature, 10-20 minutes.
Proceed immediately to electroporation.

Fig. 4 A specific example for RNP complex preparation

equimolar concentrations in a sterile PCR tube. If using both 2-part oligos in the same concentration (200 μM), then mix in equal volumes (*see* **Note 7**). The annealed 2-part gRNA will be at a final concentration of 100 μM. The total quantity of the annealed 2-part gRNA is demonstrated in Fig. 3. Incubate at 95 °C for 5 min (optional: in a thermal cycler). *Important!* Remove from heat, and allow to *gradually* cool to room temperature on the benchtop for about 10 min. Place the cooled down 2-part gRNA on ice.

3. Calculate the appropriate volume of the components of each intended complex as explained in Fig. 4 and **Note 8**. Scale up each complex's volume to the number of the desired electroporation reactions.
4. Begin complex preparation by placing the calculated PBS volume in a sterile PCR tube. Add the calculated volume of gRNA to the PBS, gently swirling the pipette tip while pipetting.
5. Lastly, take the Cas9 nuclease out of the −20 °C freezer and dilute the appropriate volume of the enzyme into the gRNA-containing-PBS, repeating the same gentle swirling motion during pipetting.
6. Briefly spin down the complexes. Incubate at room temperature for 10–20 min (optional: 25 °C in a PCR thermal cycler).

   Coordination tip: to effectively prepare the cells and the RNP complex simultaneously, skip to Subheading 3.2. Use RNP incubation time for cell centrifuge cycles.

3.4 Electroporation of Cells with CRISPR RNP Complex—Day 2

1. In advance, turn on the electroporator and select the desired electroporation program (*see* **Note 3**).
2. Follow the calculations from Table 1 to assemble the electroporation reactions. Add the relevant RNP complexes to the cells, which are already resuspended in the electroporation solution (an example can be seen in Fig. 5 and **Note 8**).
3. Next, follow the ratios described in Table 1 to add a suitable volume of the electroporation enhancer to each reaction (optional. *See* an example in Fig. 5).
4. Gently pipet up and down and spin down briefly (avoid cell pellet). Transfer each electroporation suspension to a separate cuvette (optional: use a multichannel pipette). To prevent air bubbles inside the cuvettes, it is recommended to transfer slightly less than the electroporation reaction total volume. For example, from a 26 μl suspension, 25 μl can be transferred into the cuvette. Remember to dedicate a cuvette for the mock samples, which will contain only cells.
5. Gently tap the cuvettes on the bench to eliminate any residual air bubbles. Insert the cuvettes into the electroporator and electroporate.
6. Immediately after electroporation, remove the cuvettes from the instrument. It is *a crucial step* for cell viability to add the prewarmed recovery culture media into the cuvette as quickly

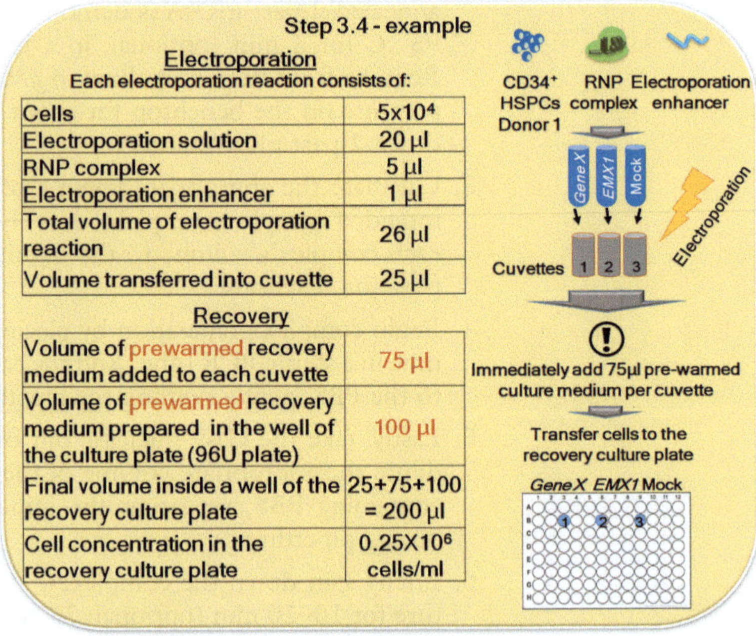

Step 3.4 - example

Electroporation

Each electroporation reaction consists of:

Cells	5x10⁴
Electroporation solution	20 μl
RNP complex	5 μl
Electroporation enhancer	1 μl
Total volume of electroporation reaction	26 μl
Volume transferred into cuvette	25 μl

Recovery

Volume of prewarmed recovery medium added to each cuvette	75 μl
Volume of prewarmed recovery medium prepared in the well of the culture plate (96U plate)	100 μl
Final volume inside a well of the recovery culture plate	25+75+100 = 200 μl
Cell concentration in the recovery culture plate	0.25X10⁶ cells/ml

Fig. 5 A specific example for an electroporation reaction assembly and for the calculation of recovery volumes

as possible! (For volume considerations consult the example in Fig. 5.)

7. Use sterile, disposable, plastic pipettes to transfer the entire volume from each cuvette to the correct recovery well in the culture plate.

8. Culture cells in complete culture medium at a density of 0.25×10^6 cells/ml in a low-oxygen incubator (37 °C, 5% CO_2, and 5% O_2).

3.5 Editing Efficiency Evaluation—Day 4–6

After 48–96 h in culture, genome-editing efficiency may be analyzed. For that purpose, extract gDNA from the cells (crude extraction is enough). Amplify the area of interest within the genome using PCR primers flanking the gRNA target site. To establish the percentage of edited sequences, compare Sanger sequencing products of the amplicons derived from the edited cells with those of mock cells by a chromatogram decomposition software such as TIDE or ICE (*see* **Note 9**). For even more precise evaluation, address NGS. Use cells in the desired downstream experiments.

4 Notes

1. The EMX1 gene gRNA specific sequence: 5-′-GAGUCCGAGCAGAAGAAGAA-3′.

The PCR primers for the amplification of the edited region: F—5′-CCATCCCCTTCTGTGAATGT-3′, R—5-′-TGCTTGTCCCTCTGTCAATG-3′.

2. Electroporation enhancer sequence:

5-′-TTAGCTCTGTTTACGTCCCAGCGGGCATGAGAGTAACAAGAGGGTGTGGTAATATTACGGTACCGAGCACTATCGATACAATATGTGTCATACGGACACG-3′.

3. It is recommended to perform an optimization experiment to test which is the most suitable program for CD34⁺ HSPCs among several candidate programs in a given electroporator. The difference between programs may be seen in the achieved editing efficiency; however, it can also manifest itself in cell viability following electroporation. For Lonza 4D-Nucleofector the DZ-100 program was found as the most suitable for $CD34^+$ HSPCs.

4. Due to the high variability among primary cells, it is recommended to examine gene editing in $CD34^+$ HSPCs based on at least three independent donors.

5. As few as 1×10^4 cells can be used in one cuvette (well) of the 20 μl Nucleocuvette™ Strip [10] while maintaining the same reaction conditions, however, it is less favorable to do so due to some cell death during the electroporation. Check the recommendations of the chosen electroporator's manufacturer and perform the necessary adaptations.

6. The PBS wash is intended to remove any residual traces of RNases, thereby preventing quick degradation of the critical CRISPR RNA components prior to the electroporation. It also facilitates the transfer of cells into a microcentrifuge tube in which the cell pellet will be clearly visible, allowing for maximal removal of excess fluid before reconstitution in an electroporation solution.

7. For better accuracy, avoid annealing less than 3 μl of a total volume of crRNA and tracrRNA. Excess annealed 2-part gRNA can be stored at −20 °C after use.

8. The RNP concentration is defined as the concentration of Cas9 in the final electroporation reaction volume. In the example given in Figs. 4 and 5, the RNP concentration is 4 μM, since the 104 pmol of Cas9 is diluted in a final electroporation reaction volume of 26 μl. Subsaturation conditions, unique for each guide type and genomic locus, can be determined by a titration experiment of the RNP concentration in the range of 0.5–4 μM.

9. TIDE software [4] can be found in https://tide.deskgen.com/

.

ICE software [11] can be found in https://ice.synthego.com/#/.

References

1. Booth C, Gaspar HB, Thrasher AJ (2016) Treating immunodeficiency through HSC gene therapy. Trends Mol Med 22 (4):317–327. https://doi.org/10.1016/j.molmed.2016.02.002
2. Mohty R, Brissot E, Battipaglia G, Ruggeri A, Sestili S, Mediavilla C, Belhocine R, Dulery R, Mohty M, Malard F (2018) CD34(+)-selected stem cell "Boost" for poor graft function after allogeneic hematopoietic stem cell transplantation. Curr Res Transl Med. https://doi.org/10.1016/j.retram.2018.12.003
3. Maeder ML, Gersbach CA (2016) Genome-editing technologies for gene and cell therapy. Mol Ther 24(3):430–446. https://doi.org/10.1038/mt.2016.10
4. Brinkman EK, Chen T, Amendola M, van Steensel B (2014) Easy quantitative assessment of genome editing by sequence trace decomposition. Nucleic Acids Res 42(22):e168. https://doi.org/10.1093/nar/gku936
5. Adli M (2018) The CRISPR tool kit for genome editing and beyond. Nat Commun 9 (1):1911. https://doi.org/10.1038/s41467-018-04252-2
6. Hendel A, Bak RO, Clark JT, Kennedy AB, Ryan DE, Roy S, Steinfeld I, Lunstad BD, Kaiser RJ, Wilkens AB, Bacchetta R, Tsalenko A, Dellinger D, Bruhn L, Porteus MH (2015) Chemically modified guide RNAs enhance CRISPR-Cas genome editing in human primary cells. Nat Biotechnol 33 (9):985–989. https://doi.org/10.1038/nbt.3290
7. Dever DP, Bak RO, Reinisch A, Camarena J, Washington G, Nicolas CE, Pavel-Dinu M, Saxena N, Wilkens AB, Mantri S, Uchida N, Hendel A, Narla A, Majeti R, Weinberg KI, Porteus MH (2016) CRISPR/Cas9 beta-globin gene targeting in human haematopoietic stem cells. Nature 539(7629):384–389. https://doi.org/10.1038/nature20134
8. Basila M, Kelley ML, Smith AVB (2017) Minimal 2′-O-methyl phosphorothioate linkage modification pattern of synthetic guide RNAs for increased stability and efficient CRISPR-Cas9 gene editing avoiding cellular toxicity. PLoS One 12(11):e0188593. https://doi.org/10.1371/journal.pone.0188593
9. Jacobi AM, Rettig GR, Turk R, Collingwood MA, Zeiner SA, Quadros RM, Harms DW, Bonthuis PJ, Gregg C, Ohtsuka M, Gurumurthy CB, Behlke MA (2017) Simplified CRISPR tools for efficient genome editing and streamlined protocols for their delivery into mammalian cells and mouse zygotes. Methods 121-122:16–28. https://doi.org/10.1016/j.ymeth.2017.03.021
10. Bak RO, Dever DP, Porteus MH (2018) CRISPR/Cas9 genome editing in human hematopoietic stem cells. Nat Protoc 13 (2):358–376. https://doi.org/10.1038/nprot.2017.143
11. Timothy Hsiau TM, Waite K, Yang J, Kelso R, Holden K, Stoner R (2018) Inference of CRISPR Edits from Sanger trace data. bioRxiv

Chapter 4

Gene Disruption Using Chemically Modified CRISPR-Cpf1 RNA

Moira A. McMahon and Meghdad Rahdar

Abstract

CRISPR-based gene editing in mammalian cells is a powerful research tool which has demonstrated efficient site-specific gene modifications and is showing promise as a therapeutic for patients with genetic diseases. Multiple different CRISPR systems have been identified, each with its own target DNA recognition sequence, expanding the editable mammalian genome. It is also now appreciated that chemically modified nucleic acids can substitute for unmodified nucleotides in guide RNAs, providing protection from exonuclease degradation and improving gene editing efficiency. CRISPR-Cpf1 unlike CRISPR-Cas9, has a substantially lower propensity for off-target genomic cleavage, making it a preferred gene editing system for many applications. Here we provide a detailed protocol for use of CRISPR-Cpf1 and chemically modified guide RNAs in cell lines, outlining the steps from designing guide RNAs to a target gene of interest, delivery and expression in cells, and analysis of gene editing events.

Key words CRISPR-Cpf1, Guide RNA, Chemically modified nucleic acid, Gene editing, Synthetic CRISPR RNA, Surveyor nuclease assay, Cas12a

1 Introduction

Clustered regularly interspaced short palindromic repeats (CRISPR) and CRISPR-associated proteins (Cas) identified in bacteria, and that function as a bacterial adaptive immunity, have been engineered for mammalian genome editing [1–7]. In mammalian cells, CRISPR systems induce site-specific DNA double-stranded breaks which are typically repaired by nonhomologous end joining (NHEJ) or homologous recombination (HR). Mutagenic NHEJ introduces nucleotide insertions or deletions which can result in gene inactivation due to disruption of protein-encoding reading frames. High fidelity HR results in precise nucleotide changes derived from the sister chromatid or a user defined repair template containing homologous sequence. Cas9 from *Streptococcus pyogenes* (SpCas9), was the first CRISPR-Cas used for genome editing in mammalian cells [5–7]. SpCas9 complexes with a short CRISPR

Tudor A. Fulga et al. (eds.), *CRISPR Guide RNA Design: Methods and Protocols*, Methods in Molecular Biology, vol. 2162, https://doi.org/10.1007/978-1-0716-0687-2_4, © Springer Science+Business Media, LLC, part of Springer Nature 2021

RNA (crRNA) which provides DNA specificity and a second longer structurally important trans-activating crRNA (tracrRNA) [5]. For many mammalian genome editing applications, these two RNAs have been fused and engineered into a single guide RNA (sgRNA) [5, 6]. SpCas9 recognizes target DNA in context of an "NGG" protospacer adjacent motif (PAM) and induces a blunt ended DNA double stranded break [5, 6]. It has been shown to be highly efficient at inducing a double stranded break at the intended genomic locus but also suffers from off-target gene editing [8–12]. Another CRISPR-Cas, CRISPR from Prevotella and Francisella 1 (Cpf1 or Cas12a), namely, Acidaminococcus sp. (AsCpf1), has also been engineered for mammalian genome editing [3]. Unlike SpCas9, AsCpf1 utilizes a single crRNA (Fig. 1a) that recognizes target sequence adjacent to a 5′ TTTV PAM and produces a staggered end DNA double stranded break. AsCpf1 also has an advantage over SpCas9 in that little to no off-target gene editing has been reported [13, 14]. But Cpf1 has been shown to have an indiscriminate single-stranded DNase activity following target engagement [15]. This newly described activity is being adapted as a molecular diagnostic, while the extent and consequences of this activity in mammalian cells still require further investigation.

Several groups have explored the use of chemical modifications of crRNA for increasing genome editing efficiency and for developing a potential CRISPR therapeutic [16–21]. Chemical modifications increase metabolic stability and affinity with complementary nucleic acid targets [22–24]. For example, phosphorothioate backbone substitution (PS) (Fig. 1b) protects RNA from nuclease degradation and increases binding affinity to serum proteins aiding in broad tissue distribution and cellular uptake in the absence of formulations [24]. Chemical modifications to the ribose sugar including 2′-O-methyl (2′-O-Me), 2′-fluoro (2′-F), and S-constrained ethyl (cEt) (Fig. 1b), also protect RNA from endo- and exonuclease-mediated degradation in cells while also increasing affinity with complementary nucleic acid targets [25, 26]. Our group, as well as others have shown that the crRNA from the SpCas9 system is amenable to substitution with chemically modified nucleotides and can be codelivered to cells in a dish or in vivo with an SpCas9 expression plasmid, adeno-associated virus encoded SpCas9, SpCas9 mRNA, SpCas9 protein, or packaged in lipid nanoparticles [16–18, 20, 21]. These synthetic crRNAs (scrRNAs) mediate similar or enhanced gene editing activity compared to unmodified RNA and in many cases have reduced off-target activity [17, 21]. We and others have extensively tested nucleotide positions in the crRNA of CRISPR-AsCpf1 and demonstrated that it too can tolerate substitution with chemically modified nucleotides although not to the same extent as SpCas9 crRNA [18, 19]. AsCpf1 scrRNAs in many cases enhance genome editing compared to the natural system.

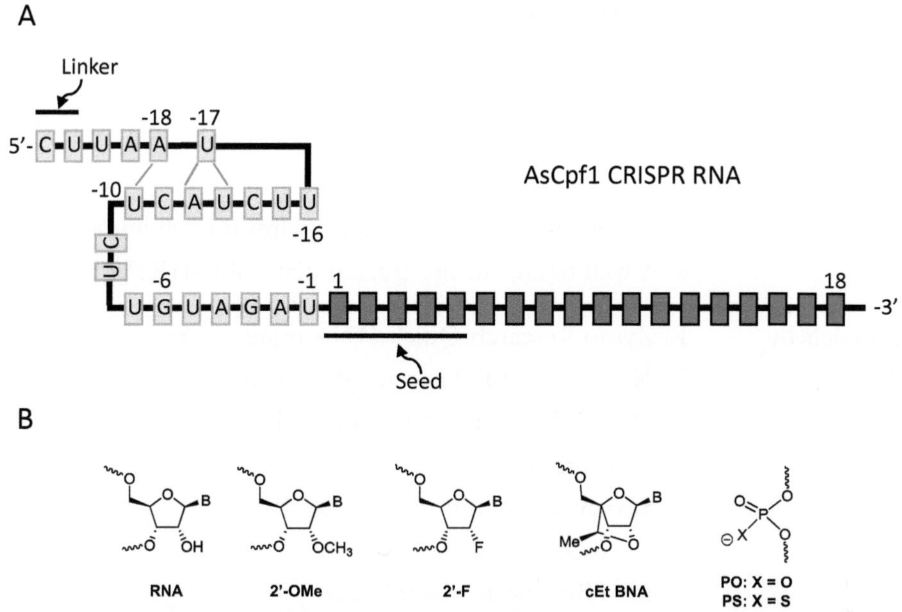

Fig. 1 Illustration of CRISPR-AsCpf1 RNA. (**a**) Structure of AsCpf1 40-mer crRNA where the constant 5′ handle nucleotide sequence is indicated and the variable DNA sequence recognition region is represented by dark gray boxes. The seed sequence (underlined) is required for DNA recognition. (**b**) Chemically modified nucleotide structures that can be substituted at various positions in crRNA. Phosphorothioate (PS), 2′-O-Methyl (2′-O-Me), 2′-Fluoro (2′-F) or S-constrained ethyl (cEt)

Here we provide a step by step procedure for use of CRISPR-Cpf1 and chemically modified crRNAs in cell lines, outlining the steps from initially designing crRNAs to a target gene of interest, through analysis of gene editing events in mammalian cells.

2 Materials

2.1 Molecular Biology and scrRNA Synthesis

1. Plasmid: Human codon-optimized AsCpf1 expression plasmid, pcDNA3.1(−) or available through Addgene (plasmid 69982).
2. AsCpf1 scrRNAs, PAGE-purified.
3. NEB Stable Competent *E. coli*.
4. LB broth base.
5. Ampicillin stock solution (100 μg/ml).
6. Zymo Research ZymoPURE-Express midiprep kit.

2.2 Tissue Culture and Transfection

1. HEK293T cells.
2. Complete DMEM medium, high-glucose DMEM supplemented with 10% fetal bovine serum and 1% antibiotic/antimycotic, HEPES, and sodium pyruvate.

3. TrypLE Express.
4. DPBS (-$CaCl_2$, -$MgCl_2$).
5. Lipofectamine 3000.
6. RNAiMAX.
7. Opti-MEM.
8. 10 cm tissue culture treated plates for cell line maintenance.
9. 6 well tissue culture treated plates for scrRNA transfection.

2.3 scrRNA Activity Confirmation

1. Zymo Research Quick-DNA miniprep kit.
2. NanoDrop 2000 spectrophotometer.
3. Q5 High-Fidelity 2× Master Mix.
4. PCR primers.
5. Surveyor Mutation Detection Kit.
6. Gel loading dye, 6× purple.
7. 5% TBE polyacrylamide gel.
8. SYBR Gold Nucleic Acid Gel Stain (10,000× concentrate in DMSO).
9. Chemidoc XRS imaging system.

2.4 Isolation and Positive Confirmation of Edited Cells

1. Cell sorting buffer: 25 mM HEPES, 1% FBS (or BSA), 2 mM EDTA in PBS.
2. 5 ml polystyrene round bottom tube with cell-strainer cap 12 × 75 mm style (Falcon).
3. 96-well tissue culture plates.
4. 96-well quick-DNA isolation kit.
5. Zero Blunt TOPO PCR cloning kit.
6. Solution basin.
7. ZR-96 DNA clean and concentrator kit.

3 Methods

3.1 crRNA Design and Synthesis to Target Gene of Interest

1. Several online tools are available for designing AsCpf1 crRNAs and for identifying potential off-target sites. We suggest using the lab page from Dr. Feng Zhang (https://zlab.bio/guide-design-resources) to gain access to available guide design resources including Benchling and CRISPOR. For all the design pages, be sure to select the appropriate design (AsCpf1 and 5′TTTV PAM). Input sequence can either be by gene name or locus or a 100–200 base pair user-defined segment of the target gene in proximity to area of desired editing. Rank crRNAs according to their predicted on/off-target effects and

choose 2–3 crRNAs with the highest on-target and lowest off-targets to test activity.

2. scrRNAs can be synthesized on a solid-phase DNA/RNA synthesizer. Alternatively, custom or commercially available chemically modified crRNAs can be ordered through Integrated DNA Technologies or TriLink BioTechnologies.
3. Resuspend scrRNAs in nuclease-free water to a final concentration of 100 μM (up to 300 μM).
4. Store aliquots of scrRNAs at −80 °C until ready to use to protect from degradation with repeated freeze–thaw cycles.

3.2 Delivery of AsCpf1 and scrRNAs to Cell Lines

1. Culture HEK293 cells in a 10 cm tissue culture dish in complete DMEM medium at 37 °C and 5% CO_2 according to the manufacturer's recommendation.
2. 24 h before transfection, aspirate culture medium and wash cells with 5 ml of DPBS by gently adding to the side of the culture dish. Remove DPBS and add 2 ml of TrypLE to the side of the dish and incubate for 2–3 min at 37 °C, until the cells detach from the plate. Add 8 ml of prewarmed complete DMEM to the plate and pipet the cells up and down to dissociate and create a single cell suspension.
3. Transfer the appropriate volume of cells to a 50 ml conical tube and bring up the total volume to 12 ml with fresh media and plate 2 ml/well of a 6-well plate. Incubate overnight (*see* **Note 1**).
4. 20–24 h after seeding, transfect AsCpfl expression plasmid into cells of interest using Lipofectamine 3000. Prepare the DNA reagent mix in a 1.7 ml tube centrifuge tube and vortex briefly to mix according to Table 1.
5. Prepare the Lipofectamine 3000 mix according to Table 2 and vortex briefly to mix.
6. Add the DNA reagent mix to the Lipofectamine 3000 mix and vortex briefly to mix.
7. Incubate at room temperature for 5 min then add 250 μl of the combined mix dropwise to each well of the 6-well plate and incubate overnight.
8. 6 h after DNA transfection, remove culture media and replace with 2 ml prewarmed complete DMEM.
9. 24 h after DNA transfection, prepare scrRNAs for transfection using RNAiMAX according to Table 3.
10. Add the RNA reagent mix to the RNAiMAX mix and vortex briefly to mix.
11. Incubate at room temperature for 5 min then add 300 μl of the combined mix dropwise to each well of the 6-well plate.

Table 1
Components of DNA mixture for transfection

Component	Amount
Opti-MEM medium	125 μl
DNA	2 μg
P3000 reagent	4 μl

Table 2
Components of Lipofectamine 3000 mixture for transfection

Component	Amount
Opti-MEM medium	125 μl
Lipofectamine 3000	3.75 μl

Table 3
RNAiMax components for transfection

Component	Amount
Opti-MEM medium	150 μl
scrRNA	3 μl
Opti-MEM medium	150 μl
RNAiMax	9 μl

12. [Optional] 6 h after RNA transfection, replace culture medium with 2 ml prewarmed complete DMEM and incubate for an additional 24–48 h (Fig. 2a) (*see* **Note 2**).

3.3 Validation of CRISPR Activity Using Surveyor Mutation Detection Kit

1. 24–48 h after RNA transfection, process entire well for gDNA using the Zymo Research quick gDNA isolation kit following the manufacturer's protocol (if this is the final step) or maintain 10% of cells in culture for Subheadings 3.4 and 3.5 and process remaining 90% for gDNA.
2. Measure gDNA concentration using a NanoDrop 2000 spectrophotometer and store at −20 °C until ready for use (and up to 6 months).
3. Design PCR primers that sit asymmetrically around the AsCpf1 binding site and that amplify a product between 500 and 1000 base pairs using the 2× Q5 master mix with the reaction conditions and thermocycler program according to Table 4. Use the NEB Tm calculator for Q5 enzyme (http://

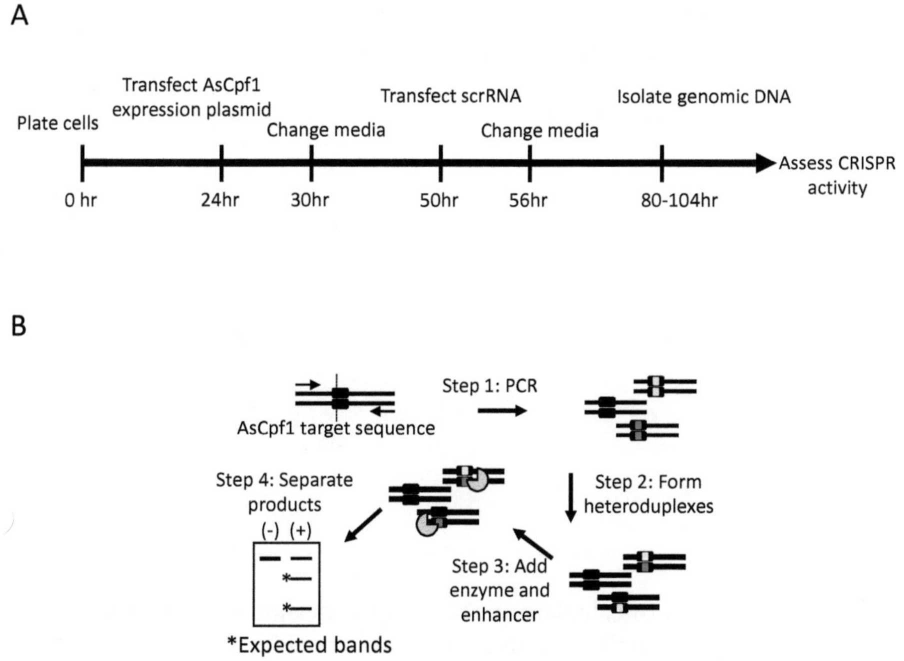

Fig. 2 Experimental timeline for validating gene disruption using AsCpf1 system. (**a**) Schematic outlines the experimental timeline from plating cells at time 0 h through assessment of CRISPR activity at 80–104 h following start of experiment. (**b**) Schematic outlines steps for assessing CRISPR activity where genomic DNA from transfected cells is purified, amplified by PCR, slowly melted and reannealed to form heteroduplexes and then incubated with an enzyme that recognizes DNA mismatches before separating products by gel electrophoresis to visualize products

Table 4
PCR reaction components and cycling conditions

Component	Amount			
2× Q5 High-Fidelity Master Mix	25 μl			
10 μM Forward Primer	2.5 μl			
10 μM Reverse Primer	2.5 μl			
gDNA	50–200 ng			
Water	Up to 50 μl			
Cycle number	Denature	Anneal	Extend	Final
1	98 °C, 30 s			
2–33	98 °C, 30 s	Variable, 20 s	72 °C, 20 s	
34			72 °C, 2 min	
35				10 °C, infinite

Table 5
Surveyor assay annealing conditions

Cycle number	Temperature	Time
1	95	10 min
2	85	1 min
3	75	1 min
4	65	1 min
5	55	1 min
6	45	1 min
7	35	1 min
8	25	1 min
9	10	Infinite

tmcalculator.neb.com) for designing compatible primers with similar melting temperatures (Fig. 2b) (*see* **Note 3**).

4. Remove 12 μl of PCR reaction into new tube (should be at least 200–400 ng DNA).
5. Melt and reanneal product using a thermocycler and the conditions in Table 5.
6. Add 1 μl of enhancer and 1 μl nuclease per reaction (*see* **Note 4**).
7. Incubate at 42 °C for 60 min.
8. Add 1.6 μl of stop solution from mutation detection kit (1/10th volume).
9. Add 3.2 μl of 6× DNA loading dye.
10. Run 15 μl on a 5% TBE polyacrylamide gel at 125 V for 40 min.
11. Incubate gel in TBE containing 1× SYBR gold on shaker for 20 min.
12. Quantify bands in each lane from gel using Image Lab software or equivalent using the following indel formula: indel (%) = $100 \times [1 - (1 - \text{fraction\_cut})^{0.5}]$. Fraction_cut = (sum of the intensity of the cleaved product)/(sum of the intensity of cleaved product plus undigested product).

3.4 Isolation of Single-Cell Clones by FACS

1. 24–48 h following confirmation of gene editing from Subheading 3.3, detach cells from plate as in Subheading 3.2, and transfer to a 50 ml falcon tube.
2. Spin cells at 500 × *g* for 5 min at 4 °C.

3. Remove the supernatant and resuspend cells in FACS buffer to a concentration of ~0.5–1 × 10^6 cells/ml.
4. Pass cells through a 5 ml polystyrene round bottom tube with cell-strainer cap and store on ice until ready to sort.
5. Prepare four to six 96-well plates by adding 100 μl prewarmed DMEM to each well using a multichannel pipette.
6. Sort single cells directly into prepared 96-well plates at 1 cell per well and allow to grow for 2–3 weeks.
7. 72 h after sorting, add an additional 50 μl of prewarmed DMEM to each well.
8. Monitor cells every 2–3 days and replace half the medium with fresh medium once a week.
9. Mark wells with colony growth and combine colonies across the four to six 96-well plates into individual wells of a single 96-well plate 2–3 weeks after sorting by removing media, adding 50 μl of TrypLE, incubating for 2–3 min at 37 °C and resuspending with 100 μl of DMEM and transferring 150 μl to a new well of a 96-well plate.
10. When majority of wells are confluent, ~2–4 days, passage the cells again into two replicate plates, one plate receiving 90% of the cells for PCR screening and the second plate 10% of the cells to culture gene edited cells from. Incubate overnight.

3.5 Identification of Gene Edited Clones

1. Remove PCR screening plate from incubator, remove lid, flip plate, and tap onto paper towel or gauze to remove media.
2. Add 100 μl of DPBS per well using a 12-channel pipette to wash.
3. Flip plate and tap out excess buffer.
4. Follow 96-well quick gDNA isolation kit protocol to extract gDNA from cells.
5. PCR: Set up PCR mastermix according to the following table and add 40 μl directly to each well of a 96-well PCR plate on ice (*see* **Note 5**).
6. Add 10 μl of gDNA and perform the PCR reaction according to Table 4.
7. Clean and concentrate the PCR products according to the manufacturer's protocol (96-well DNA clean and concentrate kit).
8. Directly sequence the PCR product using same primers used for amplification.
9. Analyze the sequencing res μlts by visualization of associated chromatograms and aligning in DNA analysis software such as

Snapgene or online tools such as TIDE (https://tide.deskgen.com).

10. For further confirmation of complete or partial gene knockout, 2 μl of remaining PCR product can be TOPO cloned (according to manufacturer's protocol) and >10 individual colonies sequenced per gene edited clone to validate sequence changes.
11. Final confirmation should be performed by Western blot analysis using the parental cell line as a control for gene expression.

4 Notes

1. You will want to seed cells so that they are 50–70% confluent on day of transfection. For HEK 293 cells this is roughly 200,000–300,000 cells per well. 2 ml of cells is optimal if starting with cells in a 10 cm dish that are approximately 90% confluent.
2. Viability in some cell types may be improved with a change of media especially when incubating for an additional 48 h.
3. Optimize primers and DNA quantity using HEK 293 cell genomic DNA (gDNA) and use increasing amounts of DNA from 50 ng to 200 ng. It is critical to have a clean PCR product with a single band to avoid extra purification steps. To confirm quality of the amplicon, run 10% of the PCR product on a 1% wt/vol agarose gel and visualize on a gel imaging system.
4. Enhancer and nuclease can be premixed together and add 2 μl of mixture added per reaction. Be sure to keep all reagents on ice.
5. Total PCR reaction volume can be adjusted to as little as 12.5 μl. Test smaller reaction volumes with HEK293 gDNA to confirm positive PCR before screening with smaller volumes. The same primers used for the mutation detection assay can be used here. Adjust mastermix volume according to the total number of PCR reactions required to screen clones.

References

1. Wiedenheft B, Sternberg SH, Doudna JA (2012) RNA-guided genetic silencing systems in bacteria and archaea. Nature 482 (7385):331–338. https://doi.org/10.1038/nature10886
2. Bhaya D, Davison M, Barrangou R (2011) CRISPR-Cas systems in bacteria and archaea: versatile small RNAs for adaptive defense and regulation. Annu Rev Genet 45:273–297. https://doi.org/10.1146/annurev-genet-110410-132430
3. Zetsche B, Gootenberg JS, Abudayyeh OO, Slaymaker IM, Makarova KS, Essletzbichler P, Volz SE, Joung J, van der Oost J, Regev A, Koonin EV, Zhang F (2015) Cpf1 is a single RNA-guided endonuclease of a class 2 CRISPR-Cas system. Cell 163(3):759–771. https://doi.org/10.1016/j.cell.2015.09.038
4. Shmakov S, Smargon A, Scott D, Cox D, Pyzocha N, Yan W, Abudayyeh OO, Gootenberg JS, Makarova KS, Wolf YI, Severinov K, Zhang F, Koonin EV (2017) Diversity and

evolution of class 2 CRISPR-Cas systems. Nat Rev Microbiol 15(3):169–182. https://doi.org/10.1038/nrmicro.2016.184

5. Jinek M, Chylinski K, Fonfara I, Hauer M, Doudna JA, Charpentier E (2012) A programmable dual-RNA-guided DNA endonuclease in adaptive bacterial immunity. Science 337 (6096):816–821. https://doi.org/10.1126/science.1225829
6. Cong L, Ran FA, Cox D, Lin S, Barretto R, Habib N, Hsu PD, Wu X, Jiang W, Marraffini LA, Zhang F (2013) Multiplex genome engineering using CRISPR/Cas systems. Science 339(6121):819–823. https://doi.org/10.1126/science.1231143
7. Mali P, Yang L, Esvelt KM, Aach J, Guell M, DiCarlo JE, Norville JE, Church GM (2013) RNA-guided human genome engineering via Cas9. Science 339(6121):823–826. https://doi.org/10.1126/science.1232033
8. Fu Y, Foden JA, Khayter C, Maeder ML, Reyon D, Joung JK, Sander JD (2013) High-frequency off-target mutagenesis induced by CRISPR-Cas nucleases in human cells. Nat Biotechnol 31(9):822–826. https://doi.org/10.1038/nbt.2623
9. Hsu PD, Scott DA, Weinstein JA, Ran FA, Konermann S, Agarwala V, Li Y, Fine EJ, Wu X, Shalem O, Cradick TJ, Marraffini LA, Bao G, Zhang F (2013) DNA targeting specificity of RNA-guided Cas9 nucleases. Nat Biotechnol 31(9):827–832. https://doi.org/10.1038/nbt.2647
10. Kuscu C, Arslan S, Singh R, Thorpe J, Adli M (2014) Genome-wide analysis reveals characteristics of off-target sites bound by the Cas9 endonuclease. Nat Biotechnol 32(7):677–683. https://doi.org/10.1038/nbt.2916
11. Pattanayak V, Lin S, Guilinger JP, Ma E, Doudna JA, Liu DR (2013) High-throughput profiling of off-target DNA cleavage reveals RNA-programmed Cas9 nuclease specificity. Nat Biotechnol 31(9):839–843. https://doi.org/10.1038/nbt.2673
12. Tsai SQ, Zheng Z, Nguyen NT, Liebers M, Topkar VV, Thapar V, Wyvekens N, Khayter C, Iafrate AJ, Le LP, Aryee MJ, Joung JK (2015) GUIDE-seq enables genome-wide profiling of off-target cleavage by CRISPR-Cas nucleases. Nat Biotechnol 33 (2):187–197. https://doi.org/10.1038/nbt.3117
13. Kim D, Kim J, Hur JK, Been KW, Yoon SH, Kim JS (2016) Genome-wide analysis reveals specificities of Cpf1 endonucleases in human cells. Nat Biotechnol 34(8):863–868. https://doi.org/10.1038/nbt.3609
14. Kleinstiver BP, Tsai SQ, Prew MS, Nguyen NT, Welch MM, Lopez JM, McCaw ZR, Aryee MJ, Joung JK (2016) Genome-wide specificities of CRISPR-Cas Cpf1 nucleases in human cells. Nat Biotechnol 34(8):869–874. https://doi.org/10.1038/nbt.3620
15. Chen JS, Ma E, Harrington LB, Da Costa M, Tian X, Palefsky JM, Doudna JA (2018) CRISPR-Cas12a target binding unleashes indiscriminate single-stranded DNase activity. Science 360(6387):436–439. https://doi.org/10.1126/science.aar6245
16. Finn JD, Smith AR, Patel MC, Shaw L, Youniss MR, van Heteren J, Dirstine T, Ciullo C, Lescarbeau R, Seitzer J, Shah RR, Shah A, Ling D, Growe J, Pink M, Rohde E, Wood KM, Salomon WE, Harrington WF, Dombrowski C, Strapps WR, Chang Y, Morrissey DV (2018) A single administration of CRISPR/Cas9 lipid nanoparticles achieves robust and persistent in vivo genome editing. Cell Rep 22(9):2227–2235. https://doi.org/10.1016/j.celrep.2018.02.014
17. Hendel A, Bak RO, Clark JT, Kennedy AB, Ryan DE, Roy S, Steinfeld I, Lunstad BD, Kaiser RJ, Wilkens AB, Bacchetta R, Tsalenko A, Dellinger D, Bruhn L, Porteus MH (2015) Chemically modified guide RNAs enhance CRISPR-Cas genome editing in human primary cells. Nat Biotechnol 33 (9):985–989. https://doi.org/10.1038/nbt.3290
18. Li B, Zhao W, Luo X, Zhang X, Li C, Zeng C, Dong Y (2017) Engineering CRISPR-Cpf1 crRNAs and mRNAs to maximize genome editing efficiency. Nat Biomed Eng 1(5). https://doi.org/10.1038/s41551-017-0066
19. McMahon MA, Prakash TP, Cleveland DW, Bennett CF, Rahdar M (2018) Chemically modified Cpf1-CRISPR RNAs mediate efficient genome editing in mammalian cells. Mol Ther 26(5):1228–1240. https://doi.org/10.1016/j.ymthe.2018.02.031
20. Mir A, Alterman JF, Hassler MR, Debacker AJ, Hudgens E, Echeverria D, Brodsky MH, Khvorova A, Watts JK, Sontheimer EJ (2018) Heavily and fully modified RNAs guide efficient SpyCas9-mediated genome editing. Nat Commun 9(1):2641. https://doi.org/10.1038/s41467-018-05073-z
21. Rahdar M, McMahon MA, Prakash TP, Swayze EE, Bennett CF, Cleveland DW (2015) Synthetic CRISPR RNA-Cas9-guided genome editing in human cells. Proc Natl Acad Sci U S A 112(51):E7110–E7117. https://doi.org/10.1073/pnas.1520883112
22. Braasch DA, Jensen S, Liu Y, Kaur K, Arar K, White MA, Corey DR (2003) RNA

interference in mammalian cells by chemically-modified RNA. Biochemistry 42 (26):7967–7975. https://doi.org/10.1021/bi0343774

23. Chiu YL, Rana TM (2003) siRNA function in RNAi: a chemical modification analysis. RNA 9 (9):1034–1048

24. Eckstein F (2014) Phosphorothioates, essential components of therapeutic oligonucleotides. Nucleic Acid Ther 24(6):374–387. https://doi.org/10.1089/nat.2014.0506

25. Bennett CF, Swayze EE (2010) RNA targeting therapeutics: molecular mechanisms of antisense oligonucleotides as a therapeutic platform. Annu Rev Pharmacol Toxicol 50:259–293. https://doi.org/10.1146/annurev.pharmtox.010909.105654

26. Shen X, Corey DR (2018) Chemistry, mechanism and clinical status of antisense oligonucleotides and duplex RNAs. Nucleic Acids Res 46(4):1584–1600. https://doi.org/10.1093/nar/gkx1239

Chapter 5

"Split-and-Click" sgRNA

Lapatrada Taemaitree, Arun Shivalingam, Afaf H. El-Sagheer, and Tom Brown

Abstract

CRISPR-Cas9 gene editing is dependent on a programmable single guide RNA (sgRNA) that directs Cas9 endonuclease activity. This RNA is often generated by enzymatic reactions, however the process becomes time-consuming as the number of sgRNAs increases and does not allow the incorporation of chemical modifications that can improve or expand the functionality of CRISPR. Solid-phase RNA synthesis can overcome these issues, but highly pure full-length sgRNA remains at the limits of current synthetic methods. Here, we demonstrate a "split-and-click" approach that separates the sgRNA into its two smaller components – a DNA-targeting ~20-mer RNA and a constant Cas9-binding 79-mer RNA – and chemically ligates them together to generate a biologically active sgRNA. The benefits of our approach lie in the stringent purification of the DNA-targeting 20-mer, the reduced synthesis of the constant 79-mer each time a new sgRNA is required, and the rapid access it provides to custom libraries of sgRNAs.

Key words Solid-phase RNA synthesis, CRISPR, sgRNA Libraries, CuAAC

1 Introduction

The Cas9–sgRNA ribonucleoprotein complex is directed to its target DNA by simple Watson–Crick base pairing [1]. Subsequent DNA cleavage triggers imperfect DNA repair that results in efficient site-specific genome editing [2, 3]. Central to this process is the sgRNA. Methods to generate the sgRNA often involve in vitro or in vivo enzymatic transcription, but these approaches have drawbacks. Firstly, in vitro transcription can give rise to errors at the 3′ end of RNA, and more importantly at the 5′ end [4] where DNA-targeting accuracy is determined. Secondly, both in vitro and in vivo transcription initiation require specific 5′ nucleotides (e.g., 5′-A, 5′-G, 5′-A/G, 5′-GG, 5′-GA and 5′-G for U3 [5], U6 [3], H1 [6], T7 [7, 8], SP6 [9], and T3 [10] promoters, respectively) that must be appended to the sgRNA and could influence target specificity [1, 11–14]. Finally, the entire process can be time-consuming and labor-intensive when a large number of different

Tudor A. Fulga et al. (eds.), *CRISPR Guide RNA Design: Methods and Protocols*, Methods in Molecular Biology, vol. 2162,
https://doi.org/10.1007/978-1-0716-0687-2_5,

sgRNAs are required for multiplexed applications (e.g., the ~4500, ~2300, or ~450 arrayed sgRNAs that have been used in high-content screens (HCS) examining viral infection [15], profiling single cell phenotypes [16] and studying epigenetic regulation [17] respectively).

Automated solid-phase RNA synthesis provides an alternative route to individual or pools of sgRNAs, and more importantly allows the incorporation of chemical modifications that increase sgRNA stability in cells [18–22], reduce off-target effects [23–26] and expand the functionality of CRISPR beyond genome editing [27]. Solid-phase oligonucleotide synthesis involves stepwise 3′ to 5′ coupling of nucleotides. However, each coupling reaction is imperfect and becomes progressively worse toward the 5′ end as the growing oligonucleotide chain blocks reagents reaching the reaction site. Under ideal conditions, the maximum yield of a 100-mer DNA is 36.6% assuming a constant coupling efficiency of 99% [28] (E^N where $E =$ coupling efficiency and $N =$ number of nucleotides in a sequence). For RNA, yields and purity are even lower as each RNA monomer has a 2′-OH protecting group that increases steric hindrance, reduces coupling efficiency, and introduces additional impurities via side reactions. Thus, the direct synthesis of full-length ~100-mer sgRNA is inefficient and pushes current synthetic approaches to their limit.

Here, we demonstrate a "split-and-click" approach [29] (Fig. 1a) that addresses the issues of solid-phase synthesis and provides a more flexible route to chemically modified sgRNAs. The sgRNA is split into a variable DNA-targeting ~20-mer (crRNA; CRISPR RNA) and an invariant Cas9-binding chemically modified 79-mer (tracrRNA; transactivating CRISPR RNA), synthesized with 3′-propargyl and 5′-azide modifications respectively. The two RNAs are subsequently joined to form a sgRNA using copper(I)-catalyzed azide alkyne cycloaddition (CuAAC) "click" ligation, a reaction that is well-developed in our group [30, 31]. The clicked sgRNA contains an artificial triazole backbone linkage between bases G1 and U2 (Fig. 1b) but remains functional in vitro and in cells [29]. With this strategy, the variable ~20-mer can be synthesized on-demand and in high purity while the constant 79-mer can be produced once on a large-scale for repeated use. This provides a simple and cost-effective route to a library of (or individual) sgRNAs.

2 Materials

2.1 Solid-Phase RNA Synthesis

1. Applied Biosystems 394 automated DNA/RNA synthesizer.
2. Empty twist-cap synthesis column.

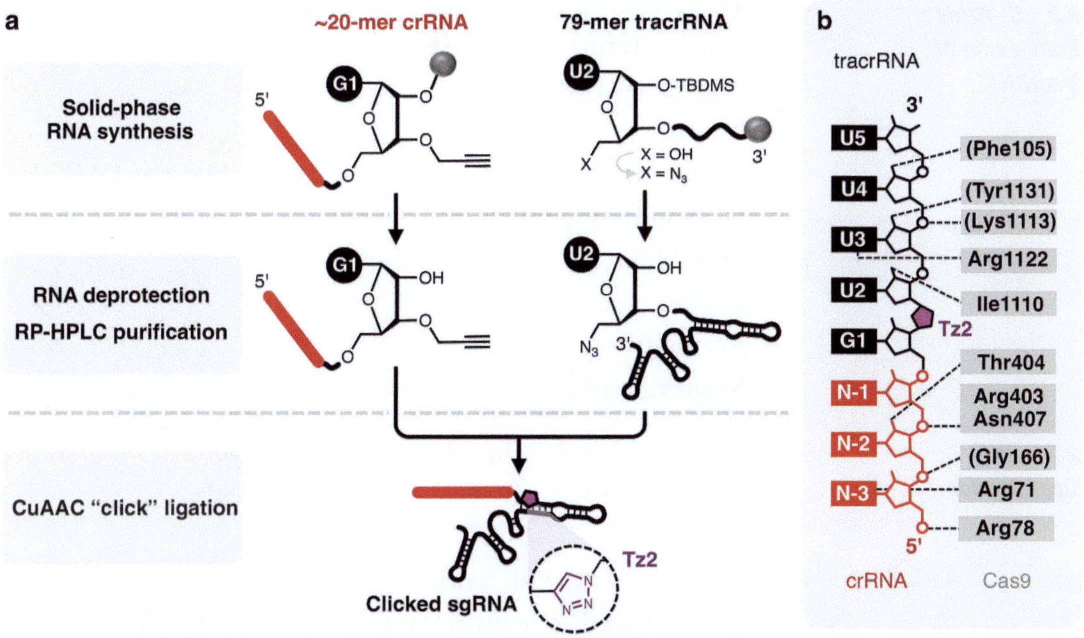

Fig. 1 Overview of the "split-and-click" strategy. (**a**) illustrates the steps involved in synthesis of the RNA precursors and CuAAC ligation to generate a sgRNA construct containing a biocompatible triazole linkage (Tz2). (**b**) demonstrates the intermolecular interactions between Cas9 and the sgRNA based on PDB: 4008 [32], and the relative position of the triazole linkage with respect to the sgRNA

3. 3′-*O*-propargyl Guanosine (*N*-ibu where ibu = isobutyryl) 2′-lcaa CPG 1000 Å.
4. U RNA SynBase™ CPG 1000/110.
5. 2′-*O*-TBDMS RNA phosphoramidites (A-tac, C-tac, G-tac, and U, where tac = *tert*-butylphenoxyacetyl).
6. 2′-OMe RNA phosphoramidites (A-pac, C-ac, G-iPr-pac, and U, where pac = phenoxyacetyl, ac = acetyl and iPr-pac = isopropylphenoxyacetyl).
7. 5′-TFA-amino-modifier C6-CE phosphoramidite.
8. Anhydrous acetonitrile.
9. Activator: 5-benzylthio-1*H*-tetrazole, 0.3 M in anhydrous acetonitrile.
10. Cap A: 5% *tert*-butylphenoxy acetic anhydride in tetrahydrofuran.
11. Cap B: 16% *N*-methylimidazole in tetrahydrofuran.
12. Oxidizer: Iodine, 0.1 M in tetrahydrofuran/pyridine/water, 77:21:2.
13. Trichloroacetic acid (TCA), 3% in dichloromethane.

2.2 5′-Azide Conversion of tracrRNA

1. Methyltriphenoxyphosphonium iodide, 0.5 M in anhydrous dimethylformamide (DMF).
2. Anhydrous DMF.
3. Argon.
4. Sodium azide solution, 100 mg in 2 mL anhydrous DMF.
5. Glass screw-top vial, 5 mL.
6. Heat block.
7. Oven.
8. Acetonitrile.
9. Syringe, 1 mL.

2.3 RNA Deprotection

1. Diethylamine, 20% in anhydrous acetonitrile.
2. Syringe, 1 mL.
3. Acetonitrile.
4. Glass screw-top vial, 5 mL.
5. Concentrated aqueous ammonia/ethanol, 3:1, v/v.
6. Round-bottom flask, 25 mL.
7. Ethanol/water, 3:1, v/v.
8. Rotary evaporator.
9. Falcon tube, 15 mL.
10. Syringe filter, Regenerated Cellulose, 0.45 μm.
11. Anhydrous dimethyl sulfoxide (DMSO).
12. Triethylamine trihydrofluoride.
13. Sodium acetate solution, 3 M pH 5.2.
14. Butanol.
15. Centrifuge.
16. Freeze dryer.
17. Nuclease-free water.
18. NAP™-10 column.

2.4 RNA Purification by Reverse-Phase (RP)-HPLC

1. Gilson HPLC system.
2. ACE® C8 column, 10 mm × 250 mm, 100 Å, 10 μm.
3. Buffer A: triethylammonium bicarbonate (TEAB), 0.1 M pH 7.5.
4. Buffer B: TEAB, 0.1 M pH 7.5 with acetonitrile, 50%, v/v.
5. Collection tube.
6. Freeze dryer.
7. Nuclease-free water.
8. UV–Vis spectrometer.

2.5 RNA Characterization by Mass Spectrometry

1. UPLC-MS Waters XEVO G2-QTOF mass spectrometer with an ACQUITY UPLC system.
2. ACQUITY UPLC BEH C18 Column, 2.1 mm × 50 mm, 130 Å, 1.7 μm.
3. Buffer A: 8.6 mM triethylamine (TEA), 200 mM hexafluoroisopropanol (HFIP) in 5% methanol/water, v/v.
4. Buffer B: 20% buffer A in methanol, v/v.
5. Mass spectrometry vial.
6. Nuclease-free water.
7. Deconvolution software MassLynx v4.1.

2.6 crRNA–tracrRNA Ligation by CuAAC

1. Triethylammonium acetate (TEAA) buffer, 2 M pH 7.
2. $MgCl_2$ solution, 100 mM.
3. DMSO.
4. Ascorbic acid solution, 125 mM (freshly prepared).
5. $CuSO_4$ and Tris(3-hydroxypropyltriazolylmethyl)amine (Cu–THPTA) solution, 250 mM in DMSO/water, 55%, v/v.
6. Nuclease-free water.
7. Argon.
8. Eppendorf tube, 0.5 mL.
9. Freeze dryer.

2.7 RNA Purification by Desalting

1. Amicon® Ultra Centrifugal Filters, 0.5 mL, 10 kDa cutoff.
2. Nuclease-free water.
3. Centrifuge.

2.8 RNA Purification by Denaturing Polyacrylamide Gel Electrophoresis (PAGE)

1. PAGE kit (W × D × H = 18 × 0.2 × 24.4 cm).
2. Acrylamide–bisacrylamide solution, 40%, 29:1.
3. Urea.
4. Ammonium persulfate solution, 10% in water.
5. *N, N, N, N′*-tetramethylethylenediamine (TEMED).
6. TBE buffer (10×): 1 M Tris base, 1 M boric acid, 20 mM EDTA, pH 8.3.
7. Nuclease-free water.
8. Sonicator.
9. Formamide.
10. Loading dye: 50% aqueous formamide, 0.01% xylene cyanol (w/v), 0.01% bromophenol blue (w/v).
11. Plastic transparent plate.
12. Fluorescent TLC plate, 20 × 20 cm.

13. Syngene G:Box Imager.
14. Blade.
15. Eppendorf tube, 0.5 mL.
16. Microcentrifuge tube, 2 mL.
17. Needle, 21 gauge.
18. RNA extraction buffer: 50 mM Tris–HCl pH 7.5, 25 mM NaCl.
19. Thermomixer.
20. NAP™-10 column.
21. Freeze dryer.

3 Methods

3.1 Solid-Phase RNA Synthesis

1. For 1 μmol synthesis, load an empty twist-cap synthesis column with solid support-bound nucleoside 3′-*O*-propargyl Guanosine (*N*-ibu) 2′-lcaa CPG 1000 Å (for crRNA synthesis) or U RNA SynBase™ CPG 1000/110 (for tracrRNA synthesis) (*see* **Note 1**).
2. Install the column on the synthesizer.
3. Prepare the 0.1 M phosphoramidite solutions by dissolving phosphoramidite monomers in anhydrous acetonitrile.
4. Install the phosphoramidite solutions including all the reagents and solvents required for each step (detritylation, coupling, capping and oxidation) on the synthesizer.
5. Perform RNA synthesis using a standard phosphoramidite cycle with the coupling time of 10 min (for all monomers).
6. Carry out detritylation at the end of the synthesis to remove the dimethoxytrityl group from the 5′ end of the RNA. Flush the column with argon for 5 min. The crRNA is subject to deprotection (*see* Subheading 3.3) while the tracrRNA undergoes 5′-OH to azide conversion (*see* Subheading 3.2) prior to deprotection.

3.2 5′-Azide Conversion of tracrRNA

1. For 1 μmol synthesis, we recommend dividing the RNA-bound solid support into four twist-cap synthesis columns (~12 mg) for efficient azide conversion in this section. The solid support-bound RNA can be recombined for deprotection in Subheading 3.3.
2. Dissolve 0.452 g methyltriphenoxyphosphonium iodide in 2 mL anhydrous DMF to make a 0.5 M solution (*see* **Note 2**).
3. Use a 1 mL syringe to take up the above iodide solution (~0.75 mL) and connect it to one end of the twist column.

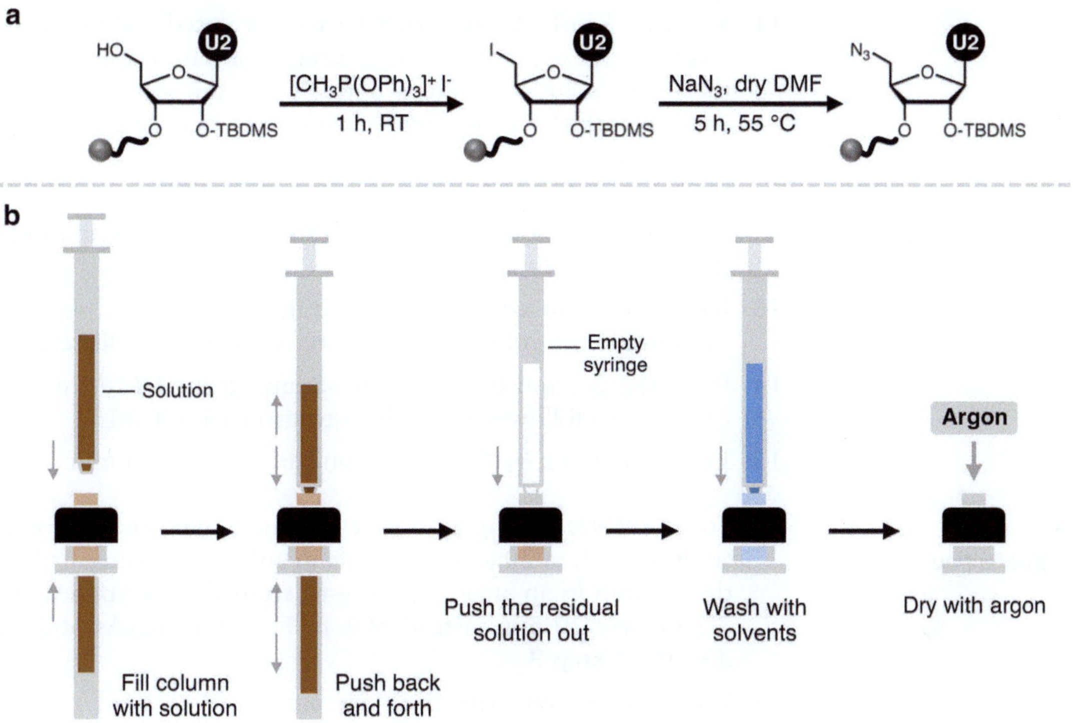

Fig. 2 Conversion of 5′-OH of tracrRNA to 5′-N_3. (**a**) shows the chemical steps involved in converting the 5′-OH of tracrRNA to a 5′-iodo intermediate prior to azide conversion. (**b**) depicts the practical manipulation of the column when treating the solid support with various reagents in (**a**)

4. Push the liquid through the column until it just reaches the end of the column.
5. Use another 1 mL syringe to take up the iodide solution (~0.75 mL) and connect it to the free end of column.
6. Push the solution through the column back and forth every 15 min for 1 h at room temperature. Ensure the solution covers the solid support and avoid bubbles within the column.
7. Discard the iodide solution and flush out residual solution from inside the column with an empty syringe.
8. Wash the column by passing fresh anhydrous DMF through (3 x 1 mL of DMF).
9. Dry the column by flushing argon through it for 5 min. The general scheme for **steps 3–9** is shown in Fig. 2.
10. Prepare a sodium azide solution by suspending 100 mg sodium azide (*see* **Note 3**) in 2 mL anhydrous DMF in a loosely sealed glass vial, heating for 10 min at 70 °C before letting it cool to room temperature. Allow the solid to settle at the bottom of the vial.

11. Use two 1 mL syringes to take the saturated sodium azide solution and apply it to the column in an analogous way to the iodide solution in **steps 3–5**. Ensure the solid support is covered with the sodium azide solution and no air bubbles are in the column.
12. Place the column (with syringes connected) in the oven for 5 h at 55 °C. During this period, push the solution back and forth through the column every hour.
13. Discard sodium azide solution and flush out residual solution from inside the column with an empty syringe (*see* **Note 4**).
14. Wash the column by passing fresh anhydrous DMF through (3 × 1 mL) followed by fresh acetonitrile (3 × 1 mL).
15. Dry the column by flushing argon through it for 5 min.

3.3 RNA Deprotection

1. For the RNA bearing a primary aliphatic amine at the 5′ end, apply 1.5 mL of 20% diethylamine in anhydrous acetonitrile to the column in an analogous way to **steps 3–7** in Subheading 3.2 but for 20 min instead of 1 h. For other RNAs proceed directly to **step 3**.
2. Wash the column with fresh acetonitrile (3 × 1 mL) before flushing the column with argon for 5 min.
3. Disassemble the column and transfer the solid support-bound RNA into a glass vial.
4. Add 3 mL of concentrated aqueous ammonia/ethanol (3:1, v/v) into the vial and quickly close the screw cap to avoid excessive release of the ammonia gas.
5. After incubation for 2 h at 55 °C, let the ammonia solution cool to room temperature or place in a −20 °C freezer until cooled (*see* **Note 5**).
6. To avoid taking up the solid support, filter the solution containing the partially deprotected RNA through a Regenerated Cellulose syringe filter (0.45 μm) into a 25 mL round-bottom flask. To increase RNA recovery, wash the solid support by adding 1 mL of ethanol/water (3:1, v/v), vortexing and filtering the solution into the same round-bottom flask.
7. Remove ammonia, ethanol, and water in vacuo using a rotary evaporator. Stop the process when the liquid turns cloudy but before dryness (*see* **Note 6**).
8. Transfer the liquid into the 15 mL Falcon tube. Add a few drops of ethanol/water (3:1, v/v) to the round-bottom flask to dissolve the residual RNA and transfer it to the same Falcon tube.
9. Freeze-dry the RNA overnight.

10. Add 300 μL anhydrous DMSO to the dried RNA. Vortex it well.
11. Add 300 μL triethylamine trihydrofluoride (*CARE!*, *see* **Note 7**). Vortex it well (*see* **Note 8**).
12. Incubate for 2.5 h at 65 °C before cooling the solution to room temperature.
13. Precipitate the fully deprotected RNA (*see* **Note 9**) by adding 50 μL 3 M sodium acetate solution and 3 mL butanol. Mix well.
14. Leave the tube for 1 h at −80 °C.
15. Centrifuge at 12,000 × *g* for 30 min at 4 °C to pellet the RNA.
16. Discard the triethylamine trihydrofluoride containing supernatant (*see* **Note 10**).
17. Wash the RNA pellet by adding 750 μL ethanol/water (3:1, v/v) and vortex well.
18. Centrifuge at 12,000 × *g* for 30 min at 4 °C and discard the supernatant.
19. Repeat the wash **steps 17** and **18** once more.
20. Freeze-dry the RNA pellet to remove residual ethanol.
21. Dissolve the pellet in 1 mL nuclease-free water.
22. Equilibrate a disposable NAP™-10 column by filling it with nuclease-free water and allowing it to completely empty. Repeat this step three times.
23. Add the RNA solution onto the NAP™-10 column and allow the solution to flow through. The RNA is now inside the column.
24. Elute the desalted RNA from the column into a 2 mL microcentrifuge tube by adding 1.5 mL nuclease-free water. The RNA is now ready for further purification.

3.4 RNA Purification by Reverse-Phase (RP) HPLC

1. On the Gilson HPLC system equipped with an ACE® C8 column, set a gradient of 15–30% buffer B over 17.5 min for the ~20-mer 3′-alkynyl crRNA or a gradient of 20–30% buffer B over 17.5 min for the 79-mer 5′-azido tracrRNA. The flow rate should be set to 4 mL/min. The RNA will be separated from impurities (e.g., truncated RNA and incompletely deprotected RNA) based on the difference in hydrophobicity.
2. Select a UV detection wavelength between 260 and 298 nm (*see* **Note 11** for precise wavelength selection).
3. Equilibrate the column by running injecting water as a "mock" sample using the defined gradient. For highly structured RNA (i.e., 2′-OMe-containing tracrRNA), also equilibrate the HPLC column temperature to 55 °C to improve separation.

4. Inject the RNA sample and start the run.
5. Collect separate fractions for each peak including shoulder peaks.
6. Freeze-dry the HPLC fractions to remove TEAB buffer and acetonitrile, which are both volatile.
7. Dissolve the dry RNA in nuclease-free water and measure the concentration using a UV–Vis spectrometer.
8. Determine the desired pure RNA fraction by UPLC-MS.

3.5 RNA Characterization by Mass Spectrometry

1. For each RNA fraction, prepare a 20 μM solution of the RNA in water (10 μL) and add to a mass spectrometry vial.
2. On the UPLC-MS Waters XEVO G2-QTOF mass spectrometer (ESI negative mode) equipped with an ACQUITY UPLC BEH C18 column, set up a gradient of buffer B from 0–70% over 7.5 min with the column heated to 50 °C. The flow rate is 0.2 mL/min.

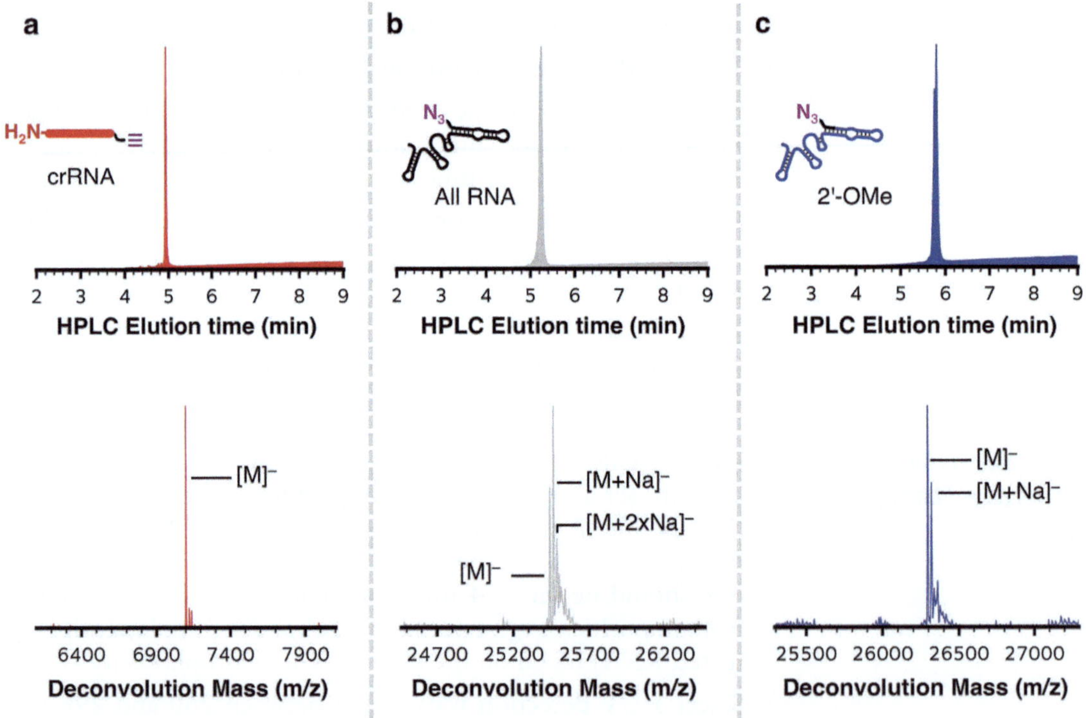

Fig. 3 Representative HPLC chromatograms and mass spectra for HPLC-purified crRNA with 3′-alkyne and 5′-azido tracrRNA. (**a**) 21-mer *EMX1*-targeting crRNA $[M]^-$ expected mass: 7100; found mass: 7102. (**b**) 79-mer tracrRNA (all RNA) $[M]^-$ expected mass: 25436; found mass: 25440. (**c**) 79-mer tracrRNA (2′-OMe) $[M]^-$ expected mass: 26295; found mass: 26295. Note that sodium ion adducts are also detected for tracrRNAs, which may appear as a broad or bifurcated peak on HPLC due to RNA secondary structure

3. Equilibrate the column by injecting water as a mock sample using the defined gradient.
4. Inject 2 μL of sample and start the run.
5. The RNA should elute at 4.8–4.9 min for ~20-mer crRNA or 5.2–5.3 min for 79-mer tracrRNA. The raw data is then deconvoluted using the MassLynx v4.1 software (Fig. 3).

3.6 crRNA–tracrRNA Ligation by CuAAC

1. Freeze-dry 750 pmol 3′-alkynyl crRNA and 750 pmol 5′-azide tracrRNA in separate tubes. Add 1 μL of nuclease-free water to dissolve each RNA. Flick the tubes and centrifuge to ensure that the RNA is dissolved.
2. In a 0.5 mL Eppendorf tube, mix the two RNAs (the ratio of crRNA/tracrRNA is 1.5:1, *see* **Note 12**) and add 1 μL 2 M TEAA buffer, 0.5 μL 100 mM $MgCl_2$ solution, 5 μL DMSO, and 1 μL 125 mM ascorbic acid solution in sequential order. Mix well.
3. In a separate 0.5 mL Eppendorf tube containing 10 μL of water, adjust argon flow to a level that allows the solution to be displaced *gently* around in the tube. Keep this level of argon constant for **steps 4** and **5**.
4. Flush argon over the mixture, moving the RNA solution in the tube for 2 min. This minimizes oxygen ingress into the solution which can oxide reactive Cu^{1+} to inactive Cu^{2+} in **step 5**.
5. Add 0.5 μL 250 mM Cu–THPTA solution under argon. Mix by moving the RNA solution in the tube using argon. The total reaction volume is 10 μL.
6. Incubate the reaction for 1 h at room temperature. The general scheme for **steps 2–6** is shown in Fig. 4.

3.7 RNA Purification by Desalting

1. Add 490 μL nuclease-free water to the reaction and transfer to the Amicon® Ultra Centrifugal Filter device inserted into a collection tube.
2. Centrifuge with a fixed angle rotor at 14,000 × *g* for 10 min.
3. Discard the flow-through in the collection tube (~35 μL should remain in the filter device).
4. Add 465 μL nuclease-free water to the filter. Mix well by pipetting or inverting the tube.
5. Repeat **steps 2–4** five times.
6. After the final round of centrifugation, recover the desalted RNA in the chamber of the filter device by placing the filter device upside down in a clean collection tube.
7. Centrifuge at 1000 × *g* for 2 min. The desalted sample can now be used as crude clicked sgRNA (*see* **Note 13**) with slightly reduced activity in vitro due to residual 79-mer starting material or can be subjected to further purification by denaturing PAGE.

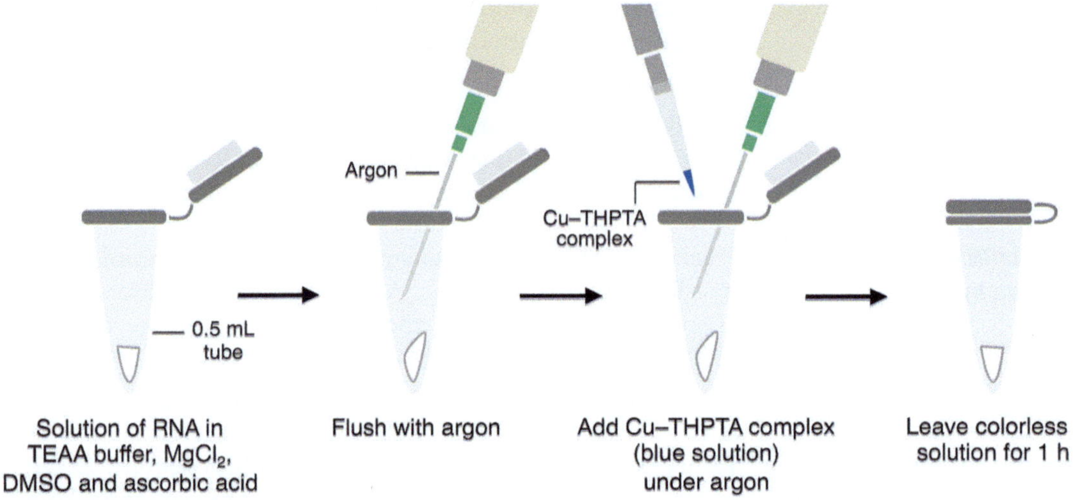

Fig. 4 Illustration of how to perform CuAAC click ligation. The components required for the reaction are added in the order of RNA, H_2O, TEAA buffer, $MgCl_2$, DMSO, and ascorbic acid. The mixture is flushed with argon for 2 min and the Cu–THPTA solution (blue) is added under argon. The Cu–THPTA solution becomes colorless after reacting with ascorbic acid. The reaction is incubated for 1 h at room temperature

3.8 RNA Purification by Denaturing Polyacrylamide Gel Electrophoresis (PAGE)

1. Prepare a 10% acrylamide gel containing 7 M urea gel by adding 17.5 mL 40% acrylamide–bisacrylamide (29:1), 22.5 mL water, and 7 mL TBE (10×) to 29.4 g urea.
2. Sonicate the gel solution until the urea is fully dissolved. Alternatively heat the sample in a water bath to dissolve the urea.
3. Add 560 μL 10% APS. Mix well.
4. Add 56 μL TEMED. Mix quickly (~30 s) and immediately pour between two glass plates.
5. Insert the comb and ensure there are no air bubbles inside the gel or at the wells.
6. Let the gel polymerize for 1 h before removing the comb and filling the upper and lower buffer chambers with 1× TBE.
7. Prerun the gel for 30 min at 20 W to equilibrate the gel and increase its temperature to aid oligonucleotide denaturation.
8. After prerunning, disassemble the electrophoresis apparatus and flush each well with 1× TBE to remove gel debris and urea.
9. Mix each sample with an equal volume of formamide and load into the bottom of the well. Also load 20 μL gel loading dye into the empty well to help track the RNA migration. For a 10% acrylamide denaturing gel, the migration of xylene cyanol and bromophenol blue dyes correspond to 55-nt and 12-nt RNA respectively.

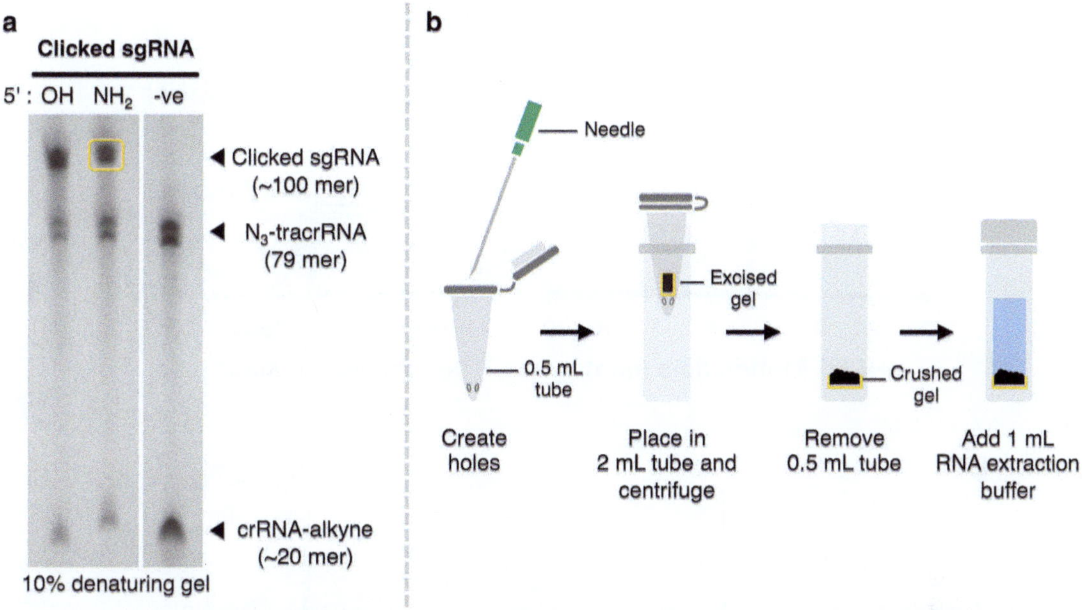

Fig. 5 Purification of clicked sgRNA by denaturing PAGE. (**a**) is an example of a clicked sgRNA purified on a 10% denaturing PAGE gel. The band at the top of the gel is the desired sgRNA (~100 mer, with and without NH_2 at 5′ end) while the other bands correspond to starting materials (~20-mer crRNA and 79-mer tracrRNA). –ve indicates a negative control in which click ligation is not performed. After ligation, minimal 79-mer (limiting component) should remain if CuAAC ligation is efficient. (**b**) shows the gel extraction step that involves excising the clicked sgRNA band (yellow box in (**a**)) by a blade (*CARE!*), crushing of the gel via centrifugation, soaking gel pieces in RNA extraction buffer and incubating overnight to extract the RNA from the gel

10. Reassemble all parts and start the run for 2h at 20 W or until the xylene cyanol (top) dye reaches the middle of the gel.
11. Disassemble the glass plates, carefully transfer the gel to the plastic transparent plate and place the gel on a fluorescent TLC plate (20 × 20 cm).
12. Visualize the RNA under a short-wavelength UV light (240–300 nm) using a Syngene G:Box Imager (Fig. 5a).
13. Excise the clicked sgRNA band using a blade (*CARE!*) and place in a 0.5 mL V-shaped Eppendorf tube with holes created at the bottom of the tube using a 21 gauge needle (*CARE!*, *see* Fig. 5b for illustrative example). Place this tube in a 2 mL flat bottom microcentrifuge tube and spin at 14,000 × *g* for 5 min.
14. Add 1 mL of RNA extraction buffer and incubate overnight at 37 °C on a thermomixer with 900 rpm shaking.
15. Equilibrate a NAP™-10 column by filling it with nuclease-free water and allowing it to completely empty. Repeat this step three times.

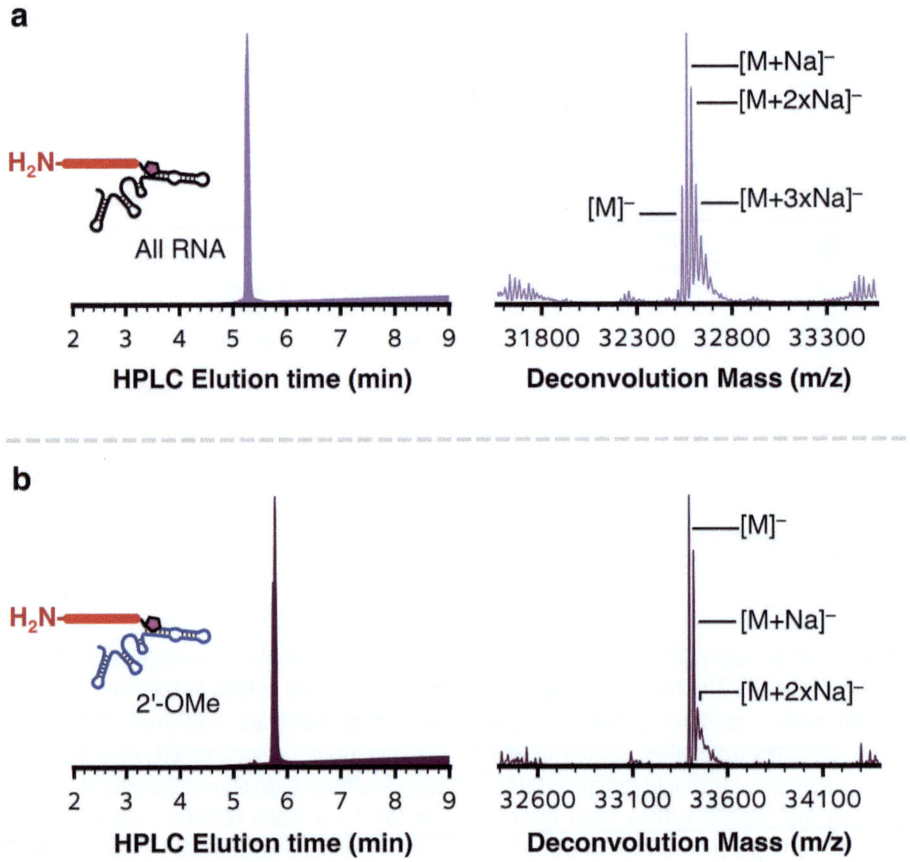

Fig. 6 Representative HPLC chromatograms and mass spectra for clicked ~20–79 sgRNAs. (**a**) Clicked sgRNA-NH_2 (targeting *EMX1* in cells) $[M]^-$ expected mass: 32536; found mass: 32540. (**b**) Clicked sgRNA-OMe-NH_2 (targeting *EMX1* in cells) $[M]^-$ expected mass: 33395; found mass: 33397. Note that sodium ion adducts are also detected and that the sgRNA may appear as a broad or bifurcated peak on HPLC due to RNA secondary structure

16. Add the RNA solution (1 mL) including the crushed gel onto the NAP™-10 column and allow the solution to flow through. The RNA is now inside the column.
17. Elute the desalted RNA from the column into a clean 2 mL microcentrifuge tube by adding 1.5 mL nuclease-free water (*see* **Notes 14** and **15**).
18. Freeze-dry the sample and analyze the purified clicked sgRNA by mass spectrometry (Fig. 6). The clicked sgRNA is now ready for use.

4 Notes

1. For 1 μmol synthesis, the amount of solid support to be packed is calculated by dividing 1 μmol by the loading capacity of the

solid support. For example, if the loading capacity of 3′-*O*-propargyl Guanosine (*N*-ibu) 2′-lcaa CPG 1000 Å is 40.6 μmol/g, 24.6 mg should be used for a 1 μmol synthesis.

2. Avoid exposure of methyltriphenoxyphosphonium iodide to the atmosphere. The solid should be a dark yellow powder. If the reagent becomes a dark brown liquid, do not use it, and reorder the reagent.
3. Sodium azide is highly toxic. Appropriate safety precautions should be taken when handling it. Do not use a metal spatula to measure sodium azide as this can form metal azides which can result in shock-sensitive explosions. Use a plastic or ceramic spatula instead.
4. To destroy sodium azide [33], place an aqueous solution of sodium azide (maximum 5% w/v) in a three-neck flask containing a stirrer bar in a fume hood. Ensure the sample is stirring before the addition of a 20% aqueous solution of sodium nitrite to the sodium azide solution. Note that toxic nitric oxide gas is evolved. The final ratio of sodium nitrite to sodium azide should be 1.5:1. Next, add a 20% aqueous solution of sulfuric acid dropwise until the mixture is slightly acidic. The order of addition in these steps is *important*. When gas evolution stops, a blue starch-iodine paper test confirms decomposition of excess nitrite. If the mixture does not contain heavy metals and the pH is 6–10, it can be disposed down the drain. Otherwise dispose of as recommended by your safety department. If at any point uncertain, please consult your safety department.
5. Opening the screw cap while the vial is still hot is hazardous; it causes rapid pressure release that expels the oligonucleotide containing solution from the vial.
6. 2′-*O*-TBDMS protected RNA is hydrophobic. Therefore, upon concentrating the solution, it precipitates and forms a cloudy solution. If the sample is taken to dryness, redissolving the RNA is very challenging and results in significant sample loss. Ethanol/water (3:1, v/v) is added to help dissolve protected RNA to ensure the maximum recovery of RNA when transferring it from one container to another.
7. Triethylamine trihydrofluoride is highly hazardous. Appropriate safety precautions should be taken when handling it. We recommend keeping the calcium gluconate gel on hand as first aid treatment in case of skin exposure to the reagent.
8. If there is still solid present, it should fully dissolve upon heating in **step 12**. Gently shake during the first 10–15 min to assist the process.

9. Deprotected RNA is highly sensitive to nuclease degradation. Gloves must be worn. Sterile containers and nuclease-free water must be used to avoid degradation.
10. Collect the triethylamine trihydrofluoride containing supernatant into a closed polyethylene container clearly labelled with the chemical composition. Plasticware (e.g., Falcon tubes and pipette tips) that comes in contact with triethylamine trihydrofluoride should also be collected in a plastic bag. Dispose of the triethylamine trihydrofluoride waste as recommended by your safety department.
11. Measure the UV–Vis absorption spectrum of the sample, diluting as necessary to ensure the spectrometer is not saturated. Next, correct the absorbance reading for dilution (e.g., for 1 in 10 dilution multiply all absorbance values by 10) and determine the wavelength that gives ~1 absorbance if only one main product is expected (i.e., crRNA) or ~5 (or higher) absorbance if multiple peaks are anticipated (i.e., tracrRNA). Ensure the detection wavelength is not above 298 nm or below 260 nm.
12. A library of clicked sgRNAs can be generated in one-pot ligation reaction. Mix appropriate crRNAs together but maintain a ratio of combined crRNAs to tracrRNA of 1.5:1. For example, a library of 6 sgRNAs requires 0.25× of each crRNA (6 crRNAs) and 1× of tracrRNA. Other components in the reaction remain unchanged.
13. If the clicked sgRNA construct is used without further purification (as crude), we recommend carrying out an analytical denaturing PAGE gel to check that click ligation has been successful.
14. The level of desalting required is dependent on the size of the band cut from the PAGE gel. To ensure complete desalting, we recommend further desalting purification as described in Subheading 3.7.
15. In the case of larger scale sgRNA click ligation, do not load more than 1 nmol of RNA per gel lane. If more than 3 gel bands must be combined, it is more convenient to place the excised gel in a 50 mL Falcon tube and crush the gel with a clean spatula. Add 15 mL of RNA extraction buffer to the Falcon tube and incubate overnight at 37 °C on a thermomixer with 900 rpm shaking. Remove the gel pieces by passing the solution through a syringe with the plunger removed and cotton wool placed inside. Collect the filtrate, which contains the RNA, into a 50 mL round bottom flask. To maximize recovery, add 5 mL nuclease-free water to the syringe containing the gel pieces and wool, and collect the filtrate into the same round-bottom flask. Next, concentrate the solution to dryness using a rotary evaporator ensuring the sample is not

heated above 50 °C. Redissolve the white solid (mixture of RNA and salt) in 1.5 mL nuclease-free water. Next, equilibrate two NAP™-25 columns by filling them with nuclease-free water and allowing them to completely empty. Repeat this step three times. Add 1.5 mL of the RNA solution onto one of the equilibrated NAP™-25 columns and allow the solution to flow through. The RNA is now inside the column. Add 1 mL nuclease-free water and allow it to flow through. The RNA is still inside the column. Now elute and collect the RNA into a new 15 mL Falcon tube using 2.5 mL nuclease-free water. Apply the eluted RNA solution onto the second equilibrated NAP™-25 column and allow the solution to flow through. The RNA is now in the column. Elute and collect the desalted RNA into a 15 mL Falcon tube by adding 3.5 mL of nuclease-free water. Freeze-dry the sample and resuspend in the desired volume of nuclease-free water. The clicked sgRNA is now ready for analysis by mass spectrometry.

Acknowledgments

This work was supported by UK BBSRC grants BB/J001694/2 (Extending the boundaries of nucleic acid chemistry), BB/M025624/1 (Next-generation DNA synthesis), BB/R008655/1 (New and versatile chemical approaches for the synthesis of mRNA and tRNA), UK EPSRC grant EP/S019944/1 (CRISPR Chemistry). We also thank ATDBio Ltd. for supporting A.H.E.S. and A.S. as well as the Royal Thai Government for supporting L.T.

References

1. Jinek M, Chylinski K, Fonfara I et al (2012) A programmable dual-RNA-guided DNA endonuclease in adaptive bacterial immunity. Science 337:816–821
2. Cong L, Ran F, Cox D et al (2013) Multiplex genome engineering using CRISPR/Cas systems. Science 339:819–823
3. Mali P, Yang L, Esvelt KM et al (2013) RNA-guided human genome engineering via Cas9. Science 339:823–826
4. Helm M, Brulé H, Giegé R, Florentz C (1999) More mistakes by T7 RNA polymerase at the 5′ ends of in vitro-transcribed RNAs. RNA 5:618–621
5. Shan Q, Wang Y, Li J et al (2013) Targeted genome modification of crop plants using a CRISPR-Cas system. Nat Biotechnol 31:686–688
6. Ranganathan V, Wahlin K, Maruotti J, Zack DJ (2014) Expansion of the CRISPR-Cas9 genome targeting space through the use of H1 promoter-expressed guide RNAs. Nat Commun 5:4516
7. Milligan JF, Groebe DR, Witherell GW, Uhlenbeck OC (1987) Oligoribonucleotide synthesis using T7 RNA polymerase and synthetic DNA templates. Nucleic Acids Res 15:8783–8798
8. Pleiss JA, Derrick ML, Uhlenbeck OC (1998) T7 RNA polymerase produces 5′ end heterogeneity during in vitro transcription from certain templates. RNA 4:1313–1317
9. Melton DA, Krieg PA, Rebagliati MR et al (1984) Efficient in vitro synthesis of biologically active RNA and RNA hybridization probes from plasmids containing a

bacteriophage SP6 promoter. Nucleic Acids Res 12:7035–7056

10. Adhya S, Basu S, Sarkar P, Maitra U (1981) Location, function, and nucleotide sequence of a promoter for bacteriophage T3 RNA polymerase. Proc Natl Acad Sci U S A 78:147–151
11. Cho SW, Kim S, Kim Y et al (2014) Analysis of off-target effects of CRISPR/Cas-derived RNA-guided endonucleases and nickases. Genome Res 24:132–141
12. Kim D, Bae S, Park J et al (2015) Digenome-seq: Genome-wide profiling of CRISPR-Cas9 off-target effects in human cells. Nat Methods 12:237–243
13. Kim E, Koo T, Park SW et al (2017) In vivo genome editing with a small Cas9 orthologue derived from Campylobacter jejuni. Nat Commun 8:14500
14. Hwang WY, Fu Y, Reyon D et al (2013) Efficient in vivo genome editing using RNA-guided nucleases. Nat Biotechnol 31:227–229
15. Kim HS, Lee K, Kim SJ et al (2018) Arrayed CRISPR screen with image-based assay reliably uncovers host genes required for coxsackievirus infection. Genome Res 28:859–868
16. de Groot R, Lüthi J, Lindsay H et al (2018) Large-scale image-based profiling of single-cell phenotypes in arrayed CRISPR-Cas9 gene perturbation screens. Mol Syst Biol 14:e8064
17. Henser-Brownhill T, Monserrat J, Scaffidi P (2017) Generation of an arrayed CRISPR-Cas9 library targeting epigenetic regulators: from high-content screens to in vivo assays. Epigenetics 12:1065–1075
18. Hendel A, Bak RO, Clark JT et al (2015) Chemically modified guide RNAs enhance CRISPR-Cas genome editing in human primary cells. Nat Biotechnol 33:985–989
19. Yin H, Song CQ, Suresh S et al (2017) Structure-guided chemical modification of guide RNA enables potent non-viral in vivo genome editing. Nat Biotechnol 35:1179–1187
20. Rahdar M, McMahon MA, Prakash TP et al (2015) Synthetic CRISPR RNA-Cas9–guided genome editing in human cells. Proc Natl Acad Sci 112:E7110–E7117
21. Mir A, Alterman JF, Hassler MR et al (2018) Heavily and fully modified RNAs guide efficient SpyCas9-mediated genome editing. bioRxiv Biochem 9:2641
22. Basila M, Kelley ML, Smith AVB (2017) Minimal 2'-O-methyl phosphorothioate linkage modification pattern of synthetic guide RNAs for increased stability and efficient CRISPR-Cas9 gene editing avoiding cellular toxicity. PLoS One 12:e0188593
23. Rueda FO, Bista M, Newton MD et al (2017) Mapping the sugar dependency for rational generation of a DNA-RNA hybrid-guided Cas9 endonuclease. Nat Commun 8:1610
24. Yin H, Song CQ, Suresh S et al (2018) Partial DNA-guided Cas9 enables genome editing with reduced off-target activity. Nat Chem Biol 14:311–316
25. Cromwell CR, Sung K, Park J et al (2018) Incorporation of bridged nucleic acids into CRISPR RNAs improves Cas9 endonuclease specificity. Nat Commun 9:1448
26. Ryan DE, Taussig D, Steinfeld I et al (2018) Improving CRISPR-Cas specificity with chemical modifications in single-guide RNAs. Nucleic Acids Res 46:792–803
27. Lee K, Mackley VA, Rao A et al (2017) Synthetically modified guide RNA and donor DNA are a versatile platform for CRISPR-Cas9 engineering. elife 6:e25312
28. Shivalingam A, Brown T (2016) Synthesis of chemically modified DNA. Biochem Soc Trans 44:709–715
29. Taemaitree L, Shivalingam A, El-Sagheer AH, Brown T (2019) An artificial triazole backbone linkage provides a split-and-click strategy to bioactive chemically modified CRISPR sgRNA. Nat Commun 10:1610
30. El-Sagheer AH, Brown T (2010) New strategy for the synthesis of chemically modified RNA constructs exemplified by hairpin and hammerhead ribozymes. Proc Natl Acad Sci 107:15329–15334
31. El-Sagheer AH, Sanzone AP, Gao R et al (2011) Biocompatible artificial DNA linker that is read through by DNA polymerases and is functional in Escherichia coli. Proc Natl Acad Sci 108:11338–11343
32. Nishimasu H, Ran FA, Hsu PD et al (2014) Crystal structure of Cas9 in complex with guide RNA and target DNA. Cell 156:935–949
33. National Research Council (2011) Prudent practices in the laboratory: handling and management of chemical hazards. The National Academies Press, Washington, DC

Chapter 6

Chimeric DNA–RNA Guide RNA Designs

Shuhan Lu, Ying Zhang, and Hao Yin

Abstract

CRISPR-associated nuclease (Cas) has been widely applied to modify the genomes of various cell types. As RNA-guided endonucleases, Cas enzymes can target different genomic sequences simply by changing the guide sequence of the CRISPR RNA (crRNA) or single guide RNA (sgRNA). Recent studies have demonstrated that DNA-RNA chimeric crRNA or sgRNA can efficiently guide the Cas9 protein for genome editing with reduced off-target effects. This chapter aims to describe a procedure for using chimeric RNA to modify the genomes of mammalian cells.

Key words CRISPR, Chemical modification, Synthetic sgRNA, Chimeric guide RNA, Off-target, Genome editing

1 Introduction

Clustered regularly interspaced short palindromic repeat (CRISPR)/CRISPR-associated nucleases (Cas) was originally identified as a prokaryotic immune defense system to eliminate foreign DNA and RNA [1–5]. The two classes of CRISPR systems include the Class 1 systems that utilizes multiple effector proteins to eliminate evading nucleic acids and the Class 2 systems that uses a single endonuclease [1–5]. CRISPR RNA (crRNA) guides Cas proteins to recognize and cut specific sequences via base pairing the crRNA guide sequences and target sequences [1–5]. A protospacer adjacent motif (PAM) sequence next to the target sequence is necessary to activate Cas proteins [1–5]. Among various Cas species, the *Streptococcus pyogenes* Cas9 (SpCas9) is the most widely used Cas endonuclease in both biomedical research and clinical studies [5]. SpCas9 uses a crRNA and a trans-activating CRISPR RNA (tracrRNA), both of which can be merged into a single guide RNA (sgRNA) [1–5]. Other Cas proteins, such as Cas12a (also known as Cpf1) require only a crRNA for efficient genome editing [6].

Plasmid transfection and viral infection are the most common delivery methods for CRISPR/Cas in basic research [7]. Although

Tudor A. Fulga et al. (eds.), *CRISPR Guide RNA Design: Methods and Protocols*, Methods in Molecular Biology, vol. 2162, https://doi.org/10.1007/978-1-0716-0687-2_6, © Springer Science+Business Media, LLC, part of Springer Nature 2021

such methods are effective, the persistent expression from DNA may result in higher off-target effects than delivery of the Cas9 protein/sgRNA complex (as Ribonucleoprotein, RNP) or RNA only (as Cas9 mRNA and sgRNA) [7]. Despite the much progress in Cas9 protein engineering to reduce off-target effects, the transient expression of CRISPR-Cas is the key to minimize off-target mutagenesis [8–14].

Although sgRNA and crRNA/tracrRNA can be obtained via in vitro transcription (IVT), these unmodified RNA are sensitive to degradation and may trigger innate immune responses [15–17]. Synthetic RNA with chemical modifications is more stable and has lower immunogenicity than unmodified IVT RNA [7]; however, the relatively high cost of synthesis limits extensive application. Because DNA-DNA duplex binding is less stable than RNA-DNA interactions, DNA-predominated guide sequences may lower the binding affinity between sgRNA and target DNA and thus increase specificity [11]. Partial replacement of the RNA nucleotides with DNA at the 5′ of the guide sequences (up to 10 nucleotides) maintains the genome editing activity of Cas9 in mammalian cells [11]. These findings are consistent with a publication in biorxiv [18]. In that study, 11 RNA nucleotides of a guide sequence were replaced with DNA, and the resulting chimera showed partial activity in cells [18]. The structure-guided replacement of RNA nucleotides at the 3′ of the crRNA with DNA also retains full activity in cells [11]. As a result, up to 60% of the RNA nucleotides of crRNA can be substituted with DNA nucleotides to decrease the synthesis cost by approximately 60%. The chimeric crRNA showed high on-target editing efficiency with substantially reduced off-target activities [11]. Similar to spCas9, the substitution of 8 RNA nucleotides with DNA at the 3′ of guide sequence of the crRNA for Cpf1 also retained its on-target activity in cells [11].

A minimal-RNA chimeric crRNA in which most RNA nucleotides were substituted with DNA showed comparable cleavage efficiency to unmodified RNA in vitro [19]; however, this modified crRNA did not generate substantial indels (insertions and deletions, as a result of genome editing) in mammalian cells, indicating that excessive DNA substitution compromises the genome editing activity [19, 20]. Similarly, tracrRNA can tolerate 50% or more DNA substitution and still retain the biochemical activity of crRNA-tracrRNA/Cas9 complex, but not genome editing activity in cells [19, 20].

Several of the ribose sugar 2′ hydroxyl groups of crRNA/tracrRNA/sgRNA interact with Cas9 protein to form hydrogen bonds, and disruption of these bonds might reduce the genome editing activity of the RNP [21, 22]. Indeed, while Cas9 biochemical activity in a test tube is largely retained with modifications at the 2′ hydroxyl contact positions of the sgRNA, the genome editing activity is dramatically reduced in cells [13, 19, 20]. Differences in the structural conformation between RNA and DNA may affect the binding affinity to Cas9, and thus excessive DNA substitution of crRNA or sgRNA is likely less tolerant for Cas9 protein interaction [19, 20].

A recent study identified that off-target cleavage can be reduced by incorporating 2'-O-methyl-3′-phosphonoacetate (MP) modification at two specific sites of the guide sequence of sgRNAs, likely due to destabilization of base pairing between the sgRNA and genomic sequences [23]. This method along with the studies discussed above, suggests that the chemical modification of guide sequences during the chemical synthesis of sgRNA or crRNA is a versatile tool for improving the specificity of CRISPR-Cas systems [23].

In this chapter, we introduce the chimeric crRNA or sgRNA design and synthesis, Cas9 protein expression and purification, RNP transfection and indels evaluation in details.

2 Materials

2.1 Mammalian Cell Lines

1. HEK293T, U2OS, or other cells that are easy to transfect.
2. Alternatively, Cas9-expressing stable cell lines (e.g., HEK293T cells that constitutively express Cas9) can be used.

2.2 Synthetic crRNA and sgRNA

1. These RNAs can be chemically synthesized or purchased from commercial vendors such as Integrated DNA Technologies.
2. We recommend HPLC purification of synthetic RNA. The purity of the resulting RNA should be greater than 80%, with an average of 90% purity.

2.3 TracrRNA

1. TracrRNA can be either chemically synthesized or in vitro transcribed (IVT) using T7 RNA Polymerase.

2.4 Cas9 Protein

1. Commercially available (e.g., New England Biolabs) or expressed and purified in house.

2.5 Other Reagents

1. The genomic DNA extraction solution (e.g., QuickExtract DNA Extraction Solution from Epicentre).
2. T7E1.
3. HEK 293T cell culture medium: DMEM.
4. Transfection agents: Lipofectamine RNAiMax or *CRISPR-MAX from Thermofisher.*

3 Methods

Figure 1 illustrates the detailed design of chimeric RNA.

3.1 Design of Guide Sequences

1. Use the website https://portals.broadinstitute.org/gpp/public/analysis-tools/sgrna-design. In brief, type in the gene name or the target sequence, specify the Cas9 species, PAM sequence (e.g., NGG for SpCas9) and the target editing species (human or mouse).
2. Select two to three top scored guide sequences. Exclude any guide sequences with a large number of off-target sites.

3.2 Synthesis of Chimeric RNA

1. Replace eight to ten RNA nucleotides at the 5′ of the guide sequence with DNA nucleotides. If the guide sequences have high (>50%) or low GC (<50%) content, substitute ten or eight RNA nucleotides with DNA, respectively. *See* Fig. 1 for details on how to substitute the 3′ end of the crRNA with DNA.
2. If using Integrated DNA Technologies, put the following code in "custom RNA synthesis." Select HPLC purification. Double-check the supplied HPLC data after the RNA arrives. The purity of the RNA should more than 80%.

   We use a guide sequence targeting GFP where the first 20 nucleotides from the 5′ is the guide sequence.

   crRNA

   GGGCGAGGrArGrCrUrGrUrUrCrArCrCrGrGrUrUrUrUrArGrArGrCrUrArUrGrCrUrGrUrUrUrUrG

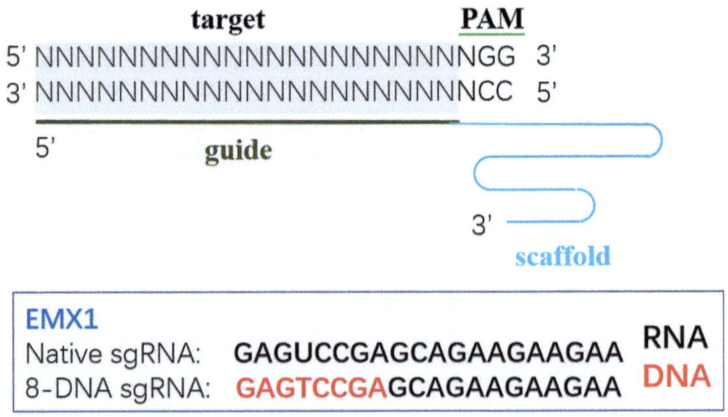

Fig. 1 The detailed design of chimeric RNA

sgRNA

GGG CGA GGrA rGrCrU rGrUrU rCrArC rCrGrG rUrUrU rUrArG rArGrC rUrArG rArArA rUrArG rCrArA rGrUrU rArArA rArUrA rArGrG rCrUrA rGrUrC rCrGrU rUrArU rCrArA rCrUrU rGrArA rArArA rGrUrG rGrCrA rCrCrG rArGrU rCrGrG rUrGrC rUrUrU rUrU

3. Dissolve RNA in RNase-free water; aliquot and store at −80 °C.

3.3 Cas9 and tracrRNA

1. Cas9 can be purchased from commercial vendors such as New England Biolabs. We also recommend expressing Cas9 in house or via a Cas9-expressing cell line.
2. To express Cas9 protein in house, we use a Cas9 expression vector with an N-terminal or C-terminal 6× His tag and a C-terminal nuclear localization sequence (NLS) (Addgene# 47327). The expression plasmid pET-28b-Cas9-His is transformed into *E. coli* BL21(DE3). Streak the transfected cells onto LB agar plates containing 50 μg/mL kanamycin and incubate at 37 °C. Pick a single colony into 5 mL of LB media containing 50 μg/mL kanamycin, and shake for 2 h. Transfer to 1 L of LB media and shake at 37 °C until OD600 reaches approximately 0.6–0.8. Lower the temperature from 37 °C to 18 °C and add 0.5 mM IPTG to induce Cas9 protein expression. Sixteen hours later, collect the cells, suspend in lysis buffer (25 mM Tris–HCl, 300 mM NaCl, and 0.25 mg/mL lysozyme) and lyse the cells by sonification or French press. Purify Cas9 protein from the cell lysate via Ni-NTA column (HisTrap HP, GE) and the size exclusion chromatography.
3. To construct a Cas9 expressing cell line, we use a lentivirus vector (e.g., lentiCas9-Blast Addgene 52962) which is optimized to express Cas9 protein in mammalian cells.
4. Although tracrRNA can be obtained via IVT, we recommend chemical synthesis of tracrRNA to enhance its performance.

   Shown is an example of synthesized tracrRNA from IDT:
   rArArArCrArGrCrArUrArGrCrArArGrUrU
   rArArArArUrArArGrGrCrUrArGrUrCrCrGrUrUrArUrCr
   ArArCrUrUrGrArArArArArGrUrGrGrCrArCrCr
   GrArGrUrCrGrGrUrGrCrUrUrUrUrUrU

3.4 Cell Culture and Transfection of Chimeric RNA

HEK293T cells are used as an example.

1. Day 1: Seed HEK293T cells at a density of 1.0×10^5 cells per well in a 24 well plate with a total DMEM medium volume of 500 μL.
2. Day 2: Incubate 125 ng of chimeric crRNA/tracrRNA or chimeric sgRNA with 500 ng of Cas9 protein in 25 μL of

Opti-MEM medium for 10 min at room temp (RT) to generate the Cas9 RNPs. Dilute 1.5 μL of Lipofectamine RNAiMax or *CRISPRMAX with* 25 μL of Opti-MEM medium and incubate for 5–10 min. *Mix these two solutions, and incubate* for 15 min before adding the mixture to the cells.

3. Day 4–5: Harvest the cells and extract the genomic DNA using the QuickExtract DNA Extraction Solution (Epicentre). To determine the efficiency of genome editing, the T7E1 assay (NEB #M0302, following the manufacturer's instruction) or Tracking of Indels by DEcomposition (TIDE, https://tide.nki.nl/) analysis can be performed to measure the indels.

3.5 The T7E1 Assay

1. Set up a standard PCR reaction using 1–100 ng of genome DNA as the template to yield amplicons containing the targeted sites. The recommended size of PCR product is 300–800 bp.
2. Denature and reanneal the PCR products to allow the formation of a heteroduplex between the PCR amplicons with and without edits. (denaturation: 95 °C, 5 min; annealing: 95–85 °C, −2 °C/s; 85–25 °C, −0.1 °C/s).
3. Add T7 Endonuclease I to the annealed PCR products and incubate at 37 °C for 20 min. Stop the reaction by adding 1.5 μL of 0.25 M EDTA. The digested DNA fragments can then be analyzed by agarose gel electrophoresis to determine the indels according to the percentage of cleaved DNA.

4 Notes

1. The general principles of substituting the 3′ end of the sgRNA (or equivalent part of the tracrRNA) for efficient genome editing in cells requires further investigation. Nevertheless, we believe that modification of the guide sequence (the 5′ of the sgRNA and crRNA for Cas9) but not the 3′ is crucial to reduce off-target effects.
2. DNA substitution can be combined with other modifications to enhance sgRNA activity.
3. The exact correlation of the GC content with the extent of DNA substitution for the guide sequences remains under investigation.
4. The Cas9 biochemical activity in vitro is different from the genome editing ability. Extensive DNA replacement of the sgRNA and tracrRNA/crRNA likely retains their biochemical activity but may disrupt their genome editing activity in cells.

References

1. Jinek M, Chylinski K, Fonfara I et al (2012) A programmable dual-RNA-guided DNA endonuclease in adaptive bacterial immunity. Science 337:816–821
2. Jinek M, East A, Cheng A et al (2013) RNA-programmed genome editing in human cells. elife 2:e00471
3. Mali P, Yang L, Esvelt KM et al (2013) RNA-guided human genome engineering via Cas9. Science 339:823–826
4. Cong L, Ran FA, Cox D et al (2013) Multiplex genome engineering using CRISPR/Cas systems. Science 339:819–823
5. Cox DB, Platt RJ, Zhang F (2015) Therapeutic genome editing: prospects and challenges. Nat Methods 21:121–131
6. Zetsche B, Gootenberg JS, Abudayyeh OO et al (2015) Cpf1 is a single RNA-guided endonuclease of a class 2 CRISPR-Cas system. Cell 163:759–771
7. Yin H, Kauffman KJ, Anderson DG (2017) Delivery technologies for genome editing. Nat Rev Drug Discov 16:387–399
8. Slaymaker IM, Gao LY, Zetsche B et al (2016) Rationally engineered Cas9 nucleases with improved specificity. Science 351:84–88
9. Kleinstiver BP, Pattanayak V, Prew MS et al (2016) High-fidelity CRISPR-Cas9 nucleases with no detectable genome wide off target effects. Nature 529:490–495
10. Chen JS, Dagdas YS, Kleinstiver BP et al (2017) Enhanced proofreading governs CRISPR-Cas9 targeting accuracy. Nature 550:407–410
11. Yin H, Song CQ, Suresh S et al (2018) Partial DNA-guided Cas9 enables genome editing with reduced off-target activity. Nat Chem Biol 14:311–316
12. Hendel A, Bak RO, Clark JT et al (2015) Chemically modified guide RNAs enhance CRISPR-Cas genome editing in human primary cells. Nat Biotechnol 33:985–989
13. Yin H, Song CQ, Suresh S et al (2017) Structure-guided chemical modification of guide RNA enables potent non-viral in vivo genome editing. Nat Biotechnol 35:1179–1187
14. Koo T, Lee J, Kim JS (2015) Measuring and reducing off-target activities of programmable nucleases including CRISPR-Cas9. Mol Cells 38:475–481
15. Kim S, Koo T, Jee HG et al (2018) CRISPR RNAs trigger innate immune responses in human cells. Genome Res 28:367–373
16. Wienert B, Shin J, Zelin E et al (2018) In vitro-transcribed guide RNAs trigger an innate immune response via the RIG-I pathway. PLoS Biol 16:e2005840
17. Mu W, Tang N, Cheng C et al (2019) In vitro transcribed sgRNA causes cell death by inducing interferon release. Protein Cell 10 (6):461–465
18. Jakimo N, Chatterjee P, Jacobson JM (2017) Chimeric CRISPR guides enhance Cas9 target specificity. bioRxiv:147686
19. Rueda FO, Bista M, Newton MD et al (2017) Mapping the sugar dependency for rational generation of a DNA-RNA hybrid-guided Cas9 endonuclease. Nat Commun 8:1610
20. Schofield A, O'Reilly D, Malek-Adamian E et al (2018) Extensive CRISPR RNA modification reveals chemical compatibility and structure-activity relationships for Cas9 biochemical activity. Nucleic Acids Res 47:546–558
21. Jiang F, Zhou K, Ma L et al (2015) A Cas9-- guide RNA complex preorganized for target DNA recognition. Science 348:1477–1481
22. Nishimasu H, Ran FA, Hsu PD et al (2014) Crystal structure of Cas9 in complex with guide RNA and target DNA. Cell 156:935–949
23. Ryan DE, Taussig D, Steinfeld I et al (2018) Improving CRISPR-Cas specificity with chemical modifications in single-guide RNAs. Nucleic Acids Res 46:792–803

Part III

Expanding the CRISPR Toolbox

Chapter 7

Harnessing tRNA for Processing Ability and Promoter Activity

David J. H. F. Knapp and Tudor A. Fulga

Abstract

Transfer RNA (tRNA) and their associated production and processing machinery can be coopted as a versatile tool for the production of guide RNAs (gRNAs) for Cas9-based genome engineering. Using different tRNA variants enables the production of gRNAs at a variety of steady state levels. Furthermore, engineered tRNAs can be used to process gRNAs from Pol-II transcripts, thus enabling spatial/temporal control of gRNA expression. Here we describe the design, cloning, and testing of tRNA scaffolds for both Pol-III-driven expression of different levels of gRNAs, and for processing gRNAs from Pol-II transcripts.

Key words tRNA, Tissue-specific, RNA Polymerase-II, CRISPR, Cas9, Guide RNA, Genome editing

1 Introduction

Guide RNAs (gRNAs) are responsible for the precise genomic targeting of the CRISPR/Cas9 editing system. Production of gRNAs is generally accomplished using the strong, constitutive U6 RNA polymerase III (Pol-III) promoter [1, 2]. This allows for the production of gRNAs with nuclear localization. For uses requiring intermediate or low levels of gRNAs, however, Pol-III promoters of differing strength would be required. Additionally, if multiple gRNAs are required simultaneously, the use of repetitive copies of U6 promoters creates a risk of recombination, and very quickly leads to size limitations in the ensuing constructs as these promoters are relatively long. Meanwhile, tissue specific or stimulus regulated gRNA expression, require the use of RNA polymerase II (Pol-II) promoters. This poses an added challenge, as Pol-II transcripts are quickly capped, polyadenylated, and exported from the nucleus. Consequently, gRNAs produced from these promoters must be excised from the surrounding transcript in order to have functional activity [3, 4].

Tudor A. Fulga et al. (eds.), *CRISPR Guide RNA Design: Methods and Protocols*, Methods in Molecular Biology, vol. 2162, https://doi.org/10.1007/978-1-0716-0687-2_7,

Transfer RNAs (tRNAs) can address a number of the limitations associate with other Pol-III-mediated gRNA production methods. tRNAs are a diverse and ubiquitous class of RNAs which contain a compact Pol-III promoter and are capable of recruiting endogenous RNase P and Z to allow precise processing of their 5′ and 3′ ends [5]. These tRNAs are substantially more compact than the canonically used U6 promoter. Furthermore, there are many tRNA variants in different species, providing the means to mitigate recombination risks in cases where multiple gRNAs are required. Various parts of the tRNA machinery are also highly conserved, making this a highly adaptable platform for many model systems. Indeed, we have shown that even tRNAs from rice (*Oryza sativa*) retain much of their processing and promoter activity in human cells, thus demonstrating cross-species compatibility [4].

tRNAs have been applied for gRNA production both directly as Pol-III promoters [4, 6, 7] and for processing multiple gRNA out of longer transcripts [8–10]. We have created a number of mutant variants of human tRNAs which have an array of Pol-III promoter strengths and processing capabilities. In addition, we have demonstrated that several of these variants have negligible intrinsic Pol-III promoter activity while retaining sufficient processing ability for gRNA excision from Pol-II transcripts [4]. Here we provide step-by-step descriptions for cloning, molecular characterization, and functional testing of tRNA mediated gRNA production. This includes direct Pol-III-mediated promoter activity of a tRNA with varying promoter strengths, as well as variants for processing gRNAs out of transcripts produced from Pol-II promoters.

2 Materials

2.1 Bacterial Transformation

1. Subcloning Efficiency™ DH5α™ Competent Cells.
2. LB Broth.
3. S.O.C. medium.
4. Ampicillin.

2.2 Cloning

1. Qiagen MinElute PCR purification columns (Qiagen).
2. Qiaquick PCR purification Kit (Qiagen).
3. QIAquick Gel Extraction Kit (Qiagen).
4. Qiaprep Spin Miniprep columns (Qiagen).
5. Antarctic phosphatase.

6. Phusion® High-Fidelity PCR Master Mix with GC Buffer (New England BioLabs).
7. T4 polynucleotide kinase.
8. FastDigest BpiI.
9. BsmBI.
10. Nuclease-free H_2O.
11. Ultrapure Agarose.
12. Tris-acetate-EDTA (TAE) buffer.
13. NanoDrop 2000 spectrophotometer (Thermo Fisher Scientific).
14. Thermocycler.

2.3 Transfection and Construct Testing

1. HEK-293T cells (ATCC-CRL-11268).
2. Dulbecco's Modified Eagle's Medium (DMEM).
3. Fetal bovine serum (FBS).
4. Polyethylenimine (PEI).
5. Opti-MEM™ (Gibco).
6. Trypsin–EDTA.
7. T4 RNA Ligase 1 (10 U/μl).
8. 10× T4 RNA ligase buffer.
9. 10 mM ATP.
10. SUPERase In RNase Inhibitor (20 U/μl, Thermo Fisher Scientific).
11. 50% PEG8000.
12. RNA 5′ pyrophosphohydrolase (RppH, 5000 units/ml, NEB).
13. 10× Thermopol Buffer (NEB).

3 Methods

3.1 Preparing a GoldenGate Destination Vector

1. Select your vector of interest, being careful to avoid those containing BpiI and BsmBI sites as these will be required for downstream steps.
2. Should the vector contain the restriction sites, they must be removed prior to further cloning. This can be accomplished using site-directed mutagenesis kits as per the manufacturer's instructions. Alternately if these are flanked by other restriction endonuclease sites, they can be removed by restriction digest of

the flanking sites followed by replacing the center portion with a pair of annealed oligos lacking the BpiI/BsmBI site.

3. Select a pair of unique restriction sites in the desired region which leave noncompatible four base pair "sticky" ends. For Pol-II expression vectors this should be somewhere in the transcribed region of the gene (i.e., between the promoter and polyadenylation signal). Ensure that there are at least 5–10 bp between the respective recognition sequences (*see* **Note 1**).
4. Design an oligonucleotide pair with sticky ends compatible to those of the vector. Examples of various scenarios are shown below. "Sticky" ends from the vector are shown in bold while the complementary bases added to the oligonucleotide insert are shown as bold and underlined. Note that this first example will destroy the flanking restriction sites.

5′...TCGCGTTGAC **CATG**CAACATGTCTTCGAATTCGAAGACTAGGAC**CCGG** CCGCTTCGAG... 3′
3′...AGCGCAACTG**GTAC** GTTGTACAGAAGCTTAAGCTTCTGATCCTG **GGCC**GGCGAAGCTC... 5′

Additional bases (bold italics) could be added to retain flanking restriction sites (such that the initial restriction site is recreated following annealing of the sticky ends) should this be desired:

5′...TCGCGTTGAC **CATG*G***CAACATGTCTTCGAATTCGAAGACTAGGAC***GG*****CCGG** CCGCTTCGAG... 3′
3′...AGCGCAACTG**GTAC** ***C***GTTGTACAGAAGCTTAAGCTTCTGATCCTG***CC*** **GGCC**GGCGAAGCTC... 5′

The strand to which the sticky ends must be added will depend on the orientation of the specific restriction enzymes used. Examples are shown below:

5′...TCGCGTTGAC**GTAC** CAACATGTCTTCGAATTCGAAGACTAGGAC**CCGG** CCGCTTCGAG... 3′
3′...AGCGCAACTG **CATG**GTTGTACAGAAGCTTAAGCTTCTGATCCTG **GGCC**GGCGAAGCTC... 5′

5′...TCGCGTTGAC **CATG**CAACATGTCTTCGAATTCGAAGACTAGGAC **GGCC**CCGCTTCGAG... 3′
3′...AGCGCAACTG**GTAC** GTTGTACAGAAGCTTAAGCTTCTGATCCTG**CCGG** GGCGAAGCTC... 5′

5′...TCGCGTTGAC**GTAC** CAACATGTCTTCGAATTCGAAGACTAGGAC **GGCC**CCGCTTCGAG... 3′
3′...AGCGCAACTG **CATG**GTTGTACAGAAGCTTAAGCTTCTGATCCTG**CCGG** GGCGAAGCTC... 5′

5. Digest 1 μg of the vector with the flanking restriction enzymes either together (if they use compatible buffers/temperatures), or sequentially. Digestion conditions can be determined using

online tools such as the RE Digest tool in the NEBCloner online app (http://nebcloner.neb.com/#!/redigest).

6. While the vector is being digested phosphorylate and anneal the oligos designed in **step 4** as follows:

100 μM Forward Oligo	1 μl
100 μM Reverse Oligo	1 μl
10× T4 Ligase Buffer	1 μl
T4 PNK	0.5 μl
Nuclease Free H_2O	6.5 μl

Incubate at 37 °C for 30 min, 95 °C for 5 min, then bring to 25 °C at a rate of 0.1 °C/s. Store on ice until ligation.

7. Run the digested vector on a 1% agarose gel and extract the expected band using a QIAquick Gel Extraction Kit as per the manufacturer's instructions, but substitute the standard columns for a MinElute PCR purification column and elute in 15 μl Elution Buffer. Store digested destination plasmid on ice until ligation (*see* **Notes 2** and **3**).

8. Make a 1:50 dilution of annealed, phosphorylated oligo in water (1 μl of annealing reaction + 49 μl water).

9. Set up a ligation reaction as follows:

Vector	1 μl
T4 DNA ligase Buffer	1 μl
1:50 oligo dilution	1 μl
H_2O	6.75 μl
T4 DNA ligase	0.25 μl

At the same time set up a control reaction with water in place of the oligo dilution.

10. Mix gently and incubate at room temperature (20–25 °C) for 10 min.

11. Stop the reaction by incubating at 65 °C for 10 min.

12. Place the reactions on ice for 2 min.

13. Transform 0.25–1 μl into 10 μl Subcloning Efficiency™ DH5α™ Competent Cells or equivalent chemically competent bacteria as per the manufacturer's protocol.

14. Plate onto relevant antibiotic selection plates for the backbone used (generally Ampicillin) and incubate overnight at 37 °C.

15. The following day there should be at least 4× more colonies on the plate with insert (generally 50–100+ colonies).

16. Pick two to three colonies into 5 ml LB with vector-relevant antibiotic for each of the insert plates. Incubate at 37 °C with shaking overnight.
17. Extract plasmids with Qiaprep Spin Miniprep kit or equivalent and quantify DNA using a NanoDrop 2000 spectrophotometer or equivalent method.
18. Take 1 μl of each plasmid and add to a reaction mix containing 0.2 μl FastDigest BpiI, 1 μl FastDigest buffer, 7.8 μl H_2O and incubate at 37 °C for 15 min to 1 h.
19. Following digestion run diagnostic digests alongside an equivalent amount of undigested vector on a 1% Agarose gel. A sharp band at the expected vector size is expected as opposed to multiple smaller or larger bands in the undigested case.
20. This represent the GoldenGate acceptor vector which will be used in downstream steps to receive tRNA flanked gRNAs.

3.2 Using tRNA as a pol-III Promoter

1. Select the tRNA with the desired levels of functional gRNA production and order this sequence as an oligonucleotide from your preferred supplier with a CAAC overhang added to the 5′ end of the forward oligo and remove the final (3′) four bases from it. The reverse oligonucleotide should be the reverse complement of the sequence from Table 1.
2. Phosphorylate and anneal oligonucleotides for the desired tRNA from Table 1.

100 μM Forward Oligo	1 μl
100 μM Reverse Oligo	1 μl
10× T4 Ligase Buffer	1 μl
T4 PNK	0.5 μl
Nuclease Free H_2O	6.5 μl

Incubate at 37 °C for 30 min, 95 °C for 5 min, then bring to 25 °C at a rate of 0.1 °C/s. Store on ice until ligation.

(a) For standard sgRNA, phosphorylate and anneal the following oligonucleotide pairs as above:

"BsmBI-plch-sgRNA1-p1-F" + "BsmBI-plch-sgRNA1-p1-R".

"BsmBI-plch-sgRNA1-p1-notRNA-F" + "BsmBI-plch-sgRNA1-p1-R".

"BsmBI-plch-sgRNA1-polIIIterm-p2-F" + "BsmBI-plch-sgRNA1-polIIIterm-p2-R".

(*see* **Note 4**).

Table 1

Tested tRNA sequences and their properties. Pol-III promoter strength (in the absence of a Pol-II promoter) is shown as a % of U6 (normalized to dCas9 as a transfection control). 3′ Processing ability and overall processing ability represent the % correctly processed band using the circularization assays in the absence and presence of a Pol-II promoter respectively. Functional activity shows the reporter assay readouts as a % of U6 in the absence of a Pol-II promoter. Finally, the $\log_{10}$(ON/OFF) ratio shows the log fold change between the with and without Pol-II promoter conditions from reporter assays. All values represent means from [4]

tRNA	Sequence	Promoter Strength	3′ Processing	Functional activity	Overall processing	ON/ OFF
Dm tRNA^{Gly}	GGGCTTTGAGTGTGTGTAGACATCAAGCATCGGTGG TTCAGTGGTAGAATGCTCGCCTGCCACGCGGGCGG CCCGGGTTCGATTCCCGGCCGATGCA	35	86	114	ND	ND
Hs tRNA^{Gly}	atcggtGCATGGGTGGTTCAGTGGTAGAATTCTCGCCTG CCACGCGGGAGGCCCGGGTTCGATTCCCGGCCCA	21	89	109	58	0.04
Hs tRNA^{Gln}	GGTTCCATGGTGTAATGGTTAGCACTCTGGACTCTGAA TCCAGCGATCCGAGTTCAAATCTCGGTGGAACCT	15	72	109	ND	ND
Hs tRNA^{Pro}	GGCTCGTTGGTCTAGGGGTATGATTCTCGCTTAGGG TGCGAGAGGTCCCGGGTTCAAATCCCGGACGAGCCC	21	90	96	58	0.06
Os tRNA^{Gly}	AACAAAGCACCAGTGGTCTAGTGGTGGAATAGTACCCTG CCACGGTACAGACCCGGGTTCGATTCCCGGCTGG	12	42	67	ND	ND
Hs tRNA^{Pro} G16C/ A56T	GGCTCGTTGGTCTAGCGGTATGATTCTCGCTTAGGGTGCG AGAGGTCCCGGGTTCTAATCCCGGACGAGCCC	16	30	64	ND	ND
Hs tRNA^{Pro} A58T/ T59A	GGCTCGTTGGTCTAGGGGTATGATTCTCGCTTAGGGTGCG AGAGGTCCCGGGTTCAATACCCGGACGAGCCC	17	16	54	ND	ND
Hs tRNA^{Pro} A56T/ T59C	GGCTCGTTGGTCTAGGGGTATGATTCTCGCTTAGGGTGCG AGAGGTCCCGGGTTCTAACCCCGGACGAGCCC	18	17	51	ND	ND
Hs tRNA^{Pro} G17A/ A20G	GGCTCGTTGGTCTAGGAGTGTGATTCTCGCTTAGGGTGCG AGAGGTCCCGGGTTCAAATCCCGGACGAGCCC	15	33	30	13	0.05
Hs tRNA^{Pro} G9C/A56T	GGCTCGTTCGTCTAGGGGTATGATTCTCGCTTAGGGTGCG AGAGGTCCCGGGTTCTAATCCCGGACGAGCCC	13	19	26	13	0.05

(continued)

Table 1 (continued)

tRNA	Sequence	Promoter Strength	3′ Processing	Functional activity	Overall processing	ON/OFF
Hs tRNA^{Pro} G18T/C55G	GGCTCGTTGGTCTAGGGTTATGATTCTCGCTTAGGGTGCG AGAGGTCCCGGGTTGAAATCCCGGACGAGCCC	9	13	23	12	−0.01
Hs tRNA^{Pro} G17T/C55A	GGCTCGTTGGTCTAGGTGTATGATTCTCGCTTAGGGTGCG AGAGGTCCCGGGTTAAAATCCCGGACGAGCCC	7	30	17	13	0.01
Hs tRNA^{Pro} T54C/A58C	GGCTCGTTGGTCTAGGGGTATGATTCTCGCTTAGGGTGCG AGAGGTCCCGGGTCCAACTCCCGGACGAGCCC	7	22	8	9	−0.03
Hs ΔtRNA^{Pro} G9C/A56T	GGCTCGTTCGGAGGTCCCGGGTTCTAATCCCGGACGAG CCC	3	75	3	12	0.12
Hs ΔtRNA^{Gly}	atcggtGCATGGGTGGAGGCCCGGGTTCGATTCCCGG CCCATGCA	ND	ND	2	ND	0.47
Hs ΔtRNA^{Pro}	GGCTCGTTGGGAGGTCCCGGGTTCAAATCCCGGACGAG CCC	2	69	1	18	0.46
Hs ΔtRNA^{Pro} C55G	GGCTCGTTGGGAGGTCCCGGGTTGAAATCCCGGACGAG CCC	1	ND	0	7	1.26
Hs ΔtRNA^{Gly} C55G	atcggtGCATGGGTGGAGGCCCGGGTTGGATTCCCGG CCCATGCA	ND	ND	0	ND	0.87
Hs ΔtRNA^{Pro} C55A	GGCTCGTTGGGAGGTCCCGGGTTAAAATCCCGGACGAG CCC	0	ND	0	3	0.61
Hs ΔtRNA^{Pro} T54C/A58C	GGCTCGTTGGGAGGTCCCGGGTCCAACTCCCGGACGAG CCC	0	ND	0	2	0.25

ND not done

Add 1 μl each of the "BsmBI-plch-sgRNA1-p1-F" + "BsmBI-plch-sgRNA1-p1-R" mix and "BsmBI-plch-sgRNA1-polIIIterm-p2-F" + "BsmBI-plch-sgRNA1-polIIIterm-p2-R" mix to 48 μl H_2O for the samples. Add 1 μl each of "BsmBI-plch-sgRNA1-p1-notRNA-F" + "BsmBI-plch-sgRNA1-p1-R" mix and "BsmBI-plch-sgRNA1-polIIIterm-p2-F" + "BsmBI-plch-sgRNA1-polIIIterm-p2-R" mix to 48 μl H_2O for no tRNA control.

(b) For activation experiments with the SAM system [11], perform the following amplifications, using either the "BpiIGCCC-BsmBI-plch-sgRNA_MS2-F" + "BpiIG-GAC-sgRNA_MS2-R" primers for the sample, or the "BpiICAAC-BsmBI-plch-sgRNA_MS2-F" + "BpiIG-GAC-sgRNA_MS2-R" primers for the no tRNA control (each primer sequence is listed in Table 2):

Nuclease-free H_2O	to 30 μl total
Addgene plasmid 61424	1 ng
10 μM sgRNA primer mix	1.5 μl
Phusion® High-Fidelity PCR Master Mix with GC Buffer	15 μl

98 °C	60 s	
98 °C 67 °C 72 °C	10 s 30 s 10 s	40×
72 °C	5 min	

(*see* **Note 5**)

Run PCR products and digested destination plasmid on a 1% agarose gel and extract the expected bands using a QIAquick Gel Extraction Kit as per the manufacturer's instructions with a 30 μl elution volume.

Quantify purified PCR products using a NanoDrop 2000 Spectrophotometer.

3. Assemble and run GoldenGate reactions as follows:

Table 2
Oligonucleotide pairs for annealing and amplification

Name	Sequence	Name	Sequence
BpiIGCCC-BsmBI-plch-sgRNA_MS2-F	AAGGAGAAGACTAGCCCG GAGACGGACGTCTCCGTT TTAGAGCTAGGCCAACATG	BpiIGGAC-sgRNA_MS2-R	AAGGAGAAGACAAG TCCAAAAAAGCACCGACTCGGTGCCA
BpiICAAC-BsmBI-plch-sgRNA_MS2-F	AAGGAGAAGACTACAACGGAGACGGACGTC TCCGTTTTAGAGCTAGGCCAACATG	BpiIGGAC-sgRNA_MS2-R	AAGGAGAAGACAAG TCCAAAAAAGCACCGACTCGGTGCCA
BpiIGCCC-BsmBI-plch-sgRNA_MS2-F	AAGGAGAAGACTAGCCCGGAGACGGACGTC TCCGTTTTAGAGCTAGGCCAACATG	BpiIAACC-sgRNA_MS2-R	AAGGAGAAGACAAGGTTGCACCGACTCGG TGCCA
BsmBI-plch-sgRNA1-p1-F	GCCCGGAGACGGGAATTCACGTCTCCG TTTTAGAGCTAGAAATAGCAAGTTAAAA TA	BsmBI-plch-sgRNA1-p1-R	GCCTTATTTTAACTTGCTATTTCTAGCTC TAAAACGGAGACGTGAATTCCCGTCTCC
BsmBI-plch-sgRNA1-p1-notRNA-F	CAACGGAGACGGGAATTCACGTCTCCG TTTTAGAGCTAGAAATAGCAAGTTAAAA TA	BsmBI-plch-sgRNA1-p1-R	GCCTTATTTTAACTTGCTATTTCTAGCTC TAAAACGGAGACGTGAATTCCCGTCTCC
BsmBI-plch-sgRNA1-polIIIterm-p2-F	AGGCTAGTCCGTTATCAACTTGAAAAAG TGGCACCGAGTCGGTGCTTTTTT	BsmBI-plch-sgRNA1-polIIIterm-p2-R	GTCCAAAAAAGCACCGACTCGGTGCCAC TTTTTCAAGTTGATAACGGACTA
BsmBI-plch-sgRNA1-p2-F	AGGCTAGTCCGTTATCAACTTGAAAAAG TGGCACCGAGTCGGTGC	BsmBI-plch-sgRNA1-p2-R	GGTTGCACCGACTCGGTGCCAC TTTTTCAAGTTGATAACGGACTA
BpiI-CAAC-5p-tRNApro-F	AAGGAGAAGACTACAACGGCTCG TTGGGAGGTC	BpiI-GCCC-5p-tRNApro-R	AAGGAGAAGACAAGGGCTCGTCCGGGA TTT
BpiI-14bpBuf-3p-deltRNApro-F	AAGGAGAAGACAAAACCAGTTTGTGTCGGC TCGTTGGGAGGT	3p-del-tRNApro-R	AAGGAGAAGACTAGTCCCGGGCTCG TCCGGGA
cRT-sgRNA_nest_F	TAGCAAGTTAAAATAAGGCTAGT	cRT-CTS2_nest_R	CTAGCTCTAAAACAGGACAGT
cRT-sgRNA1_v2_F	GGCTAGTCCGTTATCAACT	cRT-CTS2_R	AGGACAGTACTCCGACTTAC
CTS2_qPCR-F	AGTCGGAGTACTGTCCTGTT	sgRNA_qPCR_common-R	GGACTAGCCTTATTTTAACTTGCT
dCas9_qPCR-F	CCCAAGAGGAACAGCGATAAG	dCas9_qPCR-R	CCACCACCAGCACAGAATAG

(a) Samples (using the sgRNA from **step 2** for samples),

Vector	50 ng
tRNA	~5 ng
sgRNA	~12 ng for SAM-sgRNA or 1 μL of 1:50 sgRNA oligos
10× T4 DNA Ligase Buffer	1 μl
T4 DNA Ligase	0.25 μl
BpiI FastDigest	0.25 μl
H_2O	to 10 μl

(b) No tRNA control (using the no tRNA sgRNA from **step 2**).

Vector	50 ng
sgRNA	~12 ng for SAM-sgRNA or 1 μl of 1:50 sgRNA oligos
10× T4 DNA Ligase Buffer	1 μl
T4 DNA Ligase	0.25 μl
BpiI FastDigest	0.25 μl
H_2O	to 10 μl

37 °C 16 °C	5 min 5 min	30×
37 °C 50 °C 80 °C	30 min 5 min 5 min	

(*see* **Notes 6** and 7).

4. Transform into Subcloning Efficiency™ DH5α™ Competent Cells or equivalent chemically competent bacteria as per the manufacturer's protocol. We generally use 10 μl competent bacteria with 0.5–1 μl of ligation product.
5. Plate onto relevant antibiotic selection plates for the backbone used (generally Ampicillin) and incubate overnight at 37 °C.
6. Pick two to three colonies into 5 ml LB with vector-relevant antibiotic each. Incubate at 37 °C with shaking overnight.
7. Extract plasmids with QIAprep Spin Miniprep kit or equivalent and quantify on a NanoDrop 2000 spectrophotometer or by other desired method.

8. Take 1 μl of product from each for diagnostic digests. Although exact digests will be vector dependent, BsmBI can be used to verify gRNA insertion and SmaI can be used to verify tRNA insertion for the $tRNA^{Pro}$ variants. Note, these do not provide a useful double digest due to close proximity and differing temperature requirements.
9. Following digestion run reactions alongside an equivalent amount of undigested vector in a 1% Agarose gel. Banding pattern can then be compared to expected patterns (this can be predicted with a variety of software including SnapGene and Benchling) based on the enzyme combination used. For single digests a sharp band at the expected size is expected as opposed to multiple smaller or larger bands in the undigested case.
10. Vectors which show correct digestion patterns should then be validated by Sanger sequencing to ensure that no point mutations are present using a primer located ~100 bases outside of the region of insertion. We often use pBR322_ori-F: CACCTCTGACTTGAGCGTCG, which is located in the pBR322 origin of replication (present in most of our vector backbones).
11. Sequence validated vectors are now ready for downstream cloning of spacer sequences.

3.3 Using tRNA for gRNA Processing from Pol-II Transcripts

1. Select a 5′ and a 3′ tRNA variant with the required levels of processing and promoter activity from Table 1.
2. Design primer sets to amplify the 5′ and 3′ tRNA adding BpiI sites with appropriate 4 bp overhangs as illustrated in Fig. 1. Example 5′ ("BpiI-CAAC-5p-tRNApro-F" + "BpiI-GCCC-5p-tRNApro-R") and 3′ ("BpiI-14bpBuf-3p-deltRNApro-F" + "3p-del-tRNApro-R") primers for the $\Delta tRNA^{Pro}$ are shown in Table 2 (*see* **Note 8**).
3. Order the template 5′ and 3′ tRNA sequences (only need forward, no overhangs) as an oligonucleotide from your preferred supplier along with the primers designed in **step 2**.
4. Amplify the 5′ and 3′ tRNA from oligonucleotide templates as follows:

Nuclease-free H_2O	12.5 μl
100 nM oligonucleotide template	1 μl
10 μM Primer mix	1.5 μl
Phusion® High-Fidelity PCR Master Mix with GC Buffer	15 μl

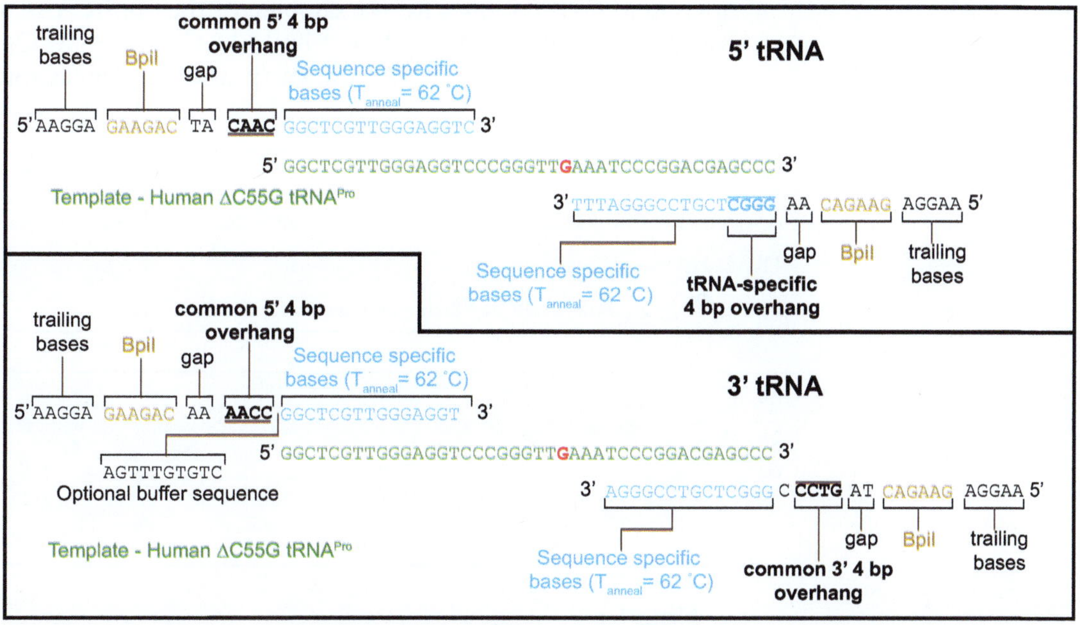

Fig. 1 Primer design for 5′ and 3′ tRNA amplification

98 °C	60 s	
98 °C 62 °C 72 °C	10 s 30 s 10 s	40×
72 °C	5 min	

(a) For standard sgRNA, phosphorylate and anneal the following oligonucleotide pairs as above:

"BsmBI-plch-sgRNA1-p1-F" + "BsmBI-plch-sgRNA1-p1-R".

"BsmBI-plch-sgRNA1-p1-notRNA-F" + "BsmBI-plch-sgRNA1-p1-R".

"BsmBI-plch-sgRNA1-p2-F" + "BsmBI-plch-sgRNA1-p2-R".

Add 1 μl each of the "BsmBI-plch-sgRNA1-p1-F" + "BsmBI-plch-sgRNA1-p1-R" mix and "BsmBI-plch-sgRNA1-p2-F" + "BsmBI-plch-sgRNA1-p2-R" mix to 48 μl H_2O for the samples. Add 1 μl each of "BsmBI-plch-sgRNA1-p1-notRNA-F" + "BsmBI-plch-sgRNA1-p1-R" mix and "BsmBI-plch-sgRNA1-p2-F" + "BsmBI-plch-sgRNA1-p2-R" mix to 48 μl H_2O for no tRNA control.

(b) For activation experiments with the SAM system [11], perform the following amplifications, using either the "BpiIGCCC-BsmBI-plch-sgRNA_MS2-F" + "BpiIAACC-sgRNA_MS2-R" primers for the samples, or the "BpiICAAC-BsmBI-plch-sgRNA_MS2-F" + "BpiIAACC-sgRNA_MS2-R" primers for the no tRNA control (primer sequences for each are in Table 2):

Nuclease-free H_2O	to 30 μl total
Addgene plasmid 61424	1 ng
10 μM sgRNA primer mix	1.5 μl
Phusion® High-Fidelity PCR Master Mix with GC Buffer	15 μl

98 °C	60 s	
98 °C 67 °C 72 °C	10 s 30 s 10 s	40×
72 °C	5 min	

(*see* **Note 5**)

5. Run PCR products and digested destination plasmid on a 1% agarose gel and extract the expected bands using a QIAquick Gel Extraction Kit as per the manufacturer's instructions with a 30 μl elution volume.
6. Quantify purified PCR products using a NanoDrop™ 2000 Spectrophotometer.
7. Assemble and run the following GoldenGate reactions:
   (a) Samples (with the sample sgRNA from **step 4**),

Vector	50 ng
5′ tRNA	~5 ng
sgRNA	~12 ng for SAM-sgRNA or 1 μl of 1:50 sgRNA oligos
3′ tRNA	~5 ng
10× T4 DNA Ligase Buffer	1 μl
T4 DNA Ligase	0.25 μl
BpiI FastDigest	0.25 μl
H_2O	to 10 μl

(b) No tRNA control (with the no tRNA sgRNA from **step 4**),

Vector	50 ng
sgRNA	~12 ng for SAM-sgRNA or 1 μl of 1:50 sgRNA oligos
10× T4 DNA Ligase Buffer	1 μl
T4 DNA Ligase	0.25 μl
BpiI FastDigest	0.25 μl
H_2O	to 10 μl

37 °C 16 °C	5 min 5 min	30×
37 °C 50 °C 80 °C	30 min 5 min 5 min	

(*see* **Notes 6** and 7)

8. Transform into Subcloning Efficiency™ DH5α™ Competent Cells or equivalent chemically competent bacteria as per the manufacturer's protocol. We generally use 10 μl competent bacteria with 0.5–1 μl of ligation product.
9. Plate onto relevant antibiotic selection plates for the backbone used (generally Ampicillin) and incubate overnight at 37 °C.
10. Pick two to three colonies into 5 ml LB with vector-relevant antibiotic each. Incubate at 37 °C with shaking overnight.
11. Extract plasmids with QIAprep Spin Miniprep kit or equivalent and quantify on a NanoDrop 2000 spectrophotometer or by other desired method.
12. Take 1 μl of product from each for diagnostic digests. Although exact digests will be vector dependent, BsmBI can be used to verify gRNA insertion, and SmaI can be used to verify tRNA insertion for the $tRNA^{Pro}$ variants (*see* **Note 9**).
13. Following digestion, run reactions alongside an equivalent amount of undigested vector on a 1% Agarose gel. Banding pattern can then be compared to expected patterns (this can be predicted with a variety of software including SnapGene and Benchling) based on the enzyme combination used. For single digests a sharp band at the expected size is expected as opposed to multiple smaller or larger bands in the undigested case.
14. Vectors which show correct digest patterns should then be validated by Sanger sequencing to ensure that no point

mutations are present using a primer located ~100 bases outside of the region of insertion.

15. Sequence validated vectors are now ready for downstream cloning of spacer sequences.

3.4 Spacer Cloning

1. Order oligonucleotides for the spacer of interest. Examples with the overlaps from several tRNAs are shown below. Additionally, order the no tRNA negative control and U6 positive control spacer (*see* **Notes 10** and **11**).

```
tRNAPro
5′...GACGA    GCCCNNNNNNNNNNNNNNNNNNNN    GTTTTAGAG... 3′
3′...CTGCTCGGG    NNNNNNNNNNNNNNNNNNNNCAAA    ATCTC... 5′

tRNAGly
5′...GCCCA    TGCANNNNNNNNNNNNNNNNNNNN    GTTTTAGAG... 3′
3′...CGGGTACGT    NNNNNNNNNNNNNNNNNNNNCAAA    ATCTC... 5′

tRNAGln
5′...GTGGA    ACCTNNNNNNNNNNNNNNNNNNNN    GTTTTAGAG... 3′
3′...CACCTTGGA    NNNNNNNNNNNNNNNNNNNNCAAA    ATCTC... 5′

tRNAGlu
5′...GTCAG    GGAANNNNNNNNNNNNNNNNNNNN    GTTTTAGAG... 3′
3′...CAGTCCCTT    NNNNNNNNNNNNNNNNNNNNCAAA    ATCTC... 5′

No tRNA negative control.
5′...CATGC    CAACNNNNNNNNNNNNNNNNNNNN    GTTTTAGAG... 3′
3′...GTACGGTTG    NNNNNNNNNNNNNNNNNNNNCAAA    ATCTC... 5′

U6 positive control.
5′...CGAAA    CACCNNNNNNNNNNNNNNNNNNNN    GTTTTAGAG... 3′
3′...GCTTTGTGG    NNNNNNNNNNNNNNNNNNNNCAAA    ATCTC... 5′
```

2. Set up a digest with the vector produced from Subheading 3.2 or Subheading 3.3 as follows:

Vector	1 μg (tRNA vector and no tRNA control)
NEBuffer 3.1	3 μl
BsmBI	1 μl
H_2O	to 30 μl

Additionally, set up digests with a standard U6 driven sgRNA containing a BbsI/BpiI (these enzymes are equivalent) placeholder sequence (For example Addgene plasmid 61424 for the SAM-sgRNA):

Vector	1 μg
10× FastDigest Buffer	3 μl
BpiI FastDigest	1 μl
H_2O	to 30 μl

(*see* **Note 12**)

3. Incubate at 55 °C for 1 h for BsmBI digests, and 37 °C for 1 h for U6 control vector and GoldenGate destination vector plasmids.
4. While the vector is being digested phosphorylate and anneal the oligos designed in **step 1** as follows:

100 μM Forward Spacer	1 μl
100 μM Reverse Spacer	1 μl
10× T4 Ligase Buffer	1 μl
T4 PNK	0.5 μl
H_2O	6.5 μl

Incubate at 37 °C for 30 min, 95 °C for 5 min, then bring to 25 °C at a rate of 0.1 °C/s. Store on ice until ligation.

5. Run the digested destination plasmid and control destination plasmids on a 1% agarose gel and extract the expected bands using a QIAquick Gel Extraction Kit as per the manufacturer's instructions substituting the standard columns for a MinElute PCR purification column. Elute in 15 μl Elution Buffer. Store digested destination plasmid on ice until ligation (*see* **Note 2**).
6. Make a 1:50 dilution of annealed, phosphorylated oligo in water (1 μl of annealing reaction + 49 μl water).
7. Set up ligation reactions for tRNA vector, no tRNA control, and U6 positive control as follows:

Vector	1 μl
T4 DNA ligase buffer	1 μl
1:50 spacer oligo dilution	1 μl
H_2O	6.75 μl
T4 DNA ligase	0.25 μl

For each vector, set up control reactions with water in place of the inserts.

8. Mix gently and incubate at room temperature (20–25 °C) for 10 min.
9. Stop the reactions by incubating at 65 °C for 10 min.
10. Place the reactions on ice for 2 min.
11. Transform 0.25–1 μl into 10 μl Subcloning Efficiency™ DH5α™ Competent Cells or equivalent chemically competent bacteria as per the manufacturer's protocol.
12. Plate onto relevant antibiotic selection plates for the backbone used (generally Ampicillin) and incubate overnight at 37 °C.
13. The following day there should be at least 4× more colonies on the plate with insert (generally 50–100+ colonies).
14. Pick two to three colonies into 5 ml LB with vector-relevant antibiotic for each of the plates with inserts. Incubate at 37 °C with shaking overnight.
15. Extract plasmids with QIAprep Spin Miniprep kit or equivalent.
16. Quantify DNA using a NanoDrop 2000 spectrophotometer or by other equivalent methods.
17. One of the clones should then be validated by Sanger sequencing to ensure that no point mutations are present in the inserted spacer using a primer located ~100 bases outside of the region of insertion.
18. Sequence validated spacers are now ready for functional testing/use.

3.5 Functional Testing of tRNA Constructs

1. Clone CTS2 (GTAAGTCGGAGTACTGTCCT) spacer [12] into the tRNA construct as well as the U6 positive control and no tRNA negative controls as in Subheading 3.4.
2. Harvest HEK293T cells (ATCC-CRL-11268) using 0.05% Trypsin-EDTA and count using a hemocytometer one night prior to transfection.
3. Plate sufficient cells per well such that the next morning cells will be at 60–80% confluence in appropriate volumes of DMEM +10% FBS (for example, 50,000–75,000 cells in 0.5 ml medium for a 24-well plate).
4. Make equimolar DNA mixtures containing dCAS9-VP64_GFP (Addgene 61422 [11]), 8xCTS2-MLP-ECFP (Addgene 55198 [3]), and the constructs to be tested. Conditions should include a U6 control, the no tRNA control, a no gRNA control (in which the gRNA plasmid is omitted entirely or replaced by a scrambled sequence), and the test condition (s). If the tRNA is being used for gRNA release from Pol-II

transcripts, wells with and without induction should be included. Additionally, wells for no transfection and single color controls should be included for both EGFP and ECFP. These could be the dCAS9-VP64_GFP alone, and dCAS-VP64_Blast (Addgene 61425 [11]) + 8xCTS2-MLP-ECFP + U6-CTS2-sgRNA.

5. Aliquot DNA mixes for each well in PCR strip tubes such that during transfection final concentration of each plasmid will be 100 pM and total DNA amount per well will be ~500 ng for a 24-well plate assuming 250 μl transfection volume (or ~1 μg for a 12-well plate assuming 500 μl transfection volume) (*see* **Note 13**).
6. Freeze strip tubes with aliquoted DNA at −20 °C for the following day (*see* **Note 14**).
7. On the day of transfection, verify that cells are at 60–80% confluence (*see* **Note 15**).
8. Remove all medium and replace with a 200 μl volume of warm DMEM +2% FBS (400 μl for a 12-well plate).
9. Return plate to the incubator.
10. Thaw plasmid mixtures and spin down briefly.
11. Aliquot PEI (3× the mass of DNA to be transfected) into new sterile 0.2 ml PCR tubes for that well (e.g., 1.5 μg per well of a 24-well plate assuming 500 ng/well).
12. Bring the volume up to 50 μl with Opti-MEM™.
13. Mix the PEI–Opti-MEM™ and add to the plasmid DNA.
14. Vortex vigorously for 10 s, then briefly spin down.
15. Let stand for 15 min at room temperature.
16. Add the PEI–DNA mixture dropwise to cells.
17. Incubate for 4 h at 37 °C and 5% CO_2.
18. Remove transfection medium and replace with 0.5 ml fresh DMEM +10% FBS (1 ml for a 12-well plate).
19. Grow cells for 1–3 days at 37 °C and 5% CO_2 (*see* **Note 16**).
20. Harvest using 0.05% trypsin–EDTA.
21. Take half of cells and extract RNA using your preferred method ensuring that the lower threshold of the column/method is 70 bp or below (methods that only retain 100 bp or more tend to preferentially lose correctly processed gRNA). The RNA will be used later for molecular validations (*see* **Notes 17** and **18**).
22. Spin down the remaining half of the cells at 300 × *g* for 5 min and remove the supernatant.

23. Resuspend cells in appropriate volumes of PBS + 2% FBS + 1 μg/ml Propidium Iodide (PI) and keep on ice in the dark.
24. Set up a flow cytometer with a 405 nm (for ECFP), 488 nm (for EGFP), and 561 nm (for PI) lasers for analysis using the untransfected and single color controls. Ensure that voltages and filter sets are appropriate to allow compensation between the ECFP and EGFP channels as some spillover will be observable (*see* **Note 19**).
25. Run samples and determine activity by examining the ECFP positivity within the viable (PI-) EGFP+ cells. Gating for EGFP should be based on the EGFP level at which ECFP positivity is observed in the U6-CTS2-sgRNA control as this is the minimum amount of dCas9-VP64 at which a cell is capable of detecting gRNAs. ECFP positivity should be set such that there are ~0.1% ECFP+ cells within the $PI^{-}EGFP^{+}$ population of the no gRNA control well. Note, the ECFP gate may have to be angled across EGFP brightness depending on machine parameters/compensation. *See* Fig. 2 for example gating.

3.6 Measurement of Total gRNA Quantity

1. Take a small aliquot of the extracted RNA for each sample from Subheading 3.5 and treat with a TURBO DNA-free™ Kit to remove residual plasmid contamination (*see***Note 20**).

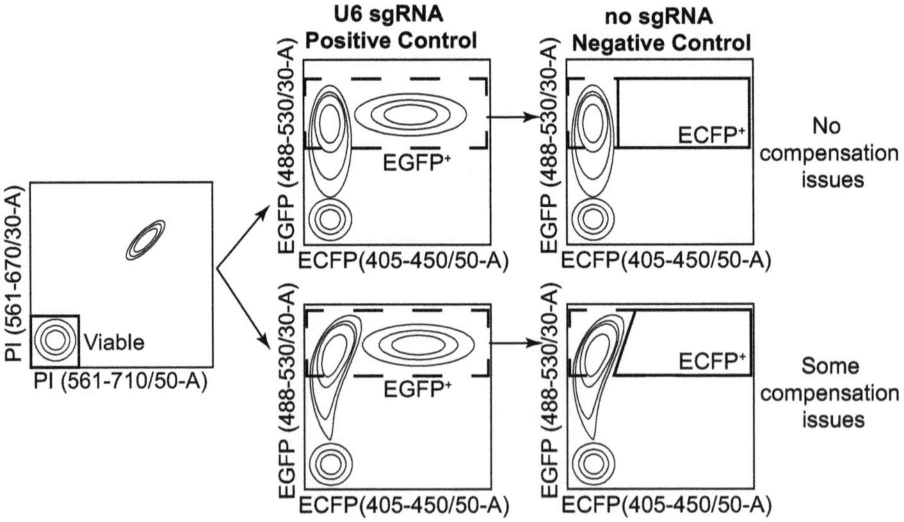

Fig. 2 Example gating hierarchy for functional gRNA activity. The viability gate is shared across controls. The channels which we generally use are listed for each detection channel in the format Laser (nm)—Bandpass filter. The control used to set each gate is listed above and expected plots are shown below without (top) and with (bottom) spillover problems

2. Perform reverse transcription using the QuantiTect® RT kit on half of the treated RNA using the supplied random primer with an incubation time of 30 min rather than 15 min at 42 °C.
3. For the remaining half of the RNA, dilute with water to an equivalent volume to the reverse transcribed RNA as a no RT control.
4. Set up triplicate qPCR reactions for each sample/no RT control with the following primer sets and conditions: "CTS2_qPCR-F" + "sgRNA_common_qPCR-R" (for total sgRNA abundance), and "dCas9_qPCR-F" + "dCas9_qPCR-R" (normalization control).

1 in 5 dilution of template	3 μl
10 μM F + R primer mix	0.25 μl
SsoAdvanced™ Universal SYBR® Green Supermix	5 μl
H_2O	1.75 μl

5. Run on a suitable qPCR system with annealing temperature 60 °C.
6. Calculate ΔCt and ΔΔCt relative to the dCas9-VP64 internal control and the U6 positive control. Limit of detection for gRNA is the ΔCt of the no RT controls.

3.7 Estimation of Processing Activity Using a Circularization-Based Amplification Assay

1. If a Pol-II promoter was used the 5′ cap must be removed prior to circularization. If not, skip to **step 5**. Note, the primers for this section are designed to work with the CTS2 spacer and standard sgRNA, if the SAM sgRNA was used then you would need to design alternative primers for the sgRNA.
2. Set up a decapping reaction as follows:

RNA	250 ng
RNA 5′ pyrophosphohydrolase (RppH, 5000 units/ml)	2.5 μl (12.5 U)
10× Thermopol Buffer (NEB)	2.5 μl
Nuclease Free H_2O	to 25 μl

3. Incubate for 1 h at 37 °C.
4. Purify the RNA by an appropriate method ensuring the minimum size cutoff is 70 bp or below (*see* **Note 21**).
5. Set up a circularization reaction as follows:

10× T4 RNA ligase buffer (NEB)	2 μl
10 mM ATP (NEB)	0.1 μl
SUPERase In RNase Inhibitor (20 U/μl, Thermo Fisher Scientific)	1 μl

(continued)

50% PEG8000 (NEB)	4 µl
T4 RNA Ligase 1 (10 U/µl, NEB)	1 µl
RNA	10 µl
H_2O	1.9 µl

6. Incubate for 4 h at room temperature.
7. Purify the RNA as done in **step 4**.
8. Perform reverse transcription using the primer "cRT-CTS2_nest_R" (Table 2) with the QuantiTect® RT kit using an incubation time of 30 min rather than 15 min at 42 °C. Other RT kits could be used but have not been tested.
9. Set up first round amplifications using the primer pair "cRT-sgRNA_nest_F" + "cRT-CTS2_nest_R" (Table 2) with the following conditions:

2× GC Phusion Master Mix	10 µl
10 µM Primer mix	1 µl
cDNA	1 µl
H_2O	8 µl

98 °C	60 s	
98 °C 58 °C 72 °C	10 s 30 s 15 s	10×
72 °C	5 min	

10. Make a one in ten dilution of the first PCR products and use this to set up a second PCR with the primers "cRT-sgRNA1_v2_F" + "cRT-CTS2_R" (Table 2) as follows:

2× GC Phusion Master Mix	10 µl
50% Sucrose +0.04% Bromophenol Blue	2 µl
10 µM Primer mix	1 µl
1:10 dilution of first PCR	1 µl
H_2O	6 µl

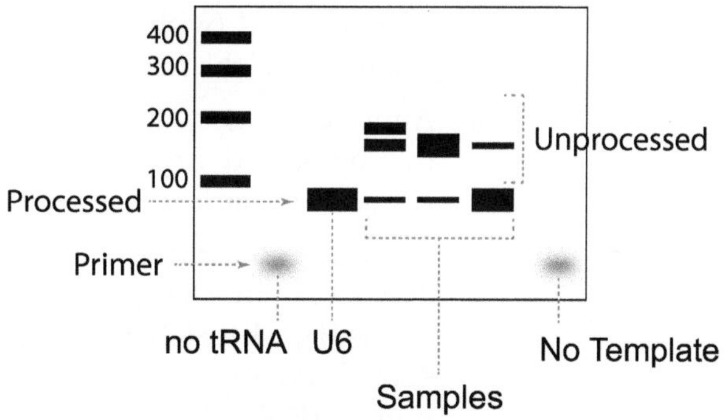

Fig. 3 Example circularization assay gel results. Symbolic band patterns expected for no tRNA control, U6 positive control, various possible sample patterns, and no template control

98 °C	60 s	
98 °C 58 °C 72 °C	10 s 30 s 15 s	15–25×
72 °C	5 min	

11. Run on a 2% agarose gel with appropriate nucleic acid dye and image on a suitable gel doc with exposure time such that no pixels are saturated.
12. Using appropriate software (e.g., GelAnalyzer2010a), remove background and quantify the pixel intensity of processed and unprocessed bands (Fig. 3).
13. Processing estimates are the raw volume of the correctly processed peak divided by the total raw volume of all peaks.

4 Notes

1. If the ultimate use requires the tRNAs to be placed on a lentivirus, ensure that they are on the reverse strand as otherwise they will process the viral genome preventing lentiviral production.
2. Vector concentration appears to matter hence the use of MinElute which allow lower elution volumes (and thus higher concentrations) than standard QIAquick columns. This is likely due to lower volume requirements and thus less contaminants transferred over from the gel purification. However, if the vector backbone is larger than 4–5 kb, standard columns

must be used (with 30 μl elution volume) as this is the maximum size of the MinElute columns.

3. Annealed oligos, digested vector and digested PCR products can be stored frozen at −20 °C and reused for additional ligations for at least 1 month.
4. These are compatible with the $tRNA^{Pro}$ variants, replace the 5′ "GCCC" overhang in "BsmBI-plch-sgRNA1-p1-F" and "BsmBI-plch-sgRNA1-p1-notRNA-F" with an appropriate 4 bp overhang for the specific tRNA of interest if not using human $tRNA^{Pro}$.
5. These primers are compatible with the $tRNA^{Pro}$ variants, replace the "GCCC" two bases downstream of the BpiI site in "BpiIGCCC-BsmBI-plch-sgRNA_MS2-F" with an appropriate 4 bp overhang for the specific tRNA of interest if not using human $tRNA^{Pro}$.
6. Reactions can be stored at 4 °C overnight or −20 °C for long-term storage prior to transformation.
7. No-template controls will often show some degree of background which does not reflect the actual reaction. As such, we generally do not setup no-template controls for GoldenGate reactions.
8. We have added a 14 bp unstructured sequence between the 3′ tRNA and the gRNA to ensure that there is no 3′ trimming of the gRNA, however, this may not be necessary and could likely be removed or replaced with other sequences of interest.
9. These do not provide a useful double digest due to close proximity and differing temperature requirements.
10. Reverse oligonucleotides for spacer cloning are common across all tRNAs as well as U6 promoted constructs. Switching between tRNA/systems thus only requires a new forward oligonucleotide.
11. For U6 controls, transcription starts at the first G in the spacer. If the spacer does not contain a G at the beginning one should be added to the start of the spacer. This is not a requirement for tRNA mediated Pol-III or Pol-II guide production.
12. Ensure that the U6 vector does not contain a Cas9 as this could interfere with downstream applications.
13. Total quantity of DNA for a 24-well plate can range up to almost 1 μg before toxicity is observed. However, reducing the amount of DNA below 500 ng/well tends to decrease cellular uptake.
14. These can be stored for up to a week if needed.

15. Ensure that the HEK293T cells are not above 80% density at time of transfection as this can severely compromise transfection efficiency.
16. Some background activation may be observed with longer incubations even in no tRNA controls. This is likely due to cryptic promoters in the vector.
17. Alternatively, these can be spun down, the supernatant removed and then snap-frozen and stored at −80 °C for extraction at a later time.
18. Ensure that the bench, and all equipment which will be in contact with the RNA is RNase clean, and be sure to wear gloves and not breath over the tubes to prevent RNA degradation.
19. Voltages for both ECFP and EGFP should be adjusted to minimize spillover of EGFP into of ECFP. This can be achieved by adjusting the EGFP channel and ECFP channel voltages using the EGFP single color control and testing various compensation parameters with it.
20. This extra DNase treatment step is critical as the plasmid DNA is not fully removed by standard RNA extraction kits (even those involving DNase treatment steps), and will severely skew qPCR results.
21. We have generally used ChargeSwitch Total RNA Cell Kit omitting the 60 °C lysis incubation and DNase treatment. We used 12.5 μl of ChargeSwitch beads per reaction and a final elution volume of 12.5 μl. We have also used phenol–chloroform extraction successfully.

References

1. Cong L, Ran FA, Cox D et al (2013) Multiplex genome engineering using CRISPR/Cas systems. Science 339:819–823
2. Mali P, Yang L, Esvelt KM, Aach J, Guell M, DiCarlo JE, Norville JE, Church GM (2013) RNA-guided human genome engineering via Cas9. Science 339:823–826
3. Nissim L, Perli SD, Fridkin A, Perez-Pinera P, Lu TK (2014) Multiplexed and programmable regulation of gene networks with an integrated RNA and CRISPR/Cas toolkit in human cells. Mol Cell 54:698–710
4. Knapp DJHF, Michaels YS, Jamilly M, Ferry QRV, Barbosa H, Milne TA, Fulga TA (2019) Decoupling tRNA promoter and processing activities enables specific Pol-II Cas9 guide RNA expression. Nat Commun 10:1490
5. Lodish H, Berk A, Zipursky SL, Matsudaira P, Baltimore D, Darnell J (2000) Molecular cell biology, 4th edn. W. H. Freeman, London
6. Mefferd AL, Kornepati AVR, Bogerd HP, Kennedy EM, Cullen BR (2015) Expression of CRISPR/Cas single guide RNAs using small tRNA promoters. RNA 21:1683–1689
7. Wei Y, Qiu Y, Chen Y, Liu G, Zhang Y, Xu L, Ding Q (2017) CRISPR/Cas9 with single guide RNA expression driven by small tRNA promoters showed reduced editing efficiency compared to a U6 promoter. RNA 23:1–5
8. Xie K, Minkenberg B, Yang Y (2015) Boosting CRISPR/Cas9 multiplex editing capability with the endogenous tRNA-processing system. Proc Natl Acad Sci U S A 112:3570–3575

9. Dong F, Xie K, Chen Y, Yang Y, Mao Y (2017) Polycistronic tRNA and CRISPR guide-RNA enables highly efficient multiplexed genome engineering in human cells. Biochem Biophys Res Commun 482:889–895
10. Qi W, Zhu T, Tian Z, Li C, Zhang W, Song R (2016) High-efficiency CRISPR/Cas9 multiplex gene editing using the glycine tRNA-processing system-based strategy in maize. BMC Biotechnol 16:58
11. Konermann S, Brigham MD, Trevino AE et al (2015) Genome-scale transcriptional activation by an engineered CRISPR-Cas9 complex. Nature 517:583–588
12. Ferry QRV, Lyutova R, Fulga TA (2017) Rational design of inducible CRISPR guide RNAs for de novo assembly of transcriptional programs. Nat Commun 8:14633

Chapter 8

Targeting Noncoding RNA Domains to Genomic Loci with CRISPR-Display: Guidelines for Designing, Building, and Testing sgRNA–ncRNA Expression Constructs

David M. Shechner

Abstract

CRISPR-Display uses the *S. pyogenes* Cas9 protein to posttranscriptionally localize noncoding RNA (ncRNA) domains to any genomic site, by directly fusing these domains to the Cas9 sgRNA cofactor. This versatile technology enables numerous applications for interrogating natural chromatin-regulatory ncRNAs, or for utilizing artificial ncRNA and ribonucleoprotein (RNP) devices at individual chromatin loci. To achieve these, a successful CRISPR-Display experiment requires that chimeric sgRNA–ncRNA fusions are stably expressed and incorporated into Cas9 complexes, and that they retain their ncRNA "cargo" domains at the targeted genomic sites. Here, I describe a workflow for designing, building, and testing such chimeric sgRNA–ncRNA expression constructs. I detail strategies for choosing expression systems and sgRNA insertion topologies, for assaying the incorporation of sgRNA–ncRNA fusions into functional Cas9 complexes, and for surveying the activities of ncRNA domains at targeted genomic loci. This establishes an initial set of "best practices" for the design and implementation of CRISPR-Display experiments.

Key words CRISPR-Display, CRISPR, Extended sgRNA, Noncoding RNA, lncRNA, RNA Aptamer, RNA Devices, lncRNA reconstitution

1 Introduction

Noncoding RNAs (ncRNAs) exhibit an astonishingly broad functional repertoire that is both essential to all kingdoms of life [1], and which has been harnessed to create an array of synthetic biological devices [2]. In eukaryotic cells, natural ncRNAs contribute to many core regulatory processes within the nucleus, including telomere maintenance, sex chromosome dosage compensation, control of local chromatin structure [3], and coordination of the genome's higher-order three-dimensional organization [4]. However, mechanistic dissection of these nuclear regulatory ncRNAs can be quite challenging. Often, the function of a nuclear-localized ncRNA can be difficult to distinguish from that of its encoding

Tudor A. Fulga et al. (eds.), *CRISPR Guide RNA Design: Methods and Protocols*, Methods in Molecular Biology, vol. 2162, https://doi.org/10.1007/978-1-0716-0687-2_8, © Springer Science+Business Media, LLC, part of Springer Nature 2021

DNA locus, or from cases where the act of transcription (but not the transcript itself) is the biologically relevant event [5]. In parallel, this compendium of natural regulatory ncRNAs has been expanded by a host of artificial RNA-based devices, including synthetic logic circuits, signal transducers, structural scaffolds, imaging systems, and catalysts [6]. In theory, applying these devices to individual genomic loci could enable powerful systems for probing and controlling genome function.

To address each of these needs—a method that directly probes the function of chromatin-regulatory ncRNAs at the transcript level, and a platform with which to deploy artificial ncRNA devices genome-wide—my colleagues and I recently developed a technology termed CRISPR-Display [7]. In CRISPR-Display ("CRISP-Disp") a nuclease-deficient mutant of the *S. pyogenes* Cas9 protein ("dCas9") is co-opted to posttranscriptionally deploy ncRNAs of interest to targeted genomic loci, by directly linking these domains to the sgRNA scaffold (Fig. 1a). CRISP-Disp can accommodate natural ncRNAs of several kilobases in length, and in some cases, can partially reconstitute the predicted regulatory activity of these ncRNAs at the targeted genomic site [7–9]. Moreover, CRISPR-Display can deploy a diverse group of artificial ncRNA domains, including aptamers, scaffolds, designed ribonucleoprotein (RNP) complexes, and large pools of random sequences [7]. This has enabled a number of novel genomic manipulations, including targeted gene activation or repression, and live-cell imaging of

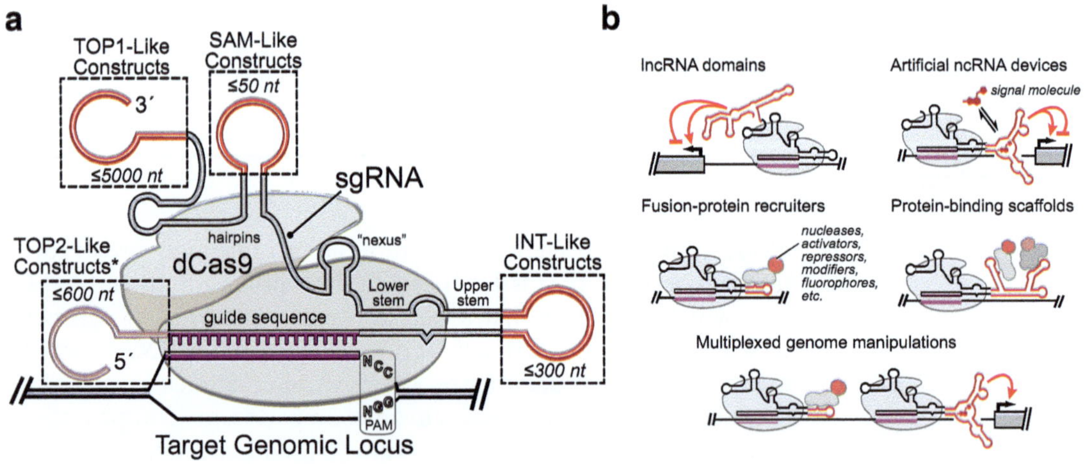

Fig. 1 Overview of CRISPR-Display. (**a**) Noncoding RNA domains can be functionally appended onto the *S. pyogenes* sgRNA at multiple points. Pertinent structural elements on the sgRNA are indicated. Cargo sizes correspond to the largest domains currently reported at each insertion point; still larger insertions may be possible. Note that concomitant insertion of cargo domains into several insertion sites greatly diminishes sgRNA functionality. Moreover, "TOP2"-like insertions are generally not tolerated under most sgRNA expression systems [7]. (**b**) Some of the RNA-based manipulations enabled by CRISPR-Display. Many such functions can be simultaneously performed at multiple loci [7, 10, 11]

individual loci, many of which can be performed simultaneously (Fig. 1b) [7, 10, 11]. Using a similar design philosophy, several groups have concomitantly developed powerful applications that use artificial RNA domains to modulate or expand Cas9 function [12–14].

Like all techniques, CRISPR-Display is constrained by methodological caveats that must be considered when designing and interpreting an experiment. These are particularly important when applying CRISP-Disp to long noncoding RNAs (lncRNAs), a large class of often low-abundance, poorly conserved transcripts, many of which are putative chromatin regulators [3, 15]. In a CRISPR-Display experiment, such ncRNAs are expressed from an atypical plasmid system (*see* Subheading 3.1, **step 4**), and in most cases at substantially higher abundance than that observed in nature [3, 7]. Thus, there are several reasons why CRISPR-Display may fail to reconstitute a particular lncRNA's activity. This activity may be intrinsically coupled to the lncRNA's biogenesis pathway, requiring *cis*-regulatory elements or assembly factors that are not recapitulated by the heterologous expression system. Overexpressing an lncRNA outside of its native context might allow it to populate misfolded states, or to assemble into spurious RNP complexes, leading to false-negative results. Alternatively, in some cases lncRNA function can be context dependent, requiring additional factors or chromatin structures at its site of action [9], thus limiting its portability to exogenous loci.

While CRISPR-Display is not a universal method for reconstituting lncRNA function, it does provide a flexible experimental platform for attempting such a reconstitution. To increase the likelihood of success, I suggest three general strategies regarding the design, implementation, and quality control stages of a CRISPR-Display lncRNA experiment, respectively. First, concerning construct design: in addition to testing the full-length lncRNA, I recommend surveying a variety of truncation constructs that comprise smaller RNA domains, or which span intervals along the lncRNA body. Many lncRNAs appear to operate as "scaffolds" that are built from discrete domains with separable functions [16]. These smaller domains may be easier to reconstitute in isolation than within the intact lncRNA, since they are less prone to misfolding and degradation and tend to be more efficiently bound by dCas9 [7]. Second, concerning lncRNA functional analysis: I suggest implementing assays that explicitly model the presumptive lncRNA activity as closely as possible. For example, transcription activation reporter systems that use minimal promoters (which require exogenous activators to recruit the entire core RNA Polymerase II machinery) may be ill suited to study lncRNAs that regulate Pol II elongation, but not initiation. Likewise, reporters driven from strong viral promoters may be irrepressible by an lncRNA that inhibits its natural targets only modestly. In general, favor assays that examine a single molecular step, such as the

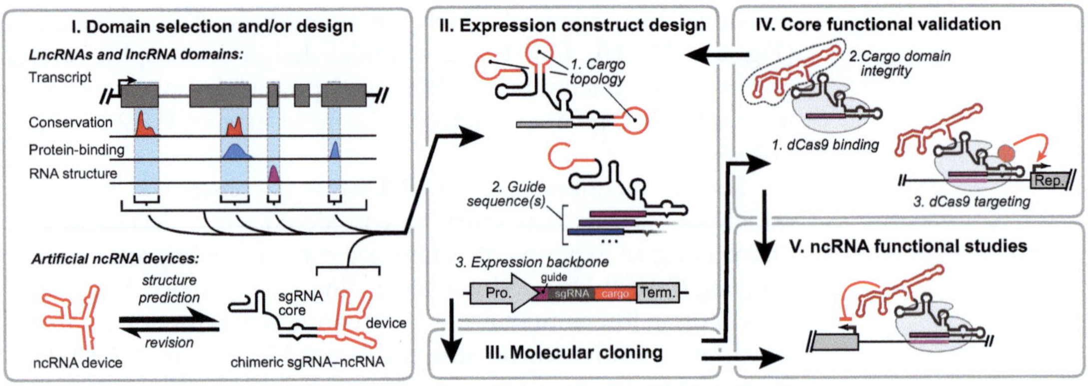

Fig. 2 Schematic summarizing the design–build–test workflow for CRISPR-Display

recruitment of a protein or the formation of a diagnostic histone modification, rather than more complex processes like transcription activation or silencing, which may be influenced by other modes of regulation [8]. Moreover, because an lncRNA's function may be context dependent, use CRISPR-Display to localize lncRNA constructs to their natural target loci (if known) before deploying them to other genomic sites, and test a series of guide RNAs sampling different positions within those loci [8, 9]. Third, regarding quality control: I recommend extensively characterizing and optimizing any novel sgRNA–ncRNA fusion construct used in a CRISPR-Display experiment, assessing its expression and stable incorporation into functional dCas9 targeting complexes. This final point is the main focus of this chapter.

Here, I present a unified experimental workflow for designing and testing CRISPR-Display constructs in cultured mammalian cells, using cargos built from both natural lncRNAs and artificial RNA devices (Fig. 2). I outline general considerations for construct design, including the choice of expression backbone, insertion topology, and cargo domain structure, and provide a molecular cloning protocol for generating these constructs. I will then detail assays that measure the incorporation of chimeric RNAs into dCas9 complexes, the integrity of RNA cargo domains in these complexes, and their targeting fidelity. Finally, I provide general guidelines for testing and utilizing ncRNA function at the targeted locus.

2 Materials

2.1 Construct Design for Natural lncRNAs

No special materials required.

2.2 Construct Design for RNA Devices

No special materials required.

2.3 Molecular Cloning

1. CMV/MASC backbone amplification primers:
   Forward:
   TAGTA**GAAGAC**TCGATTCGTCAGTAGGGTTGTAAAG.
   Reverse: TAGTA**GAAGAC**AA
   GGGTGGGTTTAGAGGAGGGATGTTGAAG.
   Each dissolved to 50 μM in nuclease-free water.
2. Gene- and domain-specific amplification primers for lncRNA constructs, designed in Subheading 3.3, each dissolved to 50 μM in nuclease-free water.
3. Oligo annealing buffer: 10 mM Tris–HCl, pH 7.5, 50 mM NaCl, in nuclease-free water.
4. Cargo-specific oligonucleotides for artificial ncRNA devices, designed in Subheading 3.3, each dissolved to 100 μM in oligo annealing buffer.
5. pCMV/MASC_(Gluc)_TOP1, and pU6_(Gluc)_INT(Gen-PurpClon) backbone plasmids (Addgene #68439 and #68433, respectively).
6. Ultra high-fidelity thermostable DNA polymerase, such as Pfu Ultra II HS, and its commercially supplied reaction buffer.
7. 10× dNTP mix: 2 mM each of dATP, dTTP, dGTP, and dCTP, in nuclease-free water.
8. Spin column-based DNA reaction cleanup and gel purification kits.
9. *BbsI* and *DpnI* restriction endonucleases, and their commercially supplied reaction buffers (e.g., NEBuffer 2.1).
10. Calf intestinal alkaline phosphatase.
11. NEB Quick Ligation Kit, or equivalent high-efficiency T4 DNA ligase, with corresponding buffer.
12. A programmable Peltier block thermocycler with heated lid.
13. Standard reagents for pouring, running, and visualizing agarose electrophoresis gels.
14. A UV-vis spectrophotometer, or equivalent.
15. Standard bacterial manipulation reagents, including competent cells, growth media, and LB agar plates supplemented with 100 μg/mL ampicillin, shaking and static 37 °C incubators.
16. Commercial endotoxin-free plasmid isolation kits, or equivalent.

2.4 RNA Incorporation and Integrity Assays

1. 0.1% (w/v) gelatin solution, in water, suitable for cell culture.
2. Cell growth medium: high-glucose Dulbecco's modified Eagle's medium (DMEM), supplemented with 10% fetal bovine serum, 1× penicillin–streptomycin, and 2 mM L-glutamine.

3. 10× Mg- and Ca-free phosphate buffered saline (PBS).
4. HEK 293T cells.
5. pEF_dCas9 plasmid (Addgene #68416). Aliquot, minimize the number of freeze-thaws per tube, and keep on ice during use.
6. "Carrier" plasmid, such as pUC19 or pNEB193.
7. Lipofection reagent, such as Lipofectamine 2000, or equivalent.
8. Methanol-free paraformaldehyde solution, 16% in water, packaged in glass ampoules.
9. Fixation solution: 0.1% formaldehyde, in 1× PBS. Prepare in a biosafety cabinet immediately before use.
10. Quenching solution: 1.25 M glycine, in 1× PBS. Sterile filter and store at room temperature for at least 6 months.
11. RIPA lysis buffer (RIPA): 50 mM Tris–HCl pH 8.0, 150 mM KCl, 5 mM EDTA, 1% (v/v) Triton X-100, 0.5% (w/v) sodium deoxycholate, 0.1% (w/v) sodium dodecyl sulfate (SDS). Add the SDS last, to avoid precipitation. Prepare under RNase-free conditions, sterile filter, and store at 4 °C for at least 1 month.
12. Native lysis buffer (NLB): 25 mM Tris–HCl pH 7.4, 150 mM KCl, 5 mM EDTA, 0.5% (v/v) Nonidet P-40. Prepare under RNase-free conditions, sterile filter, and store at 4 °C for at least 6 months.
13. 1 M Dithiothreitol (DTT), prepared in nuclease-free water. Sterile filter, aliquot and store at −20 °C for at least 1 year.
14. "Halt™ 100× EDTA-free proteinase inhibitor cocktail" (Pierce), or equivalent. The Pierce 100× formulation contains AEBSF·HCl (100 mM), aprotinin (80 μM), bestatin (5 mM), E-64 (1.5 mM), leupeptin (2 mM), and pepstatin A (1 mM).
15. Commercial RNase inhibitors, such as RNaseOUT.
16. RIPA(+) buffer: RIPA, supplemented with RNaseOUT (0.1 U/mL, final), DTT (0.5 mM, final), and EDTA-free proteinase inhibitors (1×, final). Prepare on ice, immediately prior to use.
17. RNase-Zap, or equivalent cleaning solution (e.g., 0.1 N NaOH, 1% SDS, in nuclease-free water).
18. NLB(+) buffer: NLB, supplemented with RNaseOUT (0.1 U/mL, final), DTT (0.5 mM, final), and EDTA-free proteinase inhibitors (1×, final). Prepare on ice, immediately prior to use.
19. Rabbit anti-FLAG antibody, prepared, aliquoted and stored as per the manufacturer's instructions.
20. Normal Rabbit IgG, prepared, aliquoted, and stored as per the manufacturer's instructions.

21. Protein-G Magnetic beads.
22. 3× Reverse-Crosslinking (RCL) Buffer: 3× PBS, 6% (v/v) N-lauroyl sarcosine, 30 mM EDTA, 15 mM DTT. Prepare fresh under RNase-free conditions.
23. Proteinase K Solution, 20 mg/mL, in 50 mM Tris–HCl pH 8.0, 3 mM $CaCl_2$, 50% (v/v) Glycerol. Store at −20 °C.
24. RNase-free Solid Phase Reversible Immobilization (SPRI)-type magnetic beads.
25. 70% (v/v) ethanol, in nuclease-free water. Prepare fresh.
26. RNase-free DNase (such as RQ1, or equivalent) and commercially supplied reaction buffer.
27. Random DNA hexamers, 50 ng/μL, in nuclease-free water.
28. Reverse transcriptase (SuperScript III, or equivalent), and its commercially supplied reaction buffer.
29. sgRNA core domain qPCR primers, dissolved in nuclease-free water to 50 μM.
    sgRNA_Fwd: CTCGTATTCGCAGCATAGCAA
    sgRNA_Rev: TTCAAGTTGATAACGGACTAGCCT
30. GAPDH qPCR primers, dissolved in nuclease-free water to 50 μM.
    GAPDH_Fwd: TTCGACAGTCAGCCGCATCTTCTT.
    GAPDH_Rev: GCCCAATACGACCAAATCCGTTGA
31. Gene-specific qPCR primers, designed in Subheading 3.4, **step 1**, each dissolved to 50 μM in nuclease-free water.
32. SYBR-based, Rox-normalized real-time PCR master mix.
33. Standard mammalian tissue culture equipment, such as CO_2-controlled 37 °C incubators, biosafety cabinets, dissecting microscopes, and platform rockers.
34. At least two metal modular microcentrifuge heat blocks (20 sample, 1.5 mL tube).
35. Branson SFX250 Digital sonifier, equipped with a (1/8) inch microtip, or equivalent device, ideally mounted in a sound-proof enclosure.
36. Magnetic separation stand, equipped for 1.5 mL microcentrifuge tubes.
37. A real-time PCR thermocycler capable of measuring Rox-normalized SYBR a fluorescence.

2.5 dCas9 Targeting Assays

1. HEK 293 T cells and requisite cell culture supplies (**items 1–4** in Subheading 2.4).
2. pEF_dCas9 plasmid (Addgene #68416). Aliquot, minimize the number of freeze-thaws per tube, and keep on ice during use.

3. pEF_dCas9-VP64 plasmid (Addgene #68417). Aliquot, minimize the number of freeze-thaws per tube, and keep on ice during use.
4. pU6_(GLuc)_sgRNA plasmid (Addgene #68422).
5. *Gaussia* luciferase "reporter" plasmid pTrans_9xT7_Gluc_Rep (available from the author upon request).
6. *Cypridina* luciferase "normalizer" plasmid pTrans_9xD3B_-CLuc_Norm (Addgene #68415).
7. "Carrier" plasmid, such as pUC19 or pNEB193.
8. Lipofection reagent, such as Lipofectamine 2000, or equivalent.
9. *Gaussia* luciferase substrate and assay buffer, as in the BioLux *Gaussia* Assay kit, or equivalent.
10. *Cypridina* luciferase substrate and assay buffer, as in the BioLux *Cypridina* Assay kit, or equivalent.
11. Guanidinium thiocyanate–acid phenol–based whole-cell RNA extraction reagent, such as TRIzol.
12. Commercial spin-column-based RNA reaction cleanup kits, such as QIAgen's RNeasy kit.
13. RT Primer Mix: 25 ng/μL Random DNA hexamers; 25 μM oligo-dT primer, in nuclease-free water.
14. Reverse transcriptase (SuperScript III, or equivalent), and its commercially supplied reaction buffer.
15. GAPDH qRT-PCR primers (**item 30** from Subheading 2.4).
16. Gene-specific qPCR primers, designed in Subheading 3.5, **step 10**, each dissolved to 50 μM in nuclease-free water.
17. SYBR-based, Rox-normalized real-time PCR master mix.
18. Standard mammalian tissue culture equipment, such as CO_2-controlled 37 °C incubators, biosafety cabinets, dissecting microscopes, and platform rockers.
19. A fluorescent light microscope, capable of at least 10–20× magnification and equipped with filters for Cyan Fluorescent Protein (CFP) and Yellow Fluorescent Protein (YFP).
20. A Luminometer capable of reading 96-well plates, preferably equipped with an automated injector system.
21. Clear 96-well assay microplates, with lids. These do not need to be TC-treated or nuclease-free.
22. Black-walled 96-well assay plates. Use opaque-bottomed plates for luminometers that use a top-down optic; clear-bottomed for those using a bottom-up optic.

23. A real-time PCR thermocycler capable of simultaneously measuring SYBR and Rox fluorescence.

2.6 ncRNA Functional Assays

Special reagents will vary based on assay design.

3 Methods

A typical CRISPR-Display workflow is essentially divided into four stages (Fig. 2). First, RNA expression constructs are computationally designed, taking into consideration the RNA cargo domain of interest, the expression system used, and the genomic site to which this cargo will be targeted. Second, these constructs are cloned into mammalian expression plasmids. Third, these constructs are coexpressed in cultured mammalian cells with dCas9 and any necessary accessory factors, and subjected to quality control assays that assess their incorporation into functional CRISP-Disp complexes. Constructs failing these assays are redesigned, cloned, and tested anew; this cycle is repeated as necessary. Finally using these vetted expression constructs, RNA cargo domains are deployed to targeted genomic sites and analyzed in functional assays customized to the domain of interest. Although this abstract workflow applies to both lncRNAs and artificial ncRNA devices, the technical considerations for these two RNA classes differ at several points, as detailed below.

3.1 Design of CRISPR-Display Constructs Bearing Natural lncRNAs

Here, the variables to consider are the lncRNA of interest (including the full-length transcript, and the battery of truncation constructs to be tested), the topology with which these domains are be appended onto the sgRNA scaffold, the promoter/terminator system that drives their expression [7], and the sgRNA guide sequences used. Although the optimal construct design will differ based on the particular requirements of a given lncRNA, some general guidelines are provided here.

1. Define the lncRNA of interest (*see* **Note 1**), and identify potentially useful domains therein, considering all available evidence for RNA-level function. This might include any of the following criteria: (1) sequences with higher nucleotide-level conservation; (2) exons that are more frequently retained among spliced variants; (3) regions with disease-associated variants; (4) sequences thought to bind protein factors—based on in-house biochemical data, in consultation with public datasets (such as ENCODE, https://bit.ly/2LBxUZv), or by scanning for known protein recognition motifs [17]; (5) domains with evidence for adopting ordered structures, based on high-throughput chemical probing analyses [18] or on sequence homology. In the absence of such data, functional domains

can be experimentally identified by generating constructs that naively "tile" regular intervals along the length of the lncRNA [19].

2. Chose an insertion point on the sgRNA into which to append each domain identified above. Note that this decision may vary between domains, even those isolated from the same parental lncRNA. Currently, three robust insertion sites have been characterized (Fig. 1a): at the sgRNA 3′-terminus (termed "topology 1" or "TOP1" [7]), in the engineered loop that joins sgRNAs' guide and tracer domains (termed "internal" or "INT" [7]), and in the natural tetraloop that caps the first of the 3′-hairpins (as in "SAM"-type constructs [20]). TOP1-like constructs are the best choice for RNA cargos larger than ~300 nucleotides (*see* **Notes 2** and **3**), though these domains can be vulnerable to degradation, and care must be taken to assess their integrity (*see* Subheading 3.4). Smaller inserts at the TOP1 site are also viable [7, 11], though these smaller cargos appear to be more efficiently expressed in the INT topology [7, 11, 14]. This is particularly true for RNA domains in which the 5′ and 3′ termini are base-paired to one another, since INT-like constructs insert this terminal duplex contiguously with the sgRNA "upper stem" (Fig. 1a).
3. Design sgRNA guide sequences. CRISPR-Display does not appear to significantly influence the fidelity or generality of Cas9 targeting [7], and thus guide design does not substantially differ from that used in other CRISPR-Based applications. The choice of the target genomic locus—and the guide-binding regions within that locus—will depend upon the functional assay used (*see* Subheadings 3.5, **steps 9–15** and 3.6), though putative transcription-regulatory RNAs should be targeted upstream of the regulated gene's transcription start site (TSS) (*see* **Note 4**). Consult the literature to identify vetted guide sequences that may have already been established for the target locus. Else, use conventional guide design software (for example, the Broad sgRNA design portal: https://bit.ly/2Ajgjkq) to design three to five new guides per target locus. These can be tested in isolation, to account for proximity effects between the CRISP-Disp complex and regulated gene, or together as a group, to increase the experimental signal.
4. Select an appropriate expression system (*see* **Note 5**). Longer RNAs—particularly those in the TOP1 orientation—must be driven from a Pol II promoter; stable nuclear retention of these transcripts is further enhanced using atypical (i.e., not polyadenylation-based) 3′-processing signals. The preferred expression system for long CRISP-Disp constructs is termed "CMV/MASC," and drives expression from a canonical CMV

promoter, and terminates transcription using the triple-helical Expression and Nuclear Retention Element (ENE) from the *MALAT1* gene [7, 21, 22]. An alternative system, "CMV/3′Box," can be used when expression from the CMV/MASC backbone fails to meet quality standards (*see* Subheadings 3.4 and 3.5) This system terminates transcription using an ~80 bp element derived from the 3′-terminus of the U1 snRNA gene, which directs specialized Pol II termination without polyadenylation [23]. INT-like constructs should be expressed from a standard U6 promoter. Note that Pol III will terminate transcription at sequences of greater than five consecutive uridines, which may require modification to some cargo sequences.

3.2 Design of CRISPR-Display Constructs Bearing RNA Aptamers and Other Devices

Although the design of CRISPR-Display constructs bearing artificial RNA devices is theoretically constrained by the same variables that influence lncRNA construct design, in practice the process is substantially simpler (Fig. 3). RNA devices tend to be smaller and better characterized than are lncRNAs, thus obviating the search for functional domains. Moreover, such devices often tolerate a wide range of sequence variants, which enables more design flexibility.

1. Define the RNA devices of interest. Consult the literature to identify each device's minimal functional domain, and to define

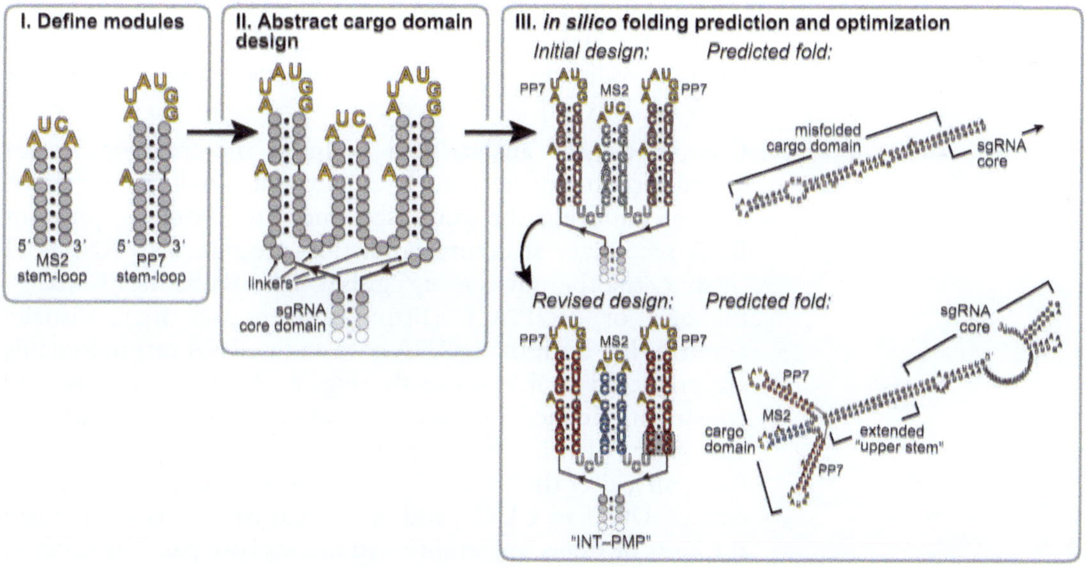

Fig. 3 Design of CRISPR-Display constructs bearing artificial ncRNA devices—here, an "INT"-like construct [7] in bearing two PP7 Stem loops bracketing an MS2 stem loop ("INT–PMP"). Gray circles can be nucleotide, as constrained by secondary structural requirements; black dots denote requisite Watson–Crick base pairs. Note that, when inserted into the sgRNA scaffold, the initial design (*top right*) is predicted to misfold. The revised design alleviates this problem by inverting two base pairs in the second PP7 stem (*gray box, lower right*). Secondary structures were predicted using RNAfold

sequences and secondary structural elements that can be altered without diminishing activity (*see* **Note 6**). Define the number of devices that will be displayed on each extended sgRNA. For example, when displaying small protein-binding stem loops, placing three to five tandem copies of the stem loop per guide RNA often significantly increases the experimental signal [7, 10, 11]. Separate each module (e.g., each stem loop) with three-nucleotide, pyrimidine-rich spacers (*see* **Note 3**), such as 5′-CUC-3′ or 5′-UCU-3′. These will be optimized below, if necessary.

2. Chose a sgRNA insertion topology. Except in rare cases (e.g., cargos longer than 300 nucleotides, or which require a free 3′-terminus for activity), INT-like constructs are the optimal choice. If the 5′- and 3′-termini of the cargo domain are base-paired to one another, insert this domain directly in-line with the sgRNA "upper stem." Else, abut the cargo domain with three-nucleotide spacers (*see* **Note 3**), as in **step 1**.
3. Design guide sequences and select an expression system. Guide design follows the identical protocol to that outlined for lncRNA-bearing constructs, above (*see* Subheading 3.1, **step 3**). If the designed constructs employ the INT topology, express them using the U6 promoter system (*see* **Note 7**). To facilitate this, remove or edit any internal sequences containing more than four contiguous uridines. TOP1-like constructs (Fig. 1a) should also be driven from the U6 system, if possible, though the stability of these domains will need to be experimentally validated (*see below*). Else, use CMV/MASC or CMV/3′ Box [7].
4. Computationally analyze and optimize the transcript's predicted secondary structure. Input the preliminary RNA design—omitting the guide sequence (*see* **Note 8**)—into an RNA secondary structure prediction program like RNAFold (http://rna.tbi.univie.ac.at/cgi-bin/RNAWebSuite/RNA-fold.cgi) or NUPACK (http://www.nupack.org). Visually confirm that both the sgRNA core and ncRNA cargo domains are predicted to fold correctly (Fig. 3, *right*). If not, use the constraints defined in **step 1** to revise the construct design: change the identity of base-pairs (e.g., change an A·U pair to a C·G pair), alter the "spacer" sequences between modules (e.g., change UCU to CUC), add additional base pairs to stabilize the ends of stems, or remove extraneous base pairs, as needed. Work iteratively and conservatively, making few changes at a time and resubmitting each revision into the prediction software. For a successful design, the RNAFold predicted Minimum Free Energy (MFE) and Centroid secondary structures will be in agreement, and the majority of bases will have high (near 1.0) base-pair probabilities and low (near 0) positional entropies (Fig. 3, *right*; *see* **Note 9**).

3.3 Cloning Chimeric sgRNA–ncRNA Expression Constructs

CRISPR-Display ncRNA expression constructs do not inherently require specialized cloning strategies, and conventional restriction enzyme-based methods, "Golden Gate" cloning, and Gibson Assembly will suffice. As examples, I here present protocols that have become the preferred techniques for cloning TOP1- and INT-like constructs in my lab. TOP1-like constructs are cloned using an approach we have termed "SheqChange," in which the plasmid backbone (here, CMV/MASC–TOP1, Addgene #68439) is amplified via an "inverted" PCR that removes the existing cargo domain and replaces with it with a pair of Type IIs restriction enzyme sites. New cargo domains are then introduced by ligation or Gibson Assembly. We routinely use a similar strategy to swap guide RNA sequences (*see* **Note 10**). INT-like constructs are cloned into a general-purpose cloning plasmid termed "INTgpc" (Addgene Plasmid #68433), by introducing the new cargo domain on a single pair of annealed synthetic oligonucleotides (*see* **Note 11**).

Cloning CMV/MASC-driven TOP1-like constructs bearing lncRNA cargos:

1. Design cloning PCR primers for each of the lncRNA constructs outlined in Subheading 3.1. These primers should append their products with Type IIs restriction sites (e.g., *BbsI*) that leave a 5′-ACCC sticky end at the 5′-terminus and a 5′-AATC sticky end at the 3′-terminus. That is, the final product should take the form: [Type IIs Site]ACCC[LncRNA domain]GATT[Type IIs Site], wherein underlined residues denote the sticky ends generated by enzymatic cleavage. In the absence of compatible Type IIs enzymes, clone inserts by Gibson Assembly, using this PCR step to add appropriate homology arms to each fragment.
2. Amplify all lncRNA cargo domains by PCR (*see* **Note 12**), and confirm their integrity by agarose gel electrophoresis. Clean up reactions using commercial spin kits, eluting into 34 μL nuclease-free water, or 10 mM Tris–HCl, pH 8.0. Avoid buffers containing EDTA.
3. "Inverse PCR" the plasmid backbone. Combine in a thin-walled, capped PCR tube: 10 ng CMV/MASC-TOP plasmid, 5 μL 10× Pfu buffer, 5 μL 10× dNTPs, 0.5 μL of each CMV/MASC amplification primer, and bring the reaction to 49 μL with nuclease-free water. Thoroughly mix by vortexing, and add 1 μL Pfu Ultra II HS polymerase. Mix by gentle flicking and inversion, and amplify in a thermocycler using the following program: (1) 95 °C–2 min, (2) 25 cycles of [95 °C–20 s; 51 °C—20 s; 72 °C—60 s], (3) 72 °C–3 min.
4. Remove 8 μL for agarose gel analysis, confirming the presence of a 3.7 kb product. Clean up the remaining 42 μL using a

commercial spin kit, eluting into 32 μL water or EDTA-free elution buffer.

5. Digest the backbone and inserts, moving all of the material remaining from **steps 2** and **3** into thin-walled, capped PCR tubes. To each insert sample, add 4 μL 10× NEBuffer 2.1 and 2 μL *BbsI* (20 U). To the backbone sample, add 4 μL 10× NEBuffer 2.1, and 2 μL each *BbsI* (20 U) and *DpnI* (50 U). Incubate at 37 °C for 2.5 h in a thermocycler. Thereafter, to the plasmid backbone only, add 2 μL (20 U) Calf Intestinal Alkaline Phosphatase, and incubate at 37 °C for one hour more.
6. Purify the digested samples. If significant off-target bands were observed in the agarose gels run in **steps 2** or **4**, then run the entire reaction on a preparative agarose gel, and purify the desired product. Clean up the samples all samples (gel purified, or not) using commercial cleanup spin kits, eluting in 32 μL water or EDTA-free elution buffer.
7. Quantify the concentration of all samples by UV-vis absorption.
8. Ligate the samples. For each construct, combine in thin-walled, capped PCR tubes: 25 ng plasmid backbone with a threefold molar excess of insert, bringing the total volume to 5 μL with nuclease-free water. Also prepare a "no-insert" control reaction with only the backbone DNA. To each sample, add 5 μL 2× Quick Ligase buffer and mix thoroughly by vortexing. Add 1 μL Quick Ligase, mix by flicking and gentle inversion, and incubate in a thermocycler at 25 °C for 15 min. Move the reactions to ice during the next step, or store at −20 °C indefinitely.
9. Transform 1 μL of each reaction into competent *E. coli* cells and select on LB-ampicillin plates. Subject several ampicillin-resistant colonies to Sanger sequencing, using standard CMV Promoter and SV40 poly(A) primers, as well as gene-specific internal primers, as needed. Ideally, for a "winning" clone, every nucleotide spanning the sgRNA and cargo domains should be confirmed in at least one high-confidence sequencing read.
10. Prepare endotoxin-free mini or maxi preparations for each construct. Store plasmids at −20 °C.
11. In parallel, generate expression constructs for minimal sgRNAs bearing each of the guide sequences that will be used in functional assays (*see* Subheading 3.6), using standard molecular cloning techniques. These constructs should be driven from the same expression backbone (here, CMV/MASC), and will serve as "no cargo" controls.

Cloning U6-driven INT-like constructs bearing ncRNA device cargos:

12. Design the cargo insert oligonucleotides. These will take the general form:

    Cargo top oligo: 5′-pCGAG[Cargo Domain]**C**-3′.

    Cargo bottom oligo: 5′-pACGA**G**[Reverse Complement of Cargo Domain]-3′.

    wherein 5′-p denotes a 5′-phosphate, underlined residues will form the sticky ends used in cloning, and the bold residues add a 3′-terminal Cytosine to the insert that restores base-pairing in the "upper stem" of the sgRNA (Fig. 1a) [7]. Dissolve each oligo in Oligo Annealing Buffer, to a final concentration of 100 μM. Oligo stocks can be stored at −20 °C indefinitely.
13. Prepare the digested INTgpc plasmid. Combine in a thin-walled, capped PCR tube: 2 μg INTgpc plasmid, 5 μL NEBuffer 2.1, and bring the reaction to 48 μL with nuclease-free water. Mix thoroughly by vortexing, and add 2 μL *BbsI* (20 U). Mix by gentle flicking, incubate at 37 °C for 2.5 h in a thermocycler, add 2 μL (20 U) Calf Intestinal Alkaline Phosphatase, and incubate at 37 °C for 1 h more. Purify the entire reaction from a preparative agarose gel, eluting into 32 μL water or EDTA-free elution buffer.
14. Anneal the cargo oligonucleotides. Aliquot 1–2 mL of Oligo Annealing Buffer into a 15 mL conical tube, and chill on ice. Separately, combine 10 μL of each cargo oligo in a single thin-walled, capped PCR tube. Heat to 95 °C in a thermocycler, and anneal by slowly lowering the temperature to 25 °C over the course of 1 h. Incubate at 25 °C for 10 min, and place the reaction on ice. This corresponds to a 50 μM annealed oligo mix. Dilute 1 μL of this stock approximately 800-fold, to a final concentration of 63 nM, in ice-cold Oligo Annealing Buffer, and mix thoroughly by vortexing. Keep this stock ice before use (*see* **Note 13**). This entire annealing and dilution step should be performed immediately prior to use.
15. Ligate the samples. In a thin-walled, capped PCR tubes, combine 25 ng plasmid backbone and 1 μL 63 nM annealed oligo mix, bringing the total volume to 5 μL with nuclease-free water. Also prepare a "no-insert" control reaction with only the backbone DNA. Ligate and transform the samples as described in **steps 8** and **9**, above. Sequence ampicillin-resistant colonies using the standard M13 Reverse primer, and prepare endotoxin-free mini or maxi preparations of "winning" clones. Store plasmids at −20 °C.
16. In parallel, generate expression constructs for minimal sgRNAs bearing each of the guide sequences that will be

used in functional assays, (*see* Subheading 3.6) using standard molecular cloning techniques. These should be driven from the same expression backbone (here, U6), and will serve as "no cargo" controls.

3.4 Assessing dCas9-Binding and ncRNA Cargo Integrity

Frequently, the CRISPR-Display expression constructs designed and cloned in Subheadings 3.1–3.3 can be immediately used in functional assays (*see* Subheading 3.6). However, it is often helpful to subject these constructs to a gamut of quality control assays that confirm their basic functionality in mammalian cells (Fig. 2). These assays aim to quantify the incorporation of hybrid sgRNA–ncRNA transcripts into dCas9 complexes, the integrity of ncRNA cargo domains in these complexes, and dCas9 targeting efficiency (Fig. 4). Constructs that are deficient in any of these essential functions can then be redesigned by modifying the sequence of the cargo domain (*see* Subheadings 3.1, **step 1** and 3.2, **step 1**), by altering linker lengths and sequences (*see* **Note 3**), by selecting a different sgRNA insertion topology or expression system (*see* Subheadings 3.1, **steps 2, 4** and 3.2, **step 2**), or can be abandoned altogether. The protocol detailed in this section interrogates the formation and integrity of CRISPR-Display complexes using formaldehyde cross-linked RNA immunoprecipitation ("fRIP") and quantitative RT-PCR (qRT-PCR), essentially as described by Hendrickson, Tenen, and colleagues (Fig. 4a–b) [24]. The assays presented in Subheading 3.5 interrogate DNA-targeting by these complexes (Fig. 4c–d; *see* **Note 14**). Both sections use cultured HEK 293T cells as a model system, but other easily transfectable cultured cell lines can be substituted (*see* **Note 5**).

1. Design qPCR primers targeting all ncRNA cargo domains. Submit cargo domain sequences to the NCBI primer design tool, with options checked as in the following link (https://bit.ly/2AoEs9q). This will filter and report any off-target transcripts in the host transcriptome that might be aberrantly targeted by each candidate primer pair. Identify primers with the fewest such off-targets, or those for which all competing amplicons are longer than 300 bp. For cargo domains that are themselves longer than 100 nt, generate several primer pairs that span lengths along the ncRNA; it may be necessary to break the cargo domain into several smaller intervals and independently submit each to the NCBI design tool (Fig. 4a). Suspend all oligos to 50 μM in nuclease-free water.
2. Prepare HEK 293T cells for transient transfection. Working in a sterile biosafety cabinet, gelatinize one 10 cm tissue culture plate for each CRISPR-Display construct being tested, as well as an additional plate for a minimal sgRNA control transfection (*see* Subheading 3.3, **steps 11** and **16**, and **Note 15**). After 1 h at room temperature, remove the gelatin solution and replace

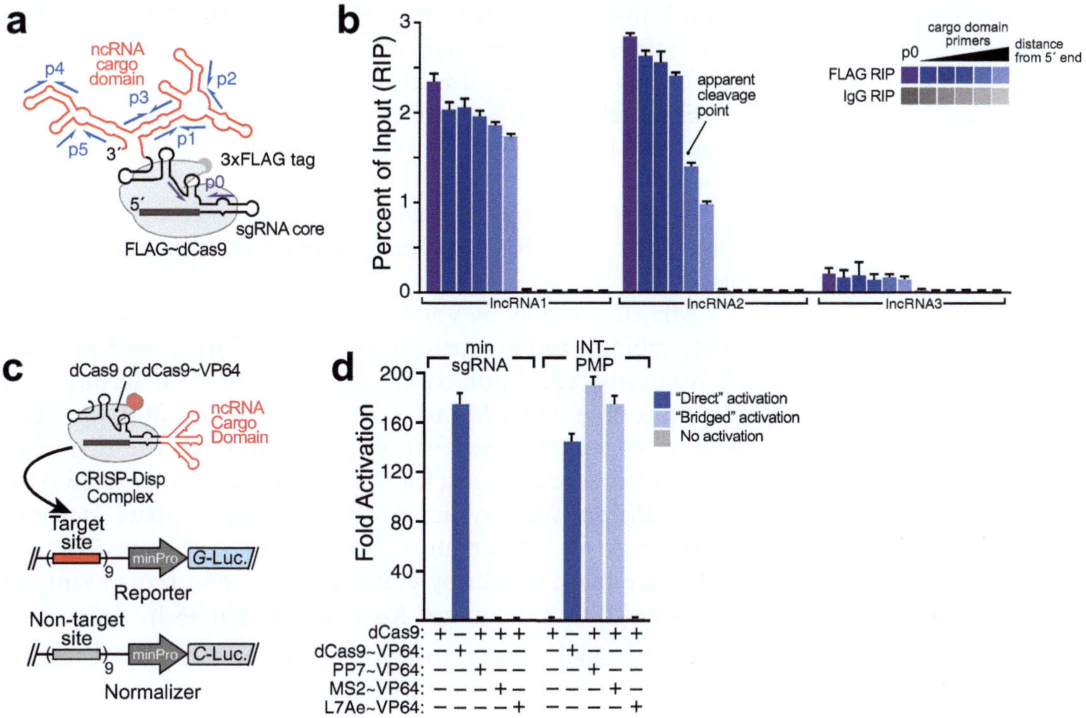

Fig. 4 CRISPR-Display quality control assays. (**a**, **b**) Assessing dCas9 binding and cargo domain integrity using formaldehyde RNA immunoprecipitation (fRIP). (**a**) Overview of the assay. CRISP-Disp RNA constructs are coexpressed with FLAG-tagged dCas9. RNA recovered from FLAG pulldown is probed by qRT-PCR, using primers that target the sgRNA core domain (p0) and various lengths along the cargo domain (here, p1—p5). (**b**) Example results. Here, "lncRNA1" appears to bind dCas9 well and retains its cargo domain in mature CRISP-Disp complexes. "lncRNA2" appears to bind dCas9 well, but fails to retain its full cargo domain, potentially indicating a hydrolysis-prone region in the construct. "lncRNA3" is weakly bound by dCas9, potentially indicating a misfolding issue. (**c**, **d**) Assessing dCas9-targeting with luciferase reporter assays, using the "INT-PMP" construct designed in Fig. 3. (**c**) Overview of the assay. CRISP-Disp RNA constructs target dCas9~VP64, a strong transcription activator, to a luciferase reporter construct (*G*-Luc, *Gaussia* Luciferase), measured relative to a nontargeted normalizer construct (*C*-Luc, *Cypridina* Luciferase). (**b**) Example data. A minimal sgRNA ("min sgRNA") strongly activates the luciferase reporter when coexpressed with dCas9~VP64, the so-called direct activation [7]. INT–PMP also exhibits strong Direct Activation, indicating efficient dCas9 targeting. This construct also supports "Bridged Activation," by recruiting VP64 fusions to its cognate PP7 and MS2 proteins, but not to L7Ae, a noncognate RNA-binding protein

with 10 mL of growth media. Meanwhile, trypsinize and count HEK 293T cells. Seed each plate with 2×10^6 cells, and incubate for 16–24 h at 37 °C, 5% CO_2, in a sterile tissue culture incubator.

3. (*Optional*) Three hours prior to transfection, replace the growth media with 10 mL per plate of prewarmed media.
4. Transfection. Thaw pEF_dCas9 plasmid, the experimental and minimal sgRNA expression plasmids cloned in Subheading 3.3, and carrier plasmid, and store on ice during use. In sterile

1.5 mL microcentrifuge tubes, prepare the following transfection mixes, each diluted with serum- and antibiotic-free DMEM to 150 μL final volume:

Pol II-driven RNA-expression constructs: 926.3 ng pEF_d-Cas9, 8.51 μg RNA expression plasmid, 2.34 μg carrier plasmid.

U6-driven RNA-expression constructs: 926.3 ng pEF_d-Cas9, 182.1 ng RNA expression plasmid, 10.67 μg carrier plasmid.

Mix thoroughly by vortexing, move tubes to a sterile biosafety cabinet, and incubate at room temperature for 5 min. To all reactions, add lipofection reagent, diluted in serum- and antibiotic-free DMEM, to a final volume of 300 μL. The volume of reagent used will vary according to the manufacturer's recommendations: for Lipofectamine 2000, use 35.3 μL per condition. Mix by gentle flicking and incubate at room temperature for 20 min more. Add the entire volume dropwise to a single plate of cells by gentle swirling, and return samples the tissue culture incubator for an additional 48 h.

5. (*Optional*) 24 h after transfection, replace the growth media with 10 mL fresh, warmed media.
6. Cross-link and harvest 48 h after transfection, remove the growth media and wash each plate twice with 10 mL room temperature 1× PBS. Immediately before use, prepare fresh fixation solution—approximately 5.5 mL per experimental condition (e.g., 55 mL for ten samples). Slowly add 5 mL fixation solution to each dish, using a serological pipette to dribble the solution down the wall, and taking care to not mechanically disturb the cells. Rock plates gently (approximately one revolution every 5 s) on a platform rocker for 10 min at room temperature. To each plate, add 0.6 mL of Quenching solution dropwise, and incubate with gentle rocking at room temperature for 5 min more. Gently aspirate the media and wash twice with 10 mL room temperature 1× PBS. Add 5 mL room temperature 1× PBS to each plate, harvest the cells the using a mechanical cell scraper, and transfer samples to ice-cold 15 mL conical tubes. Pellet samples in a swinging bucket centrifuge at 800 × *g* for 5 min at 4 °C, and aspirate the media. Gently resuspend each pellet into 1 mL of ice-cold 1× PBS by flicking and gentle pipetting, and transfer each to chilled 1.5 mL microcentrifuge tubes. Spin again at 800 × *g* for 5 min at 4 °C, and remove as much residual PBS as possible. Flash-freeze samples in liquid nitrogen and store at −80 °C (*see* **Note 15**).
7. Cell lysis. Prepare a working stock of RIPA(+) buffer in a prechilled conical tube, making at least 1.1 mL buffer per experimental sample (e.g., at least 11 mL for ten samples),

and store on ice. Thaw cross-linked cell pellets on ice; when they have fully thawed, gently flick the tube until the pellet has disaggregated into a milky fluid. Suspend each sample by gentle pipetting into 1 mL ice-cold RIPA(+) buffer, rotate end-over-end at 4 °C for 10 min, and place them on ice.

8. Sonication. Place two metal heat blocks on ice. Using lab wipers, clean the sonifier microtip with RNaseZap, rinse it several times with nuclease-free water, and dry it completely. Place the first sample tube with its cap open in the center well of a prechilled metal block. Adjust the block position such that the sonifier microtip is centered within the sample, its tip approximately 1 mm above the bottom of the tube. Secure the sample to prevent it from jumping (*see* **Note 16**). Sonicate at 10% amplitude for three 30-s intervals (0.7 s on, 1.3 s off), with 30-s resting steps between each interval. Immediately return both the sample and thermal block to ice. When processing multiple samples, alternate the metal thermal block used for each, so that the block is always freshly chilled at the start of sonication. Also rinse the microtip with nuclease-free water and dry it thoroughly between samples, so as to avoid cross-contamination. When all samples have been processed, spin them at 14,000 × *g* for 10 min at 4 °C, and decant supernatants to new, chilled 2 mL tubes, taking care not to carry over pelleted debris.
9. During the 10-min spin, prepare a working stock of NLB(+) in a prechilled conical tube, making at least 1.1 mL per sample, plus an additional 3 mL (e.g., 14 mL for 10 samples). Aliquot Magnetic Protein-G beads (28 μL bead slurry per experimental sample, e.g., 280 μL for ten samples) into a chilled 1.5 mL microcentrifuge tube. Place this tube on the magnetic stand, and decant the suspension buffer. To equilibrate the beads, wash them twice with 1 mL NLB(+), and then resuspend them to their original slurry volume (e.g., 280 μL for ten samples) in NLB(+). Store beads on ice.
10. Postsonication clarification. To each of the lysates generated in **step 8**, add 1 mL ice-cold NLB(+), and decant into an empty 3.0 mL syringe barrel equipped with a 0.45 μm filter. Position the filter outlet above a new, ice-cold 2.0 mL tube, and clarify the sample by gently inserting the plunger. Add 25 μL NLB (+)–equilibrated Protein-G beads, and rotate end-over-end for 2 hours at 4 °C. Magnetize the sample, remove 56 μL to a separate tube, and flash-freeze this tube in liquid nitrogen; this is the "input" sample used in subsequent analysis (*see* **steps 18** and **19**).
11. RNA immunoprecipitation. Divide each of the remaining clarified lysates equally into two 1.5 mL tubes (~950 μL per tube;

see **Note 15**). To one, add 6 μg Rabbit anti-FLAG antibody; to the other, add 6 μg Normal Rabbit IgG. Rock all samples end-over-end at 4 °C for 2 h. Toward the end of this incubation, equilibrate a second batch of Magnetic Protein-G beads with freshly prepared NLB(+), as described in **step 9**, but increasing the volume to 55 μL bead slurry per sample (e.g., 550 μL for ten samples). Add 50 μL equilibrated bead slurry to each sample and rock end-over-end at 4 °C for an additional hour. Thereafter, magnetize the samples, decant and discard the supernatants, and wash the beads twice with 1 mL freshly prepared, ice-cold NLB(+). Rock samples end-over-end at 4 °C for 10 min at each washing step. After the final wash, suspend beads in 56 μL nuclease-free water and proceed immediately to reverse cross-linking.

12. Reverse formaldehyde cross-links. Thaw all "input" (*see* **step 10**) samples on ice, and transfer the entire volume of each sample (56 μL) to thin-walled capped PCR tubes. To each add: 33 μL 3× RCL Buffer, 10 μL Proteinase K solution, and 1 μL RNaseOUT (0.4 U/μL, final). Incubate all samples for 1 h at 42 °C, followed by 1 h at 55 °C, in a thermocycler.

13. RNA isolation (*see* **Note 17**). Equilibrate SPRI beads to room temperature and mix thoroughly by vigorous shaking immediately before use. Prepare 50 mL 70% (v/v) ethanol, in nuclease-free water. Transfer each reverse–cross-linked RNA sample to a new 1.5 mL microcentrifuge tube, using a magnet to remove Protein-G beads from RIP samples. To each, add four volumes (400 μL) SPRI beads, and proceed with the manufacturer's protocol, eluting into 40 μL nuclease-free water. To remove residual genomic DNA, transfer samples to PCR tubes, add 5 μL RQ1 buffer and 5 μL RQ1 RNase-free DNase, and incubate at 37 °C for 30 min, in a thermocycler. Transfer samples to new 1.5 mL microcentrifuge tubes, add four volumes (200 μL) of SPRI beads, and purify RNA following the manufacturer's protocol, eluting into 20 μL nuclease-free water. Store purified RNA at −20 °C.

14. Reverse transcription. Aliquot 8 μL of each "input" RNA sample into two thin-walled, capped PCR tubes ("+RT" and "–RT"), and 8 μL of each "RIP" sample into a single PCR tube ("+RT" only). Reverse transcribe all samples using SuperScript III Reverse Transcriptase, as described in the manufacturer's instructions, priming with 1 μL 50 ng/μL random DNA hexamers and in a final volume of 20 μL. Dilute the completed reactions with 180 μL nuclease-free water (final volume: 200 μL) and store at −20 °C.

15. Quantitative PCR. Each experimental condition (i.e., minimal sgRNA control; CRISPR-Display constructs) has now

produced four samples: input(+RT), input(−RT), IgG mock IP, and FLAG IP. Working in 8-tube PCR strips, prepare qPCR mixes that pair each of these samples with all appropriate primers, that is, targeting both the sgRNA core and intervals along the ncRNA cargo domain (*see* **step 1**), as well as a GAPDH negative control. Each 50 μL qPCR mix contains 5 μL diluted cDNA, 19.4 μL nuclease-free water, 0.3 μL of each primer, and 25 μL SYBR/Rox RT-PCR Master Mix. Aliquot 10 μL of each qPCR mix into four replicate wells of a 384-well optical reaction plate, and seal the plate with microplate sealing film. Amplify reactions in a real-time thermocycler with the following cycling parameters (record SYBR and ROX fluorescence during all cycles in brackets):

Stage 1: 50 °C for 2 min; 90 °C for 10 min.

Stage 2: 95 °C for 15 s; 50 °C for 15 s; [60 °C for 1 min]. Repeat this stage 40 times.

Stage 3 (thermal melt): 95 °C for 15 s; ramp at 1 °C/s to 60 °C and hold for 1 min; [ramp to 95 °C at 0.075 °C/s and hold for 15 s].

16. Data analysis. Assess the specificity of all new qPCR primers designed in **step 1**. Inspect (−RT) conditions to confirm the absence of untemplated species, and the melting curves of all (+RT) conditions to confirm the formation of discrete, homogeneous PCR products. If necessary, analyze these products by agarose gel electrophoresis. Discard and redesign any pathological primer pairs.

For all remaining (+RT) conditions, upload Rn values (per-cycle ROX-normalized fluorescence) into "Realtime PCR Miner," (http://ewindup.info/miner/) to calculate C_T values and primer efficiencies for each reaction. Adjust C_T values to account for the fraction of the total lysate used in each condition: input samples corresponded to 56 μL out of 2 mL total lysate (2.8%); each RIP sample was ~950 μL (47.5%). Hence, the adjusted C_T values will be:

$$C_{T,\text{adj}}^{\text{input}} = C_{T,\text{obs}}^{\text{input}} + \log_{(1+\varepsilon)}(0.028)$$

$$C_{T,\text{adj}}^{\text{RIP}} = C_{T,\text{obs}}^{\text{RIP}} + \log_{(1+\varepsilon)}(0.475)$$

where ε is amplification efficiency for each primer pair, as determined by Miner. From these values, calculate the percent yield for each amplicon, in each experimental condition.

$$\text{percent yield} = 100 \times (1+\varepsilon)^{\left(C_{T,\text{adj}}^{\text{input}} - C_{T,\text{adj}}^{\text{RIP}}\right)}$$

notated as above (*see* **Note 22**).

Gauge the overall quality of the pulldown (Fig. 4b). In the minimal sgRNA transfection control, FLAG immunoprecipitation should enrich the sgRNA core domain approximately 100–1000-fold relative to the IgG control. Neither pulldown should significantly enrich GAPDH mRNA. Low specific enrichment of the sgRNA (relative to the IgG pulldown, or to GAPDH) may indicate the need for a new FLAG antibody, a revised bead-washing protocol, or both.

Quantify the incorporation of hybrid sgRNA–ncRNA transcripts into CRISPR-Display complexes. Compare the percent yield (from the FLAG IP) of each construct's sgRNA core domain to that observed in the minimal sgRNA transfection condition. Longer transcripts tend to bind dCas9 with slightly lower efficiency than does a minimal sgRNA: typically, TOP1-like constructs bearing 0.5–5 kb cargo domains bind at approximately 20–50% the efficiency of a minimal sgRNA. Consider redesigning constructs that fall below this threshold (*see* Subheadings 3.1, **steps 1–4** and 3.2, **steps 1** and **2**), or altering transfection conditions to accommodate for this deficiency.

Measure the integrity of ncRNA cargo domains in CRISPR-Display complexes. For each construct, compare the percent yield (from the FLAG IP) of the sgRNA core domain to each of the segments probed along the cargo domain. For smaller, and well-folded cargos, the recovery of the sgRNA core and cargo domains should be nearly quantitative. For longer constructs, recovery of the cargo domain may drop to 75% that of the core, particularly at regions nearing the transcript's 3′-terminus. Consider redesigning potentially degradation-prone constructs for which cargo recovery falls below this threshold.

3.5 Assessing dCas9 Targeting Fidelity

We next quantify the efficiency and fidelity with which the CRISPR-Display constructs established in Subheading 3.4 direct dCas9 to genomic loci. Our preferred assays use transcription activation as a proxy for dCas9-targeting (Fig. 4c). Briefly, using guide sequences that target a reporter gene's promoter, we coexpress CRISPR-Display RNAs with dCas9 fused to VP64, a strong transcription activator, and measure expression of the targeted gene. This section details two variants of this assay, targeting CRISP-Disp complexes to artificial luciferase reporters and to endogenous loci. We favor these assays because they are extremely sensitive, robust, and simple to perform. Moreover, because many chromatin-regulatory ncRNAs (both natural lncRNAs and artificial devices) regulate their target genes at the transcriptional level, these methods can be easily adapted to functionally interrogate the ncRNA cargo domain itself [7] (*see* Subheading 3.6). Alternative quality-control assays may be better suited in cases where the ncRNA cargo domain will eventually be deployed outside of promoter regions (*see* **Note 18**).

Measuring dCas9 targeting by luciferase reporter assays (*see* **Note 19**)

1. Prepare HEK 293T cells for transient transfection. Working in a sterile biosafety cabinet, gelatinize 12-well tissue culture plates. You will need six wells for each CRISPR-Display construct being tested, plus an additional six wells for minimal sgRNA control transfections (*see* Subheading 3.3. **steps 11** and **16**, *and* **Note 15**). Trypsinize and count HEK293T cells, and prepare a working stock: 182,000 cells/mL, in growth media, in a total volume that allows for at least 0.8 mL per experimental well. Working one 12-well plate at a time, aspirate the gelatin solution and, using a manual pipettor, immediately plate 0.7 mL (127,400 cells) of working stock per well (*see* **Note 20**). Incubate plates for 16–24 h at 37 °C, 5% CO_2, in a sterile tissue culture incubator.
2. (*Optional*) Three hours prior to transfection, replace the growth media with 0.7 mL per well of prewarmed media.
3. Transfection. Thaw pEF_dCas9-VP64 and pEF_dCas9 plasmids, the experimental and minimal sgRNA expression plasmids cloned in Subheading 3.3, and carrier plasmid, and store on ice during use. You will prepare transfection mixes for all RNA expression vectors—the minimal sgRNA, and all CRISP-Disp constructs—bound both to dCas9 and to dCas9~VP64; each mix will be transfected into biological triplicates. In sterile 1.5 mL microcentrifuge tubes, prepare the following plasmid mixes, each diluted with serum- and antibiotic-free DMEM to 36 μL final volume:

   Pol II-driven constructs: 132 ng reporter; 132 ng normalizer; 2168 ng RNA-expression construct; 236 ng dCas9 or dCas9~VP64; 332 ng carrier.

   U6-driven constructs: 132 ng reporter; 132 ng normalizer; 46.4 ng RNA-expression construct, 236 ng dCas9 or dCas9~VP64; 2453.6 ng carrier.

   If the same ncRNA cargo is being simultaneously deployed on multiple guide sequences, hold the total mass of RNA expression constructs constant at the values listed above (e.g., for four U6-driven RNAs, input 11.6 ng of each plasmid; for six Pol II-driven RNAs, input 361.3 ng of each).

   Mix thoroughly by vortexing, move tubes to a sterile biosafety cabinet, and incubate at room temperature for 5 min. To all reactions, add lipofection reagent, diluted in serum- and antibiotic-free DMEM, to a final volume of 72 μL. The volume of reagent used will vary according to the manufacturer's recommendations: for Lipofectamine 2000, use 9 μL per condition. Mix by gentle flicking and incubate at room temperature for 20 min more. To each replicate well, add 18 μL transfection mix dropwise by gentle swirling (*see* **Note 20**), and return samples the tissue culture incubator for an additional 72 h.

4. Monitor the experiment. Using a fluorescence microscope equipped with CFP and YFP filters, examine the cells each day following transfection. Successful activation by dCas9~VP64 will induce expression of CFP but not YFP; this is often most prominent in the minimal sgRNA wells. Make note of the conditions that induce the highest apparent CFP expression—this will be used to set the experimental Gain in **step 6**.
5. Harvest the sample. 60–72 h following transfection, transfer 200 μL of media from each condition into a lidded 96-well assay microplate. If not analyzing the samples immediately, seal the plate with Parafilm and store at 4 °C for up to 1 week. *Optional*: samples can also be collected at earlier time points following transfection as well, enabling kinetic analysis of gene induction. When doing so, replenish the wells with 200 μL of warm media after sample collection.
6. Measure *Gaussia* luciferase expression (*see* **Note 21**). Bring experimental samples to room temperature. Prepare working stocks of the *Gaussia* and *Cypridina* luciferase substrates in their corresponding assay buffers, as described in the manufacturer's instructions: you will need 50 μL of each stock for every well, plus additional volume for priming the luminometer injector system. Pipet 20 μL from each experimental sample into an opaque 96-well assay microplate, and pipet 20 μL from one of the high CFP-expressing conditions noted in **step 4** into an additional "Standard" well. Prime the luminometer injection system with *Gaussia* assay buffer, and confirm the optical filter will measure light at 475 nm wavelength (*see* **Note 21**).

   Manually pipet 50 μL *Gaussia* working substrate stock into the Standard well and mix by vigorous pipetting. Immediately load the plate into the luminometer, and use the instrument's control software to determine the maximum nonsaturating detector gain for the Standard sample. Set the detector gain to this value, and measure the *Gaussia* activity for all samples in the experimental series. Use the automatic injector system to inject 50 μL per well, shake for 2 s, then record and integrate the luciferase signal for 20 s.
7. Measure *Cypridina* luciferase expression. Confirm that the optical filter will read light at 465 nm, and prime the injector pump with the *Cypridina* working substrate stock. Pipet 20 μL from each experimental sample into a new opaque 96-well plate, and pipet 20 μL of the same Standard sample into an additional well. Follow the same workflow detailed in **step 6** to determine the detector gain and measure *Cypridina* expression.

8. Data analysis (Fig. 4d). For each well, calculate the ratio of *Gaussia* to *Cypridina* activities, (*Gluc/Cluc*), and for each set of triplicates calculate the mean value, <*Gluc/Cluc*>. For each ncRNA construct, including the minimal sgRNA, calculate the Fold Activation:

$$\text{Fold activation} = \frac{\left\langle \frac{GLuc}{CLuc} \right\rangle (dCas9\tilde{V}P64)}{\left\langle \frac{GLuc}{CLuc} \right\rangle (dCas9)}$$

wherein (dCas9~VP64) and (dCas9) denote conditions where the ncRNA was coexpressed with each of the corresponding Cas9 variants (*see* **Note 22**). Minimal sgRNAs driven from the U6 promoter typically induce between 200 and 600 Fold Activation, while Pol II-driven constructs typically exhibit between 50 and 200 Fold Activation. Consider redesigning constructs that fall below either of these ranges.

Measuring dCas9-targeting at endogenous loci

9. Select an endogenous target gene that will serve as transcriptional reporter, and design four to five guide sequences that target within 50 bp upstream of that gene's transcription start site (*see* Subheading 3.1, **step 3**). The choice of target gene will depend on the ncRNA being deployed; for lncRNAs this can significantly influence the experimental outcome [9]. For artificial devices, the Titin locus (*TTN*) may be a good starting point, since it can be activated over a broad dynamic range using literature-vetted guides [7]. Use the selected CRISPR guide sequences to computationally predict potential off-target genes (e.g., http://www.rgenome.net/cas-offinder/) focusing on loci with PAM-adjacent guide-binding sequences near the transcriptional start site.
10. Design qRT–PCR primers for the target and off-target genes using the NCBI design tool (https://bit.ly/2AoEs9q), as described in Subheading 3.4, **step 1**. Select primers that span exon–exon junctions, if possible.
11. Plate HEK 293T cells, as described in Subheading 3.4, **step 2**. Prepare six wells for each cargo domain, and six wells for the corresponding minimal sgRNA controls. Incubate plates for 16–24 h at 37 °C, 5% CO_2, in a sterile tissue culture incubator.
12. Transfections. Prepare the following transfection mixes, bringing the reaction volume to 36 μL with serum- and antibiotic-free DMEM:

    Pol II-driven constructs: 2168 ng pooled RNA-expression constructs; 236 ng dCas9 or dCas9~VP64; 596 ng carrier.

    U6-driven constructs: 46.4 ng pooled RNA-expression constructs, 236 ng dCas9 or dCas9~VP64; 2717.6 ng carrier.

where cargo RNA domains being deployed by multiple guides are pooled in equal ratios to the total mass listed above (e.g., for five U6-driven RNAs, input 9.28 ng of each plasmid). Continue as described in Subheading 3.4, **step 3** and incubate transfected cells for 48–72 h at 37 °C, 5% CO_2, in a sterile tissue culture incubator. *Optional:* replace the growth media 24 h following transfection.

13. Harvest total RNA. Aspirate growth media, and immediately add 350 μL TRIzol (or equivalent RNA extraction reagent) to each well. Proceed with cell lysis and organic phase separation as described in the manufacturer's protocol. Purify total RNA using commercial spin-column-based RNA cleanup kits, with on-column DNase treatment, according to the manufacturer's instructions. Quantify the RNA by UV-Vis absorption.
14. RT-qPCR. Reverse-transcribe 100–1000 ng of RNA from each sample, in 20 μL final reaction volume, following the manufacturer's instructions. Prime each reaction using 2 μL RT Primer Mix. Separately, use the RNA extracted from the minimal sgRNA·dCas9~VP64 condition to prepare a (−RT) control reaction. Dilute completed RT reactions 300-fold with nuclease-free water. Prepare qPCR mixes for all pertinent primer pairs (target, off-target, and GAPDH), and using 5 μL of each diluted RT reaction as template, as described in Subheading 3.4, **step 15**. Amplify reactions in a Real-time thermocycler using the cycling and recording parameters detailed in Subheading 3.4, **step 15**.
15. Data analysis. Assess the quality of all qPCR primer pairs by inspecting the (−RT) conditions and product melting curves. Calculate C_T values and primer efficiencies for (+RT) conditions using Realtime PCR Miner (http://ewindup.info/miner/), as described in Subheading 3.4, **step 16**.

    For each condition, calculate the fold activation of all target and off-target genes:

$$\text{Fold activation} = 2^{\left(\Delta C_{T,dCas9} - \Delta C_{T,dCas9\tilde{V}P64}\right)}$$

Wherein ΔC_T corresponds to the GAPDH-normalized gene expression for each condition: $C_{T,\text{target}} - C_{T,\text{GAPDH}}$ (*see* **Note 22**). The maximum fold activation, typically observed in the minimal sgRNA conditions, will vary significantly between loci. In our hands, *TTN* can usually be activated 400–1000-fold, while other loci may only respond 20-fold [7]. For target genes falling below this threshold, consider redesigning guide RNA sequences, or select a different target locus altogether. Finally, examine the targeting and fidelity of CRISPR-Display constructs. Compare the fold-activation of each CRISP-Disp construct to that of its corresponding minimal sgRNA, and examine the fold-activation of all off-target genes. Functional constructs typically exhibit at least 10% of the activity of their minimal sgRNA counterparts, without significant increases in off-target activation [7].

3.6 Interrogating or Utilizing ncRNA Domains by CRISPR-Display

The particular approach by which CRISPR-Display is used to interrogate natural lncRNA domains, or to utilize artificial ncRNA devices, will vary significantly based on presumed function of the ncRNAs in question. As described in the Subheading 1, care must be taken to implement an assay that explicitly probes this function as is best achievable. Often, this entails probing a single molecular interaction, or the expression of a single target gene [8, 9], but can be expanded to multifaceted manipulations [10, 11], and conceivably, to phenotypic assays. Hence, in lieu of a detailed protocol, I offer brief suggestions to aid the initial design of these experiments.

1. Transcription regulation by natural lncRNA domains (Fig. 5). Use the endogenous gene activation protocol detailed in Subheading 3.5, **steps 9–15** as a starting point for assay design. LncRNA-induced changes in gene expression tend to be quite small [7–9], and so a gamut of controls may be needed to normalize the experimental signal. An omit-dCas9 control will account for the changes in gene expression induced by overexpressing the sgRNA–lncRNA hybrid. A minimal

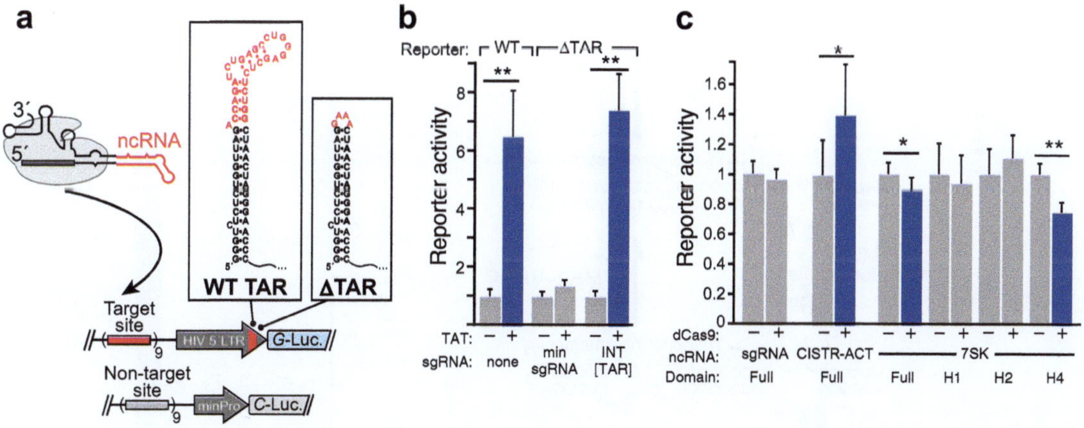

Fig. 5 Using CRISPR-Display to interrogate natural lncRNA function. (**a**) A CRISP-Disp reporter system probing regulators of RNA Pol II pausing. The HIV-1 5′-LTR encodes a structured RNA (TAR) that halts RNA Pol II transcription; binding of the viral TAT protein facilitates processive elongation. Ablation of this element ("ΔTAR") creates a constitutive transcriptional pause. (**b**) Validation of the assay. The wild-type reporter—but not the ΔTAR variant—is activated by expression of the TAT protein. Ectopically targeting the TAR RNA element upstream of the ΔTAR promoter, using CRISPR-Display, restores wild-type activity. (**c**) Identifying ncRNA domains that regulate transcription elongation. The CISTR-Act and 7SK lncRNAs were recruited to the ΔTAR reporter construct, in the absence of TAT. While CISTR-Act appeared to activate gene expression, 7SK RNA was a mild repressor. Further truncation analysis localized this repressive activity to the 7SK "helix four" domain (H4). "sgRNA" is a minimal sgRNA construct lacking an RNA cargo domain

sgRNA·dCas9 control will account for the effect of binding dCas9 to the target locus, irrespective of the cargo domain. The example presented in (Fig. 5) uses a luciferase reporter system similar to that described in Fig. 4 and Subheading 3.5, but which explicitly tests for lncRNAs that regulate RNA Pol II transcriptional pausing. Alternatively, consider alternative assays like deep sequencing or single-molecule RNA FISH, which may be better suited to measure small lncRNA-induced effects.

2. Transcription regulation by artificial ncRNA devices (Fig. 6). Either the luciferase reporter or endogenous gene activation protocols detailed in Subheading 3.5 can serve as a starting framework for assay design. Increasing the transfected mass of the RNA-expressing plasmids (up to 400 ng total, per four-well transfection premix) may significantly improve the experimental signal. For devices that are designed to recruit exogenous protein factors (e.g., the MS2 phage coat protein [7, 10, 11]), include the corresponding protein expression plasmids at 264 ng per four-well transfection premix. The example presented in (Fig. 6) uses the standard CRISP-Disp luciferase reporter system (Fig. 4c, d, [7]) to prototype an artificial

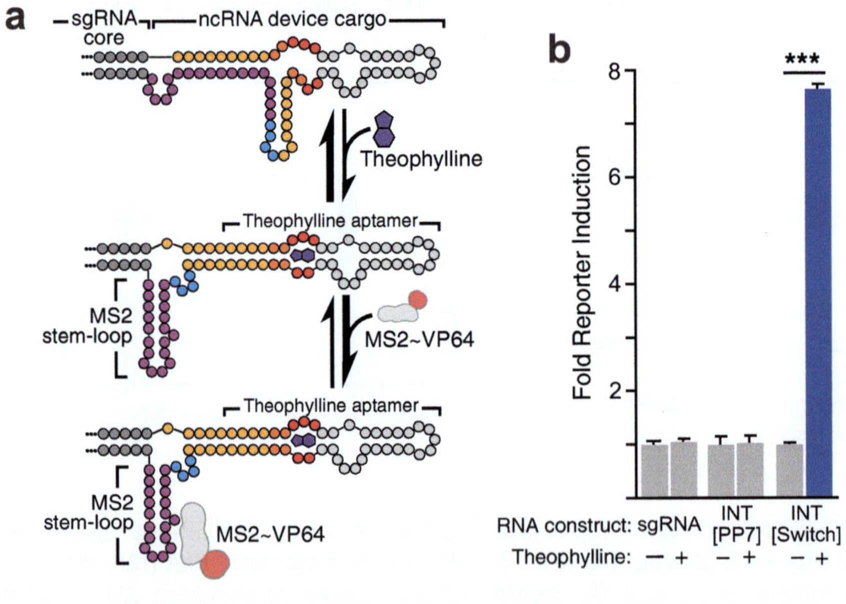

Fig. 6 CRISPR-Display with artificial RNA devices—here, a synthetic riboswitch that enables small-molecule control of CRISPR-activation. (**a**) Construct design. The riboswitch domain predominantly populates an inert conformation, but can be driven to recruit the MS2~VP64 transcription activator by the small molecule theophylline. (**b**) Proof of concept, using the luciferase reporter system outlined in Fig. 4c [7]. RNAs were coexpressed with dCas9 and MS2~VP64 for 2 days, and dosed with theophylline (+) or a vehicle control (−) for an additional day. The artificial switch RNA but not control constructs activates reporter expression nearly eightfold during theophylline treatment

riboswitch device that activates gene expression upon addition of a small-molecule activator.

3. Multiplexing functions. CRISPR-Display can be used to simultaneously perform several genomic manipulations, including gene activation, gene repression, and locus-specific imaging (Fig. 1b) [7, 10, 11]. To date, we have only attempted such complex manipulations using U6-driven RNAs. Use the following one-well transfection premix as a starting point, and optimize plasmid masses as needed: 625 ng dCas9 plasmid, 500 ng of each accessory protein-encoding plasmid, 375 ng aggregated RNA-expression plasmid per locus (e.g., if targeting two loci with three guides each, input 125 ng of each guide for a total mass of 750 ng). Additional assay details will depend on the functions being multiplexed. For imaging-based experiments, grow cells in glass-bottomed culture dishes (e.g., Lab-Tek two-chambered slides) or on glass coverslips, and image cells by confocal microscopy [10].

4 Notes

1. At the transcript level, lncRNAs tend to be substantially more heterogeneous than are mRNAs, with the average lncRNA gene producing a multitude of species that differ in start site, polyadenylation site, and splicing structure [15]. Hence, defining the "full length lncRNA" construct may require special consideration. Use available RNA-sequencing data (either generated "in house," or from public databases) to identify the most abundant transcript isoforms in the cell type of interest, or RNA domains that are common to the most highly expressed species. It may also be helpful to manually curate a target lncRNA's 5′- and 3′-termini by examining the sequencing read coverage; these can be poorly defined in public databases.
2. CRISPR-Display is currently inefficient at deploying RNA domains with exposed 5′ termini, meaning that reconstituting a hypothetical lncRNA that requires a free 5′-terminus would be challenging. Although the CMV/3′ Box expression system enables larger domains to be appended on the 5′-terminus of the sgRNA (the so-called TOP2-like constructs [7], Fig. 1a), this topology appears less robust than its TOP1- and INT-like counterparts, and is generally recommended if no other options are available.
3. In some cases, it may be fruitful to optimize the length and sequence of the linkers that connect the sgRNA core and cargo domains. I generally prefer to keep these linkers relatively short (often just three nucleotides) and unstructured (comprising mostly cytidines and uridines), while avoiding longer stretches

of poly(uridine). This is particularly important for constructs driven from U6 promoters. For Pol II-driven TOP1 constructs, I have previously used spacers of up to 25 nt in length, which provide more of a physical spacer between the cargo domain and dCas9 [7]; the cloning protocol detailed in Subheading 3.3, **steps 1–11** retains these long linkers. However, the experimenter should feel free to vary or ablate these linkers as needed. By contrast, the spacer sequences placed between modules in artificial RNA devices (Fig. 3) can have a dramatic impact on the predicted fold of the RNA domain, and should be designed carefully (*see* Subheading 3.2).

4. Guide design should take into consideration the potential confounding effects incurred simply by binding dCas9 to the target site, irrespective of its ncRNA cargo. While these effects can be controlled for by using no-cargo controls, in some cases they may overshadow the influence of the ncRNA. For example, binding dCas9 within the first 50 bp downstream of a transcription start site (TSS) may result in substantial CRISPR-interference (CRISPRi), which could mask the effects of a putative activator (or even repressor) ncRNA cargo domain.

5. To date, we have not tried stably integrating CRISPR-Display RNA expression constructs into the host genome (by lentiviral transduction, PiggyBAC insertion, etc.), and in the protocols detailed here these chimeric RNAs are expressed transiently. However, a priori there is no reason why stable expression systems would not also work.

6. In many cases, a functional RNA device can be folded from an extremely large number of unique sequence variants (Fig. 3, *left*). For example, the RNA stem-loop recognized by the bacteriophage MS2 coat protein—a common module used in artificial RNA devices [7, 10, 11]—is 19 nt long, but 14 of these residues can form any of the four Watson–Crick base pairs and retain activity. This means that there are theoretically $4^7 = 16{,}384$ sequences that perfectly satisfy the constraints of a minimal MS2 stem loop. While many such sequences may prove suboptimal (e.g., by being prone to misfolding) the large pool of remaining viable sequence variants empowers the designer with a tremendous amount of flexibility.

7. Although U6-driven INT-like constructs are extremely robust, this topology is substantially less efficient when expressed from a Pol II promoter [7]. While the U6 promoter system adds negligible additional sequence to the sgRNA transcript, all CRISPR-Display Pol II expression systems employ transcription terminator elements that add substantial (>50 nt) sequence to the sgRNA's 3′-terminus. Pol II-expressed

INT-like constructs therefore carry large exogenous RNA domains at two locations, which I suspect that this may diminish their efficacy. In support of this hypothesis, placing an HDV self-cleaving ribozyme between the sgRNA core and terminator domains—which would "trim" the sgRNA at its typical 3′-terminus—partially rescues the activity of Pol II-driven INT-like constructs (***unpublished data***). However, the activity of these HDV-trimmed INT-like constructs is still dwarfed by that of their U6-driven counterparts.

8. Guide sequences are removed during secondary structure prediction because many guides may be used over the course of an experiment, and it is nearly impossible to optimize any individual cargo domain for use with all them (at least, as predicted by RNAFold or NUPACK).

9. For cargo domains that are themselves refractory to secondary structure prediction (e.g., pseudoknotted structures, or riboswitch-like domains that stably populate multiple conformations), one can only gauge whether fusing this cargo onto the sgRNA core is predicted to perturb the folding of either domain. To do so, calculate the (potentially erroneous) predicted secondary structure of the cargo domain alone, using RNAFold or NUPACK, and compare it to the predicted fold of this domain in the sgRNA–cargo RNA chimera. Mutate the sequence as needed (altering, adding, or deleting base pairs, or changing linkers) so that the predicted fold of the chimeric RNA recapitulates that of the isolated cargo domain.

10. The SheqChange strategy detailed above can be adapted to swap guide sequences as follows. Amplify the plasmid backbone using forward and reverse primers:

    Forward: TAGTA**GAAGAC**
    AAGTTTAAGAGCTATGCTGCGAATACG.

    Reverse:
    TAGTA**GAAGAC**AAGGTGTTTCGTCCTTTCCACAAG.

    (wherein bold nucleotides denote *BbsI* recognition sites and underlined nucleotides denote *BbsI*-generated sticky ends). Digest and purify the plasmid backbone as described above, but do not treat with phosphatase. New guides are then introduced using synthetic oligonucleotides bearing the following sequences:

    Top oligo: 5′-CCAC**G**[new guide sequence]-3′.

    Bottom oligo: 5′-AAAC[reverse-complement of new guide sequence]**C**-3′.

    wherein the bold nucleotides encode the 5′-guanosine required for Pol III expression. Dissolve these guide oligos to 100 μM in Oligo Annealing Buffer, anneal and ligate as described in Subheading 3.3, **steps 14** and **15**. Note that this strategy uses the promoter and sgRNA variants as in INTgpc;

for constructs using different backbones, PCR primers may need to be modified.

11. If needed, larger cargos (up to at least 250 nucleotides [7]) can be introduced on PCR products or using tandem sets of annealed oligo duplexes. After *BbsI* digestion, the INTgpc plasmid has 5′-CTCG and 5′-TCGT sticky ends upstream and downstream of the insert site, respectively. Hence, any PCR product with the sequence 5′-[Type IIs site]CGAG[Cargo Domain]**C**TCGT[Type IIs Site]-3′ (annotated as in **step 12** of Subheading 3.3) will generate sticky ends for compatible for cloning. Alternatively, cargo domains of up to 120 nucleotides can be introduced by simultaneously inserting two adjoining pairs of synthetic oligonucleotide duplexes (a "5′-Duplex" and a "3′-Duplex"), using an approach similar to that described in **steps 12–15** of Subheading 3.3. To do this, synthesize the following four oligonucleotides:

    5′Top: 5′-CGAG[Cargo 5′-half]-3′.

    5′Bottom: 5′-pXXXXX[Reverse complement of Cargo 5′-half]-3′.

    3′Top: 5′-pYYYYY[Cargo 3′-half]**C**-3′.

    3′Bottom: 5′-ACGA**G**[Reverse complement of Cargo 3′-half]-3′.

    . . .annotated as in **step 12** of Subheading 3.3, and where XXXXX and YYYYY are a pair of internal five-nucleotide sticky ends. These sticky ends should and have a low propensity for self-pairing (e.g., avoid sequences like CCAGG) or for pairing with either sticky end on the cut backbone. Anneal each duplex in a separate tube, dilute, as described in **step 14** of Subheading 3.3. Add 1 μL of each annealed, diluted duplexes to the ligation reaction, as described in **step 15** of Subheading 3.3, adjusting the volume of nuclease-free water accordingly. This approach is not recommended for simultaneously introducing more than two insert duplexes. To clone longer cargo domains, assemble the entire domain as a PCR product, and clone as described above.

12. lncRNAs and lncRNA domains can be amplified from a variety of sources. If possible, we prefer using clones that are experimentally isolated in-house from cDNA libraries or, failing that, publically or commercially available clones generated by the same method. This ensures that the CRISP-Disp constructs we generate are derived from bona fide experimentally observed transcripts. We have also used gene synthesis and genomic PCR to isolate consensus lncRNA sequences described elsewhere in the literature [7]. Since this strategy relies on database annotations (which are not always accurate) to define the lncRNA sequence, it should be considered a route of last resort.

13. This annealing protocol aims to drive formation of the oligo duplex at a high concentration, and then to kinetically trap this duplex during the subsequent dilution step. The final ligation reaction will contain a fivefold molar excess of insert duplex (relative to the 25 ng of cut plasmid); diluting this duplex to 63 nM means that only 1 μL is needed per reaction. To ensure homogeneous mixing, I typically dilute the duplex in two steps: 1 μL of the 50 μM stock with 19 μL ice-cold Oligo Annealing Buffer, to generate a 2.5 μM stock, of which 1 μL is diluted with 38.6 μL buffer to yield the final 63.1 nM stock.

14. Theoretically, the control assays presented in Subheadings 3.4 and 3.5 do not explicitly confirm that a given CRISP-Disp construct's ncRNA cargo domain is retained at the targeted genomic locus, merely that a pool of dCas9 binds to the intact RNA (Subheading 3.4), and that a pool can be targeted to DNA (Subheading 3.5). These might correspond to two discrete pools of dCas9. However, previous work strongly suggests that passing these quality controls is sufficient to ensure that cargo domains remain intact at the targeted locus, given the diverse set of RNA-based functions that have reconstituted at individual genomic sites [7, 9, 10]. While alternative assays that might directly interrogate RNA integrity at the targeted locus are available, they are disfavored for several reasons. For example, one might use "bridged activation" assays for this purpose, by appending the 3′-end of cargo domain with motifs that recruit an exogenous transcription activator [7]. However, such activation assays tend to be distance dependent, and a placing the trans-activator at the end of a long ncRNA domain may fail to activate transcription, irrespective of RNA integrity. An alternative approach might assay if affinity purification of the ncRNA accessory domain (e.g., using antisense oligonucleotides) is sufficient to recover the targeted DNA locus. However, such methods tend to be quite technically challenging, and prone to experimental artifacts [25].

15. In ChIP- or RIP-type methods, like that detailed in Subheading 3.4, the number of cells used is often a critical variable limiting the experimental outcome. This issue manifests itself in two ways. Using too many cells will interfere with the sonication step, resulting in incomplete cell lysis and suboptimal chromatin shearing. Although lowering the cell input solves this problem, doing so too aggressively can squelch the experimental signal. In our experience, sonicating approximately 1×10^7 cells in a volume of 1 mL, and using half of the resulting lysate per pulldown condition, addresses both of these conflicting constraints [7, 24]. For experiments that probe abundant targets with strong signals, half as many cells can be lysed in the same volume. For weaker targets that might

require more input material, lyse a larger number of cells (in multiple 1 mL, 1 × 10^7 cell batches) and pool the lysates into a single sample. Using HEK 293T cells, the protocol detailed in Subheading 3.4, **steps 2–6** routinely yields approximately one million cells per 10 cm growth plate. For other cell types, I advise first performing a mock transfection in one 10 cm plate, harvesting and counting the cells prior to the formaldehyde step, and in subsequent experiments rescaling the growth as needed.

16. Aligning and securing a sample during sonication can be something of an art form, with most labs arriving at their own optimal setup empirically. In my lab, we use a soundproof enclosure that holds the microtip at in a fixed position, and use a small adjustable-height lab support jack to move the thermal block and sample into place. Our makeshift sample stabilizer consists of a linearized ~0.5 cm width rubber band and two 1.5 mL "dummy" microcentrifuge tubes. Trim the rubber band so that it is stretched taut when running along the metal thermal block's diagonal, place each end into one of the dummy tubes, and snap the lids closed.

    To position a sample, open the tube and place it in the central slot of a prechilled thermal block, and place this block on the jack platform. Use the jack to raise the sample height, and move the thermal block back and forth so as to center the microtip in the middle of the tube. Slowly adjust the height so that the tip gently touches the bottom of the tube, and then back the jack downward approximately 1 mm. Secure the sample by stretching the rubber band across the hinge that connects the tube lid and body, and place the dummy tubes in the corner slots of the metal block, diagonally opposing one another. Take care not to brush against the microtip during this step, which knock it out of alignment. Start the sonication program, and pay careful attention to the sound: a low-pitched "hum" indicates that the tip is touching the tube's bottom or wall; a frothing sound indicates that the tip is too far from the bottom. Halt the program, readjust the sample position, and try again.

17. The RNA isolation approach described in Subheading 3.4, **step 13** varies dramatically from the common practice of using organic phase extraction (e.g., TRIzol), followed by alcohol precipitation or spin-column cleanup. While these conventional methods have worked in our hands, Hendrickson and Tenen [24] observed that organic extraction rendered a sizeable fraction of fRIP samples (~5–20% of samples) recalcitrant to qRT-PCR or sequencing library preparation, presumably due to a yet-unknown contaminant. I was unable to remove

this contaminant by further organic extraction or chromatography. In contrast, this phenomenon is never observed when RNA is isolated using SPRI beads.

18. If necessary, chromatin immunoprecipitation (ChIP) can be used to assess CRISPR-Display targeting fidelity, though it is a significantly less sensitive, less quantitative, and more challenging than the assays presented in Subheading 3.5. Nonetheless, I have successfully used both FLAG-tagged dCas9 and HA-tagged proteins bound to cargo domain (e.g., HA–PP7 and HA–MS2) for ChIP (*data not shown*). Use the fRIP protocol detailed in Subheading 3.4 as a starting point, with the following modifications: increase the formaldehyde concentration in **step 6** to 1.0%, consider harsher bead-washing conditions in **step 11**, treat samples with RNase A (not DNase I) in **step 13**, and omit **step 14** altogether. If the ultimate goal is to use CRISPR-Display as a genomic imaging tool (termed CLING [10]), then this assay can itself quantify targeting fidelity. Design CLING constructs that label the adjoining regions of the same target locus in two colors (e.g., using MS2–mCherry and PP7–Venus), and quantify the colocalization of differently labeled puncta [10].
19. The luciferase reporter used in Subheading 3.6 is described in detail in [7] and is available from the author upon request. Briefly, this construct contains a cassette of nine CRISPR-targeting motifs (ATCTAGATACGACTCACTAT<u>AGG</u>, PAM underlined) spaced 13–23 base pairs from one another and 19 base pairs upstream of a minimal CMV promoter. This promoter drives expression *Gaussia* luciferase and mCerulean CFP, separated by a 2A "self-cleaving" peptide. The targeting cassette, promoter, luciferase, and fluorophore can be replaced with alternative parts, as required by the experiment.
20. Homogeneous and consistent cell plating is a critical experimental variable. Immediately after plating each well, use the pipette tip to gently swirl the cell suspension several times, using a "figure eight" pattern, before plating the next well. When all wells in a plate have been filled, gently tilt the plate back-and-forth several times in both the horizontal and vertical dimension, but take care not to move it with a circular motion—this tends to deposit the cells in rings around each well's periphery. If preparing multiple plates, gently mix the tube containing the working cell stock before starting each plate, to prevent the cell stock from settling. Similar care should be taken during transfection to ensure even distribution of the transfection mix. Use a pipette tip to swirl each well in "figure eights" immediately after addition of the transfection reagent, gently tilt the plate back-and-forth several times after

each triplicate set has been added. After all wells have received transfection mixes, rock the entire plate in "figure eights," followed by tilting in the horizontal and vertical dimension.

21. *Gaussia* and *Cypridina* luciferases metabolize different substrates and emit light at slightly different wavelengths (475 and 465 nm, respectively). However, given the technical challenges of measuring both enzymes simultaneously, the protocol outlined in Subheading 3.5 individually determines each enzyme's activity in succession, collecting light over a wide spectrum.

22. A note regarding error propagation in qRT-PCR and Luciferase assays. In all cases, we propagate uncertainties (e.g., as measured by qPCR Miner) as follows. Given $S = A + B$, or $S = A - B$:

$$\sigma_S = \sqrt{(\sigma_A)^2 + (\sigma_B)^2}$$

where σ_A and σ_B are the measurement errors of A and B, respectively. For $P = A \times B$ or $P = A/B$:

$$\sigma_P = P \times \sqrt{\left(\frac{\sigma_A}{A}\right)^2 + \left(\frac{\sigma_B}{B}\right)^2}$$

The uncertainty of other functions, $F(x)$, is estimated using the first derivative approximation:

$$\sigma_{f(x)} = \sigma_x \times f'(x)$$

Acknowledgments

I thank Yasemin Sancak for her tireless assistance in the preparation of the manuscript, David Hendrickson and Danielle Tenen for sharing their technical expertise in fRIP, John Rinn, Ezgi Hacisuleyman, and Scott Younger for their assistance in developing CRISPR-Display, and Philipp Maass for his assistance with CLING.

References

1. Cech TR, Steitz JA (2014) The noncoding RNA revolution-trashing old rules to forge new ones. Cell 157(1):77–94. https://doi.org/10.1016/j.cell.2014.03.008
2. Liang JC, Bloom RJ, Smolke CD (2011) Engineering biological systems with synthetic RNA molecules. Mol Cell 43(6):915–926. https://doi.org/10.1016/j.molcel.2011.08.023
3. Rinn JL, Chang HY (2012) Genome regulation by long noncoding RNAs. Annu Rev Biochem 81:145–166. https://doi.org/10.1146/annurev-biochem-051410-092902
4. Engreitz JM, Ollikainen N, Guttman M (2016) Long non-coding RNAs: spatial amplifiers that control nuclear structure and gene expression. Nat Rev Mol Cell Biol 17(12):756–770. https://doi.org/10.1038/nrm.2016.126
5. Bassett AR, Akhtar A, Barlow DP, Bird AP, Brockdorff N, Duboule D, Ephrussi A,

Ferguson-Smith AC, Gingeras TR, Haerty W, Higgs DR, Miska EA, Ponting CP (2014) Considerations when investigating lncRNA function in vivo. elife 3:e03058. https://doi.org/10.7554/eLife.03058
6. Kushwaha M, Rostain W, Prakash S, Duncan JN, Jaramillo A (2016) Using RNA as molecular code for programming cellular function. ACS Synth Biol 5(8):795–809. https://doi.org/10.1021/acssynbio.5b00297
7. Shechner DM, Hacisuleyman E, Younger ST, Rinn JL (2015) Multiplexable, locus-specific targeting of long RNAs with CRISPR-display. Nat Methods 12(7):664–670. https://doi.org/10.1038/nmeth.3433
8. Sigova AA, Abraham BJ, Ji X, Molinie B, Hannett NM, Guo YE, Jangi M, Giallourakis CC, Sharp PA, Young RA (2015) Transcription factor trapping by RNA in gene regulatory elements. Science 350(6263):978–981. https://doi.org/10.1126/science.aad3346
9. Luo S, Lu JY, Liu L, Yin Y, Chen C, Han X, Wu B, Xu R, Liu W, Yan P, Shao W, Lu Z, Li H, Na J, Tang F, Wang J, Zhang YE, Shen X (2016) Divergent lncRNAs regulate gene expression and lineage differentiation in pluripotent cells. Cell Stem Cell 18(5):637–652. https://doi.org/10.1016/j.stem.2016.01.024
10. Maass PG, Barutcu AR, Shechner DM, Weiner CL, Mele M, Rinn JL (2018) Spatiotemporal allele organization by allele-specific CRISPR live-cell imaging (SNP-CLING). Nat Struct Mol Biol 25(2):176–184. https://doi.org/10.1038/s41594-017-0015-3
11. Zalatan JG, Lee ME, Almeida R, Gilbert LA, Whitehead EH, La Russa M, Tsai JC, Weissman JS, Dueber JE, Qi LS, Lim WA (2015) Engineering complex synthetic transcriptional programs with CRISPR RNA scaffolds. Cell 160(1-2):339–350. https://doi.org/10.1016/j.cell.2014.11.052
12. Liu Y, Zhan Y, Chen Z, He A, Li J, Wu H, Liu L, Zhuang C, Lin J, Guo X, Zhang Q, Huang W, Cai Z (2016) Directing cellular information flow via CRISPR signal conductors. Nat Methods 13(11):938–944. https://doi.org/10.1038/nmeth.3994
13. Ferry QR, Lyutova R, Fulga TA (2017) Rational design of inducible CRISPR guide RNAs for de novo assembly of transcriptional programs. Nat Commun 8:14633. https://doi.org/10.1038/ncomms14633
14. Ma H, Tu LC, Naseri A, Chung YC, Grunwald D, Zhang S, Pederson T (2018) CRISPR-Sirius: RNA scaffolds for signal amplification in genome imaging. Nat Methods 15(11):928–931. https://doi.org/10.1038/s41592-018-0174-0
15. Mele M, Mattioli K, Mallard W, Shechner DM, Gerhardinger C, Rinn JL (2017) Chromatin environment, transcriptional regulation, and splicing distinguish lincRNAs and mRNAs. Genome Res 27(1):27–37. https://doi.org/10.1101/gr.214205.116
16. Ribeiro DM, Zanzoni A, Cipriano A, Delli Ponti R, Spinelli L, Ballarino M, Bozzoni I, Tartaglia GG, Brun C (2018) Protein complex scaffolding predicted as a prevalent function of long non-coding RNAs. Nucleic Acids Res 46(2):917–928. https://doi.org/10.1093/nar/gkx1169
17. Dominguez D, Freese P, Alexis MS, Su A, Hochman M, Palden T, Bazile C, Lambert NJ, Van Nostrand EL, Pratt GA, Yeo GW, Graveley BR, Burge CB (2018) Sequence, structure, and context preferences of human RNA binding proteins. Mol Cell 70(5):854–867.e9. https://doi.org/10.1016/j.molcel.2018.05.001
18. Kubota M, Tran C, Spitale RC (2015) Progress and challenges for chemical probing of RNA structure inside living cells. Nat Chem Biol 11(12):933–941. https://doi.org/10.1038/nchembio.1958
19. Shukla CJ, McCorkindale AL, Gerhardinger C, Korthauer KD, Cabili MN, Shechner DM, Irizarry RA, Maass PG, Rinn JL (2018) High-throughput identification of RNA nuclear enrichment sequences. EMBO J 37(6). https://doi.org/10.15252/embj.201798452
20. Konermann S, Brigham MD, Trevino AE, Joung J, Abudayyeh OO, Barcena C, Hsu PD, Habib N, Gootenberg JS, Nishimasu H, Nureki O, Zhang F (2015) Genome-scale transcriptional activation by an engineered CRISPR-Cas9 complex. Nature 517(7536):583–588. https://doi.org/10.1038/nature14136
21. Brown JA, Bulkley D, Wang J, Valenstein ML, Yario TA, Steitz TA, Steitz JA (2014) Structural insights into the stabilization of MALAT1 noncoding RNA by a bipartite triple helix. Nat Struct Mol Biol 21(7):633–640. https://doi.org/10.1038/nsmb.2844
22. Wilusz JE, JnBaptiste CK, Lu LY, Kuhn CD, Joshua-Tor L, Sharp PA (2012) A triple helix stabilizes the 3′ ends of long noncoding RNAs that lack poly(A) tails. Genes Dev 26(21):2392–2407. https://doi.org/10.1101/gad.204438.112

23. Cuello P, Boyd DC, Dye MJ, Proudfoot NJ, Murphy S (1999) Transcription of the human U2 snRNA genes continues beyond the 3′ box in vivo. EMBO J 18(10):2867–2877. https://doi.org/10.1093/emboj/18.10.2867

24. Hendrickson D, Kelley DR, Tenen D, Bernstein B, Rinn JL (2016) Widespread RNA binding by chromatin-associated proteins. Genome Biol 17:28. https://doi.org/10.1186/s13059-016-0878-3

25. Simon MD (2016) Insight into lncRNA biology using hybridization capture analyses. Biochim Biophys Acta 1859(1):121–127. https://doi.org/10.1016/j.bbagrm.2015.09.004

Chapter 9

Controlling the Activity of CRISPR Transcriptional Regulators with Inducible sgRNAs

Quentin R. V. Ferry and Tudor A. Fulga

Abstract

The type-II CRISPR-Cas9 system has been repurposed to create synthetic programmable transcriptional regulators (CRISPR-TRs). Subsequent modifications of the system now allow for spatiotemporal control of CRISPR-mediated gene activation and repression. Among these solutions, the development of inducible spacer-blocking hairpin guide RNAs (iSBH-sgRNAs) provide an easy to implement and versatile way to condition the activation of most CRISPR-TRs on the presence of a user defined inducer. In this chapter, I cover the know-how relating to the design and synthesis of iSBH-sgRNAs, as well as the implementation in mammalian cells of inducible CRISPR-TR strategies based on this technology.

Key words CRISPR, Cas9, Inducible, sgRNA, Transcription

1 Introduction

Adapted from bacteria, the *Streptococcus pyogenes* type-II CRISPR system works as an RNA-guided endoribonuclease [1]. A short single guide RNA (sgRNA) complexes with the nuclease Cas9 to form a ribonucleoprotein that specifically identifies, hybridizes with, and cleaves a double stranded DNA target complementary to the guide. Accordingly, a simple modification of the sgRNA first 20 nucleotides (nt), known as the spacer sequence, allows for the reprogramming of the system to distinct CRISPR target sites (CTS).

The system has been subsequently repurposed from programmable molecular scissors to programmable transcriptional regulators (CRISPR-TRs), that have been successfully used in multiple species to control the transcriptional output of both transgenes and endogenous genes of interest [2]. CRISPR-TRs were engineered by tethering activating or repressing effector domains (VP64, KRAB, etc.) to a nuclease deficient version of Cas9 (dCas9), which when combined with the sgRNA works as a DNA binder that can be programmed to direct the effectors to the

Tudor A. Fulga et al. (eds.), *CRISPR Guide RNA Design: Methods and Protocols*, Methods in Molecular Biology, vol. 2162, https://doi.org/10.1007/978-1-0716-0687-2_9, © Springer Science+Business Media, LLC, part of Springer Nature 2021

transcriptional start site of a gene of interest, hereby locally influencing mRNA production.

Further modifications of the system have brought inducible CRISPR-TR variants that not only let the researcher decide on what gene to activate or silence, but also allow for precise spatiotemporal control over the activity of these synthetic transcription factors. Notably, inducible approaches based on modification of the sgRNA sequence have offered a simple framework to create orthogonal responsive CRISPR-TRs at a low metabolic cost for the cell [3–7]. In these methods, the activity of CRISPR-TR is controlled by conditionally sequestering (OFF-state) or releasing (ON-state) the sgRNA spacer segment, to prevent or allow the CRISPR complex to bind on target and drive gene expression.

Researchers at the University of Oxford have shown that such inducible CRISPR-TRs can easily be implemented by appending on the 5′ end of the sgRNA an RNA sequence designed to posttranscriptionally form an inducible spacer-blocking hairpin (iSBH, Fig. 1) [3]. From 5′ to 3′, the appendage is composed of a sequence designed fully or partially complementary to the sgRNA spacer (back-fold segment), followed by a "sensor/actuator" segment encoding an RNA-cleaving unit, whose slicing is conditioned on the presence of a specific inducer. This strategy has been successfully used to create inducible sgRNAs, and consequently inducible CRISPR-TRs that specifically respond to genetically encoded (proteins) and externally delivered (antisense oligonucleotide) triggers. Additionally, Ferry et al. have shown that by multiplexing iSBH-sgRNAs expression, one can concomitantly or independently control the transcriptional output of multiple genes in a same mammalian cell [3].

Due to its modularity and ease to implement, the SBH platform represents a valuable addition to the growing synthetic biology toolkit. The advent of inducible CRISPR-TR strategies allowing for the spatiotemporal control of gene expression will certainly help answer fundamental biological questions regarding the transcriptional programs at plays during development, as well as explain the consequences of dysregulations observed in pathologies such as cancers. Furthermore, our ability to interface CRISPR-TRs within more complex gene circuits, notably coupling CRISPR-based transcription factors with synthetic receptors [8], will usher new opportunities to rewire cellular behavior for the purpose of engineering new therapeutic strategies.

In this chapter, I present guidelines, as well as a step-by-step protocol to implement iSBH-based inducible CRISPR-TRs in mammalian cells. After reviewing the general design principles of iSBH-sgRNAs, I will explain how to derive the final iSBH sequence for a preselected guide RNA, implement the system in mammalian cells, and troubleshoot these experiments.

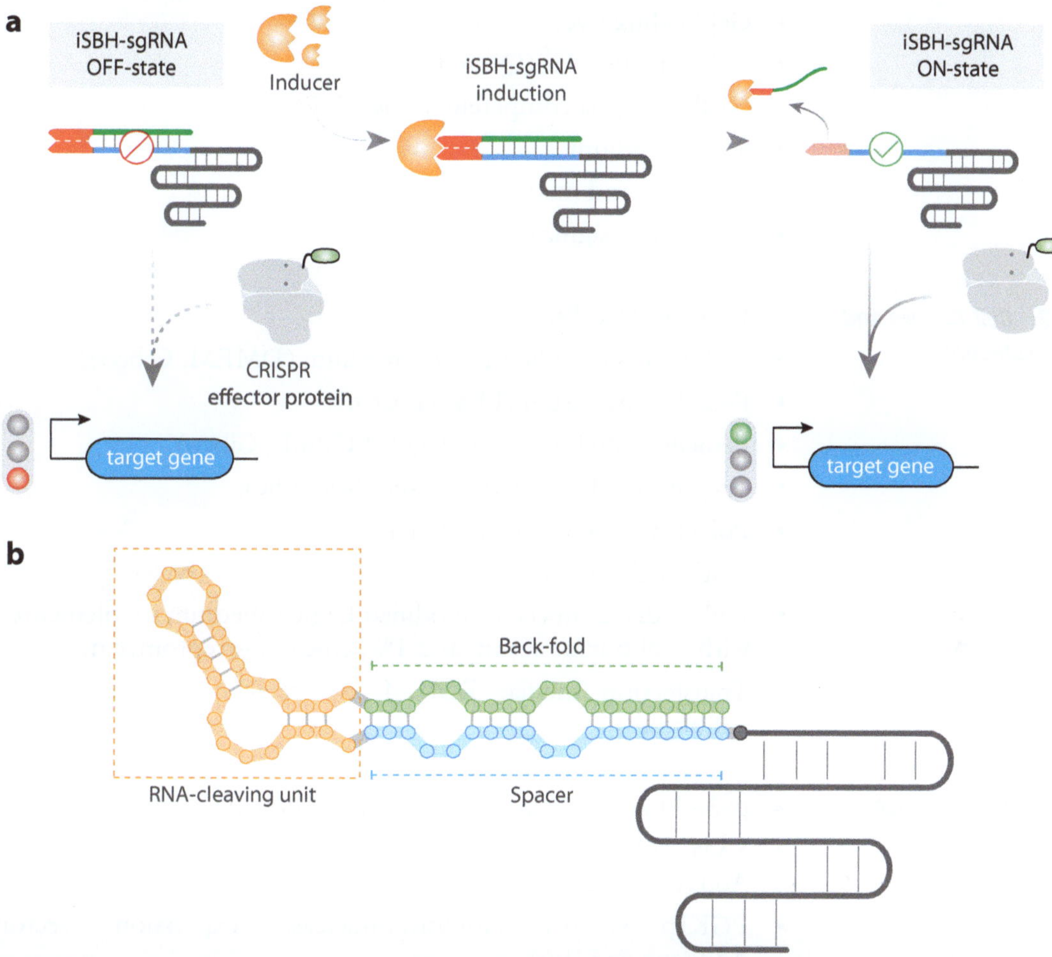

Fig. 1 iSBH methodology. (**a**) Quiescent iSBH-sgRNAs are conditionally turned on in the presence of their respective inducer. Upon activation, a functional sgRNA can direct the CRISPR effector (e.g., dCas9-VP64) to a user define gene target and regulate mRNA expression. (**b**) Canonical structure of inducible spacer-blocking hairpins showing the back-fold, cleaving unit and spacer segments

2 Materials and Software

2.1 Cloning

- Nuclease-free H_2O.
- T4 ligase and 10× T4 ligase buffer (NEB).
- T4 polynucleotide kinase (NEB).
- pDNA extraction kit (Qiagen).
- BbsI (ThermoFisher scientific) and compatible buffer.
- Antarctic phosphatase buffer (NEB) and compatible buffer.
- UltraPure™ Agarose (Thermofisher scientific).
- 1 kb DNA Ladder (NEB).

- Gel loading dye.
- Gel extraction kit (Qiagen).
- NEB® 5-alpha competent *E. coli* (NEB).
- S.O.C medium.
- LB broth and medium.
- Ampicillin (sigma).

2.2 Cell Culture and Transfection

- HEK-293T cells.
- Dulbecco's modified Eagle's medium (DMEM, Gibco).
- Fetal bovine serum (FBS, Gibco).
- Penicillin–streptomycin (10,000 U/mL, Gibco).
- Trypsin–EDTA (Thermo Fisher Scientific).
- Polyethylenimine (PEI, Sigma).
- Opti-MEM™ (Gibco).
- Full media: Dulbecco's modified Eagle's medium supplemented with fetal bovine serum and 1% penicillin–streptomycin.
- Transfection media: DMEM, 2% fetal bovine serum, no antibiotics.

2.3 Plasmid DNA

- px330 (sgRNA expressing plasmid, Addgene plasmid 42230).
- SAM sgRNA-2×MS2 vectors (sgRNA expressing plasmid, Addgene plasmid 42230).
- PGKlp-Csy4-pA (Endoribonuclease expression vector, Addgene #55196).
- pcDNA3-EGFP (Transfection positive control, Addgene #13031).
- SAM dCas9-VP64 pDNA (Addgene #61422).
- SAM MCP-p65-HSF1 (Addgene #61423).

2.4 Readout

- 1× phosphate-buffered saline (PBS).
- RNeasy Mini Kit (Qiagen).
- QuantiTect Reverse Transcription Kit (Qiagen).
- SsoAdvanced Universal SYBR Green Supermix (Bio-Rad).

2.5 Software

- iSBHfold, webtool for the automated design of iSBH-sgRNAs: http://apps.molbiol.ox.ac.uk/iSBHfold/.
- NUPACK, RNA secondary structure prediction: http://www.nupack.org/.
- Primer3m, qPCR primer pair design: http://primer3.ut.ee/.

3 Methods

3.1 Rationales for iSBH Design

See **Notes 1–3** for an overview on the design of iSBH-sgRNAs (*see* **Note 1**), as well as guidelines on how to choose a suitable inducer (*see* **Note 2**) and spacer sequence for your application (*see* **Note 3**).

3.2 Design of the iSBH Sequence

Follow guidelines in the note section to (1) design the iSBH backfold segment (*see* **Note 4**), (2) design an ASO-responsive iSBH (*see* **Note 5**), (3) achieve orthogonal and multiplexed gene activation with ASO-iSBH (*see* **Note 6**), and (4) design protein-responsive iSBH-sgRNAs for orthogonal and multiplexed gene activation (*see* **Note 7**).

3.3 Cloning of iSBH-sgRNAs

1. Select a sgRNA expressing vector (*see* **Note 8**).
2. Clone the spacer or iSBH sequence as an oligo duplex in the sgRNA expression vector (*see* **Notes 9** and **10**).
   - Design and order forward and reverse primers encoding the spacer or iSBH (*see* **Note 11**).
   - Anneal oligonucleotides and phosphorylate the oligo duplex. Set up the following reaction:

Reagent	Volume, μl
Nuclease-free H_2O	6.5
Forward primer (100 μM concentration)	1
Reverse primer (100 μM concentration)	1
10× T4 ligase buffer	1
T4 Polynucleotide Kinase	0.5

   - Incubate for 30 min at 37 °C to allow for the oligonucleotide to be phosphorylated. Heat up the reaction mix to 95 °C for 5 min before progressively cooling down to 4 °C using a 0.1 °C/s ramp (keep on ice until ligation) (*see* **Note 12**).
   - Digest, dephosphorylate, and purify the DNA backbone. Set up the following digestion mix to digest 1 μg of pDNA and incubate for 1 h at 37 °C.

Reagent	Volume, μl
Nuclease-free H_2O	15
sgRNA expressing pDNA (500 ng/μl)	2
10× enzyme buffer	2
BbsI	1

- Once the digestion completed, top up the digestion mix with the following:

Reagent	Volume, µl
Digestion mix (see previous step)	20
Antarctic phosphatase buffer	1
10× Antarctic phosphatase buffer	2.3

- Incubate for 1 h at 37 °C.
- Once the dephosphorylation completed, run the digestion mix on a 1% agarose gel and extract the linearized pDNA using a gel extraction kit. Run a 1 kb DNA Ladder alongside your digestion product to ensure that the linearized backbone has the expected size. Elute the 1 µg DNA digested in 30 µl of EB solution and keep on ice until ligation (*see* **Note 13**).
- Ligate insert and backbone together. Set up the ligation reaction below and incubate at room temperature for 1 h followed by 10 min at 65 °C to inactivate the T4 ligase (*see* **Notes 14** and **15**).

Reagent	Volume, µl
Nuclease-free H_2O	5.5
Annealed and phosphorylated oligo duplex (10 µl reaction)	1
Gel-purified pDNA backbone (30 µl elution)	2
10× T4 ligase buffer	1
T4 ligase	0.5

- Transform 1 µl of the 10 µl ligation mix into competent Dh5α bacteria. Thaw a 10–20 µl bacteria aliquot on ice, gently add ligation mix, and incubate on ice for 30 min, followed by a 30 s heat shock at 42 °C. Place back on ice for 5 min then add 50 µl of SOC medium and place the mix in a 37 °C shaker for 1 h.
- Place 25–50 µl of the transformation mix onto an LB plate containing the right selective antibiotic (ampicillin for PX330 and SAM sgRNA-2×MS2 expression vector) and incubate at 37 °C overnight.
- Take the plate out once sizable colonies are formed but do not extend the incubation time pass 16 h as single colonies might start to merge.

- Pick and grow single colonies overnight (~16 h) at 37 °C in 2 μl LB media supplemented with the right antibiotic. Extract the pDNA using a pDNA extraction kit.

3.4 Inducer Synthesis

- If using an ASO-responsive guide, order the corresponding ASO inducer as a single-stranded DNA antisense oligonucleotide with four phosphorothioate bonds on both ends. In the example below, the ASO inducer should be ordered as follows:

```
5′-AGGACAGTACAGCGAGATACGAACATCCCTAACAGTAAGTCGGAGTACTGTCCT-3′
                     ||||||||||||||||||||
                 3′-ATGCTTGTAGGGATTGTCAT-5′
```

(back-fold, sensing-loop, spacer, and ASO inducer)

5′-T*A*C*T*GTTAGGGATGTT*C*G*T*A-3′ , where the asterisks denote phosphorothioate bonds.

- Upon receiving the ASO, resuspend to 100 uM concentration and aliquot to avoid repeated freeze–thaw cycles between experiments.
- If using a protein-inducible guide, clone the corresponding endoribonuclease under any pol-II promoter (constitutive, tissue-specific, or inducible) in any vector of choice. *See* **Note 16** for codon-optimized endoribonuclease sequences).

3.5 System Delivery: Implementation in Mammalian Cells

3.5.1 Preparing Cells for Experimentations

- Thaw a fresh aliquot of human embryonic kidney (HEK293-T) cells expressing a mutant version of the SV40 large T antigen.
- Culture cells in a 75 cm^2 flask at 37 °C and 5% CO^2 in full media.
- Passage cells every 48 h in a 1:6 ratio using full media for ~7 days prior conducting the first experiments. Subsequently, the same batch of cells can be used for up to 30 passages after what a fresh aliquot has to be used to seed a new culture. When passaging, remove full media, detach the cells using 2 ml of trypsin, resuspend with 10 ml of full media to deactivate the trypsin. Finally, transfer 2 ml into a new flask and bring to a volume of 12 ml with full media.
- Once the current batch has been passaged for 7 days and is ready for experiments, test the transfection efficiency using a pDNA expressing a fluorophore (e.g., EGFP) from a constitutive promoter (e.g., pcDNA3-EGFP). Transfect 500 ng per well in 12-well plate (see below for protocol on transfection) and observe the cell layer under a fluorescence microscope 24 h posttransfection to assess transfection efficiency.

3.5.2 System Delivery

- 24 h prior to transfection, passage cells and seed a 12-well plate so as to reach 70% confluency (cell layer should cover approximately 70% of the surface area of the well) by the time of the experiment (*see* **Note 17**).
- On the day of transfection, gently remove the full media from each well without touching the cell layer and replace with the equivalent volume of transfection media. Place back the cells in the 37 °C incubator and allow them to rest for 15 min prior to pDNA delivery (see next step).
- For each well (12-well plates), prepare a transfection mix containing the following:

Reagent	Volume
Opti-MEM™	100 μl
The various pDNAs to be transfected (*see* below for example, **Note 18**)	<2 μg
Transfection reagent, polyethylenimine (PEI; Sigma-Aldrich, 1 mg/ml)	Recommended 2:3 ratio (μg DNA/ml PEI) [9]

- Vortex the mix for 10 s and allow to rest for 15 min at room temperature.
- Once the two 15 min incubations above are completed, use a pipette (200 μl tip) to deliver the transfection mix drop by drop to the cells.
- 24 h posttransfection, delicately replace the transfection media with full media (*see* **Note 19**).

3.5.3 Transfection Mixes

Below I provide examples of transfection mixes used to implement iSBH-sgRNA mediated gene regulation in mammalian cells.

- Conditional activation of *geneX* using an ASO-responsive iSBH-sgRNA and dCas9-VP64 (*see* **Notes 20** and **21**). For each of well of a 12-well plate prepare the following transfection mix:

Reagent	Volume	Note
Opti-MEM	100 μl	
iSBH-sgRNA pDNA	500 ng	Guide expression pDNA, ~5 kb. iSBH has to be designed to target *geneX* and respond to the selected ASO
dCas9m4-VP64	250 ng	
PEI	2:3 ratio	

- Allow 24 h for the system component (iSBH-sgRNA, dCas9m4-VP64) to be expressed prior to delivering the ASO trigger.
- 24 h posttransfection, replace transfection media with fresh transfection media and proceed to transfect the following transfection mix:

Reagent	Volume	Note
Opti-MEM	100 μl	
Carrier pDNA (e.g., pcDNA3.1-empty)	500 ng	Should not contain any elements that would interfere either directly with the CRISPR-TR or indirectly by altering the readout, by expressing a fluorophore for example
100 uM ASO trigger	1 μl	

- Change transfection media for full media 24 h post-ASO delivery.
- Conditional activation of *geneX* using a protein-responsive iSBH-sgRNA. For each of 12-well plate prepare the following transfection mix (*see* **Notes 22–24**):

Reagent	Volume	Note
Opti-MEM	100 μl	
iSBH-sgRNA pDNA	500 ng	Guide expression pDNA, ~5 kb. iSBH has to be designed to target *geneX* and respond to the selected ASO
dCas9m4-VP64	250 ng	
Csy4 or Cas6A expressing vector (~4.2 kb)	250 ng	
PEI	2:3 ratio	

Replace the transfection for full media 24 h posttransfection

3.6 Readout: Assessing the System's ON/OFF Performances

Here I assume that the selected target is an endogenous gene. Accordingly, I only cover the methodology to readout changes in transcriptional output via qPCR. I refer the reader aiming to use iSBH-sgRNAs to control the expression of fluorescent reporter genes (e.g., EGFP) to the original iSBH publication [3]. Details are provided on how to create a reporter construct with synthetic CRISPR promoters that can be controlled with iSBH-sgRNAs and dCas9-VP64, as well as the methodology on how to use flow cytometry analysis to assess system performances. For the remainder of this section I assume that cells have been transfected with the right construct to create the ON-state condition (+inducer), and OFF-state condition (-inducer) shown below:

ON-state, cell transfected with the following:

- One iSBH-sgRNA targeting endogenous *geneX*.
- CRISPR-TR effector protein (dCas9-VP64).
- Inducer matching the RNA-cleaving unit of the iSBH (Csy4, ASO, etc.).

OFF-state, cell transfected with the following:

- One iSBH-sgRNA targeting endogenous *geneX*.
- CRISPR-TR effector protein (e.g., dCas9-VP64).
- Decoy inducer nonmatching the RNA-cleaving unit of the iSBH (different endoribonuclease or decoy pDNA, scramble ASO, etc.)

In my experience, and using the transfection mixes detailed in the last section, changes in the transcriptional output of the targeted gene should be detectable as early as +24h postdelivery of all system component.

3.6.1 Harvest the Cells and Perform Total RNA Extraction

- Remove transfection media and wash the cell layer twice with 500 μl/well of 1× PBS (room temperature).
- Proceed with RNA extraction using RNeasy Mini Kit or similar kit and following the manufacturer's protocol.
- Nanodrop the extracted RNA samples for concentrations and keep on ice until the reverse transcription step (see below). Alternatively, samples can be stored at −20 °C for several days.

3.6.2 Convert Total RNA into Corresponding cDNA

- For each condition, reverse-transcribe 1 μg of RNA using the QuantiTect Reverse Transcription Kit or similar kit and following the manufacturer's protocol (*see* **Notes 25** and **26**).
- Keep the cDNA on ice until qPCR (see below). Alternatively, samples can be stored at −20 °C for several weeks.

3.6.3 Conduct qPCR Analysis: Primer Pair Selection and Validation

- Select qPCR primer pairs: For each gene to be quantified, use the NCBI portal to obtain the mRNA sequence of the most abundant isoform (https://www.ncbi.nlm.nih.gov/gene). Input the mRNA sequence into Primer3 (http://primer3.ut.ee/) [10] to find a suitable qPCR primer pair (*see* **Note 27**).
- Validate primer pairs using qPCR analysis: If testing primer pair for *geneX*, create serial dilutions (1, 1:10, 1:100, 1:1000) of the cDNA sample for which you expect having the greatest concentration of *geneX* mRNA. For each dilution, prepare the following qPCR reaction mix:

Reagent	Volume, μl
Nuclease-free water	12.25
SsoAdvanced Universal SYBR Green Supermix	17.5
Forward primer (10 μM)	1.75
Reverse primer (10 μM)	1.75
Reverse transcription reaction (cDNA)	3.5

- Split the content in three 10 μl technical replicates, load in a qPCR plate and run the following protocol for qPCR analysis (*see* **Note 28**):

Step 1	Heat up to 98 °C for 2 min
Step 2	98 °C 10 s
Step 3	60 °C 30 s
Step 4	Image
Go to Step 2, 39 times	
Step 5	65 °C for 5 s and then perform a melting curve analysis

3.6.4 Conduct qPCR Analysis

- Run qPCR analysis on all conditions. For each condition, prepare n qPCR reaction mixes, where n is the number of primer pairs (*see* **Note 29**):

Reagent	Volume, μl
Nuclease-free water	12.25
SsoAdvanced Universal SYBR Green Supermix	17.5
Forward primer (10 μM)	1.75
Reverse primer (10 μM)	1.75
Reverse transcription reaction (cDNA)	3.5

- Split the content in three 10 μl technical replicates in a qPCR plate and run the following protocol for qPCR analysis:

Step 1	Heat up to 98 °C for 2 min
Step 2	98 °C 10 s
Step 3	60 °C 30 s
Step 4	Image

(continued)

Go to Step 2, 39 times	
Step 5	65 °C for 5 s and then perform a melting curve analysis

- Once the *GAPDH*, *dCas9-VP64*, and *geneX* Ct values have been obtained for the ON and OFF-state conditions, use the following formula to compute the fold change in *geneX* transcript levels triggered by the presence of the inducer:

$$\text{Fold change} = \frac{2^{\left(\mathrm{Ct}_{\mathrm{GAPDH}}^{\mathrm{ON}} - \mathrm{Ct}_{\mathrm{GeneX}}^{\mathrm{ON}}\right) - \left(\mathrm{Ct}_{\mathrm{GeneX}}^{\mathrm{ON}} - \mathrm{Ct}_{\mathrm{dCas9-VP64}}^{\mathrm{ON}}\right)}}{2^{\left(\mathrm{Ct}_{\mathrm{GAPDH}}^{\mathrm{OFF}} - \mathrm{Ct}_{\mathrm{GeneX}}^{\mathrm{OFF}}\right) - \left(\mathrm{Ct}_{\mathrm{GeneX}}^{\mathrm{OFF}} - \mathrm{Ct}_{\mathrm{dCas9-VP64}}^{\mathrm{OFF}}\right)}}$$

3.7 Characterizing System Performances and Troubleshooting

See **Note 30**.

4 Notes

1. Inducible spacer-blocking hairpins can be evolved against both genetically encoded inducers and externally delivered inducers. CRISPR endoribonucleases from the Cas6 family are short proteins (~250 amino acids) that specifically recognize and cleave short cognate RNA structures. By utilizing such RNA structures as RNA-cleaving unit in the design of iSBH-sgRNAs, Ferry et al. have created silent SpCas9 guides conditionally activated by the presence of the *Pseudomonas aeruginosa* Csy4, and others that orthogonally respond to the *Streptococcus thermophilus* Cas6A endoribonuclease (Figs. 2 and 3). Given the relatively small size of Cas6 proteins, protein-responsive iSBH-sgRNAs can be used to engineer genetically encoded synthetic circuits whereby the activation of downstream target genes (CRISPR-TR activation) is conditioned on the expression of the protein inducer. For example, one could wire the activation of two synthetic receptors (e.g., GPCRs) with two distinct transcriptional programs, by modifying those receptors to release membrane-tethered Cas6 proteins only in the presence of their cognate agonist. Upon release of Csy4 for example, all iSBH-sgRNAs designed to be responsive to the endoribonuclease should be switched ON, leading to the transcriptional regulation of their corresponding targets. Meanwhile, iSBH-sgRNAs designed to respond to a different a distinct protein inducer (e.g., Cas6A) would remain silent. Conversely, triggering the Cas6A-fusion chimeric receptor should activate all Cas6A-

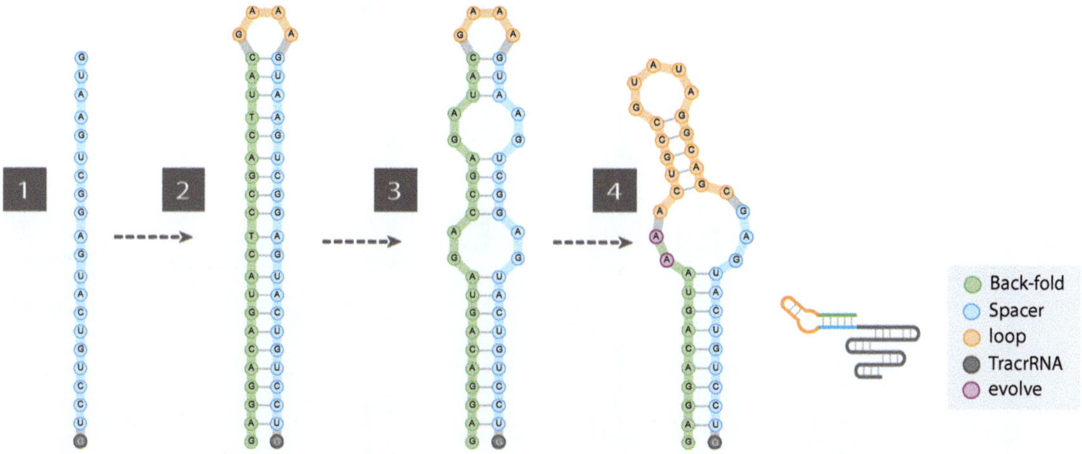

Fig. 2 Design of Csy4-responsive iSBH. The spacer sequence (1, blue) is appended with a back-fold (green) and linker (orange) segments to form the SBH$^{(0)}$ hairpin (2). Bulges are added to create SBH$^{(0B)}$ (3). Finally, the RNA-cleaving unit for Csy4 (orange) is grafted onto the seed proximal bulge of SBH$^{(0B)}$ to obtain the final Csy4-responsive iSBH sequence. The identity of the purple nucleotides is evolved by the iSBHfold software to guarantee that the bulge stays unpaired

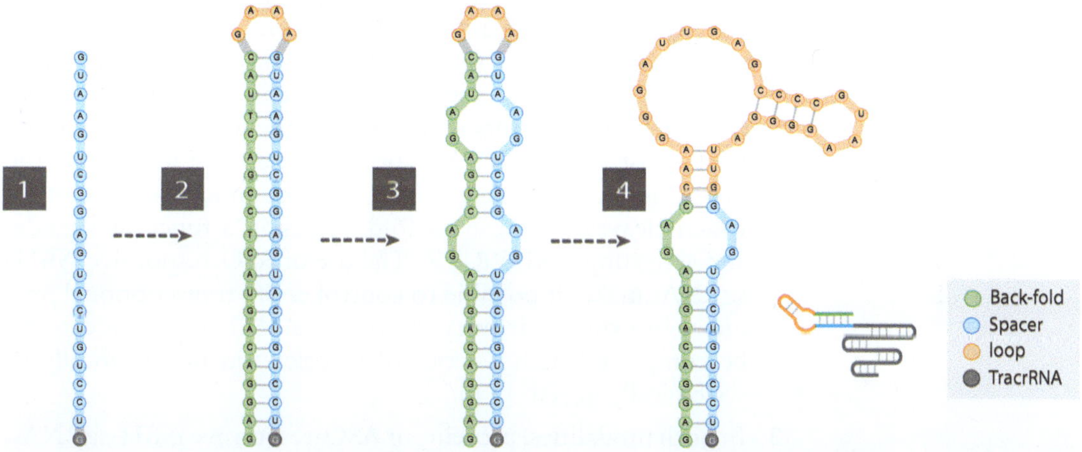

Fig. 3 Design of Cas6A-responsive iSBH. The spacer sequence (1, blue) is appended with a back-fold (green) and linker (orange) segments to form the SBH$^{(0)}$ hairpin (2). Bulges are added to create SBH$^{(0B)}$ (3). Finally, the RNA-cleaving unit for Cas6A (orange) is grafted right after the seed proximal bulge of SBH$^{(0B)}$ to obtain the final Cas6A-responsive iSBH sequence

responsive iSBH-sgRNAs and regulate their downstream target genes, without affecting Csy4-responsive guides. Both receptors, iSBH-sgRNAs, and SpCas9-TR can be encoded and delivered as plasmids to reprogram the recipient cells.

On the other hand, Ferry and colleagues have also developed iSBH-sgRNAs that specifically respond to short single stranded DNA antisense oligonucleotides (ASOs, Fig. 4). It was demonstrated that iSBH-sgRNAs with a 14-nt-long

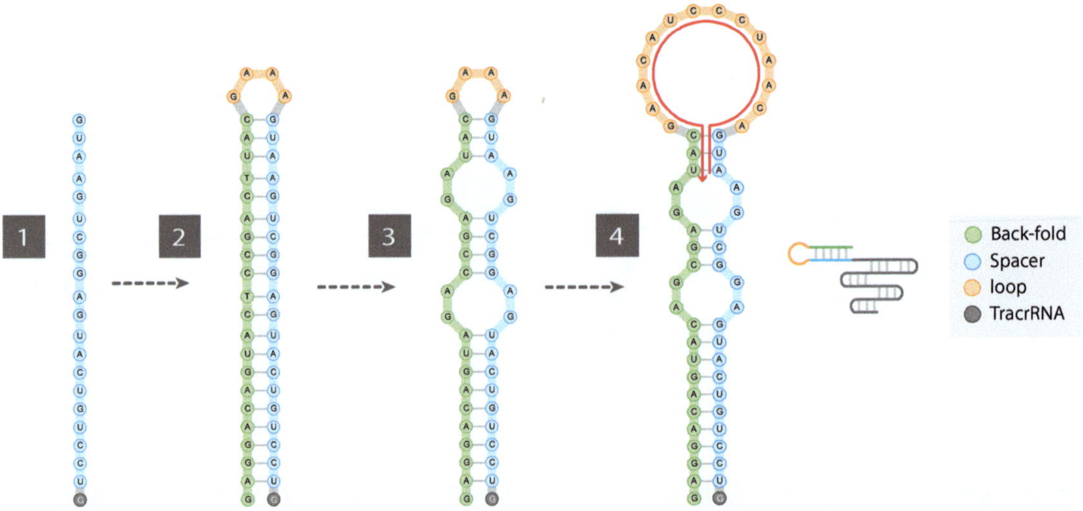

Fig. 4 Design of ASO-responsive iSBH. The spacer sequence (1, blue) is appended with a back-fold (green) and linker (orange) segments to form the SBH$^{(0)}$ hairpin (2). Bulges are added to create SBH$^{(0B)}$ (3). Finally, the SBH$^{(0B)}$ loop is replaced by a 14 nt ASO sensing-loop (orange). The ASO inducer (red) is designed to base-pair with the entire sensing-loop and three flanking nucleotides on the back-fold and on the spacer side

sensing-loop, acting as an RNA-cleaving unit, can be conditionally sliced by the delayed delivery of a sequence complementary ASO. While the mechanism of action remains to be confirmed, the authors speculate that the formation in the cell nucleus of a DNA-RNA hybrid between the iSBH and the matching ASO, engages the endoribonuclease RNase-H, which cleaves off the back-fold, releasing a functional guide, and activating CRISPR-TR. The use of ASO-responsive iSBH-sgRNAs makes it possible to control entire transcriptional programs by external delivery of a cognate ASO. Such inducers can be complexed with a series of vehicles and deliver locally or systemically in vivo [11].

2. In both modalities, protein- or ASO-responsive iSBH-sgRNAs, the modularity of the iSBH design make it possible to (a) simultaneously control multiple gene targets with a single inducers (branching), and/or (b) independently control several genes using orthogonal inducer–target pairs (orthogonal control). Indeed, the sequence of the sensing-loop being mainly independent from the choice of spacer, several sgRNAs targeting distinct genes (distinct spacer sequences) can be engineered to respond to either the same inducer (all iSBH-sgRNAs share the same sensing-loop) or different inducers (all iSBH-sgRNAs have a distinct sensing-loop). The main difference between the two modalities is (a) the type of induction, (b) the strength of the induction. As explained above, protein inducers are more advantageous when trying to design preprogrammed gene circuits, where the activation of gene targets is

conditioned on the internal state of the system. Additionally, Cas6 inducers can be used to achieve tissue-specific CRISPR-TR activity by placing the endoribonuclease in a Cre-specific cassette or expressing the protein under the control of tissue-specific promoters [12]. On the other hand, ASO inducers offer the possibility to temporally control gene expression via external interventions. Note nevertheless that protein inducers can also be combined with inducible promoters (e.g., tetracycline-responsive promoter [13]) to create inducible CRISPR-TR responsive to small molecules.

A second factor one should take into consideration when deciding on the type of inducer is the level of CRISPR-TR activation required to achieve biologically relevant gene expression of the downstream endogenous targets. In general, Ferry et al. found that protein-based induction leads to stronger CRISPR-TR activity when compared to ASO-responsive systems. This difference is primarily thought to be explained by inducer bioavailability. Indeed, in experiments using Cas6 as a trigger, the inducer was constantly being expressed from a parent plasmid; hence, the pool of inducers was constantly being replenished. On the other hand, ASO-mediated inductions involved a one-time dose delivery, which only transiently activated the iSBH-sgRNA pool.

3. The iSBH technology was primarily developed by modification of the canonical single guide RNA (sgRNA) sequence to control a dCAS9-VP64 transcriptional activator [14, 15]. Nevertheless, the methodology being solely based on small modifications of the guide 5′-end, iSBH should, in theory, be compatible with all type-II CRISPR systems using conventional 20 nt-long spacer sgRNA free of 5′ modifications. Ferry et al. notably demonstrated that the altered sgRNA 2×MS2 guides that power the synergistic activation mediator (SAM system) [16] can be rendered inducible using the iSBH methodology. As such, it is expected that iSBH will also be compatible with other new generation CRISPR-TRs such as the Suntag and VPR systems [17, 18]. Additionally, iSBH should tolerate large 3′ modification of the sgRNAs such as the ones observed in the CRISPR display system and similar strategies [19, 20]. Given that each of the aforementioned CRISPR-TR system will have different guidelines regarding spacer selection (e.g., position with regard to the target's transcription start site), I refer the reader to these original publications to choose their guide sequence.

4. Assuming that the reader has selected the following 20 nt spacer sequence:

   5′-GTAAGTCGGAGTACTGTCCT -3′ (DNA sequence).

The final sequence of the corresponding full native sgRNA will be:

5′-GTAAGTCGGAGTACTGTCCT GTTTTAGAGCTAGAAATAG-CAAGTTAAAATAAGGCTAGTCCGTTATCAACTTGAAAAAGTGG-CACCGAGTCGGTGC-3′ (DNA sequence, spacer , guide scaffold).

Silencing of this native sgRNA is achieved by appending on its 5′-end a 20 nt back-fold segment, designed by taking the reverse complement of the spacer, and linked to the guide via a neutral "GAAA" linker:

5′-GAGGACAGTACTCCGACTTACGAAAGTAAGTCGGAGTACTGTCCT GTTTTAGAGCTAGAAATAGCAAGTTAAAATAAGGCTAGTCCGT-TATCAACTTGAAAAAGTGGCACCGAGTCGGTGC-3′ (DNA sequence, back-fold , linker, spacer , guide scaffold).

Note that an extra guanosine, required for successful U6 transcription, was inserted at the 5′-end of the back-fold. Once transcribed, the back-fold/linker/spacer segment folds to form a spacer blocking hairpin where all spacer nucleotides are base-pairing with the back-fold (*see* Fig. 2). I refer to this hairpin as $SBH^{(0)}$, where (0) denotes that none of the spacer nucleotides are free to hybridize with their DNA target.

The back-fold sequence can then further be mutated to incorporate bulges in the hairpin. The addition of bulges helps lower the thermodynamics of the hairpin thus favoring strand separation postcleavage (guide activation). The size and position of the bulges must be chosen so as to guarantee that the interaction between back-fold and spacer remains strong enough to maintain the guide silent in the absence of the cognate inducer. Ferry et al. found adding two 2 nt bulges at position ~10 and ~15 to be optimal (nucleotides along the stem are numbered from seed proximal to seed distal). These bulges are referred to as seed proximal and seed distal in reference to the spacer seed sequence, which is defined as the ten 3′ most spacer nucleotides. Using as an example the sequence above, the back-fold sequence could be changed from the fully complementary AGGACAGTACT CCGACT TAC to AGGACAGTAGA CCGAGA TAC by replacing the underlined nucleotide with the complementary base (A>T, T>A, G>C, C>G) to obtain the hairpin shown in Fig. 2 ($SBH^{(0B)}$). I refer to the bulged hairpin as $SBH^{(0B)}$.

From experience, several bulge patterns are equally likely to reduce stability while maintaining complete off-state silencing. In general, I recommend maintaining full base pairing for nucleotides 1–8, which correspond to the better part of the spacer seed sequence, and nucleotides 18–20, which I refer to as hairpin collar and correspond to a nucleation site for the melting of the back-fold–spacer hybrid. Additionally, I recommend leaving at least three base-paired nucleotides between consecutive 2 nt bulges. As an example, alteration of the back-fold sequence from AGGACAGTACTCCGACTTAC to AGGACAGTTGTCCGTGTTAC might be satisfying too.

Use Nupack [21] or any other RNA secondary structure prediction algorithms when designing the hairpin to verify that the designed back-fold sequence base-pairs correctly with the spacer to yield the expected hairpin. For that purpose, one can concatenate the back-fold and spacer sequences with the GAAA linker and submit the back-fold–linker–spacer DNA sequence to the folding algorithm. In the example above, I concatenate the mutated back-fold 5′-AGGACAGTAGACCGAGATAC -3′, the linker 5′-GAAA-3′, and spacer sequence 5′-GTAAGTCGGAGTACTGTCCT -3′ to obtain the hairpin sequence:

5′-AGGACAGTAGACCGAGATACGAAAGTAAGTCGGAGTACTGTCCT-3′

Nupack predicts that the corresponding RNA sequence will fold in a hairpin with the following structure:

5′-AGGACAGTAGACCGAGATACGAAAGTAAGTCGGAGTACTGTCCT-3′

5′-((((((((..((((..(((....)))..))))..))))))))-3′

where a couple of parentheses denotes base pairing between nucleotides, while "." denotes unpaired nucleotides.

5. Once the spacer sequence is chosen and the corresponding SBH$^{(0B)}$ sequence has been derived, one can modify the design to create ASO-responsive iSBH-sgRNAs. ASO-responsive inducible hairpins are generated by replacing the default "GAAA" linker with various 14 nt ASO sensing-loop bearing sequence complementarity to a specific trigger ASO (*see* Fig. 4). Ferry et al. found that ASO-responsive iSBH-sgRNAs are best activated by 20 nt-long single stranded DNA oligonucleotides, designed to hybridize with the 14 nt ASO sensing-loop, as well as flanking 3 nt on the 3′-end of the back-fold (blue and underlined) and 3 nt on the 5′-end of the spacer (green and underlined):

```
5′-AGGACAGTACAGCGAGATACGAACATCCCTAACAGTAAGTCGGAGTACTGTCCT-3′
                    ||||||||||||||||||||
                 3′-ATGCTTGTAGGGATTGTCAT-5′
```

(back-fold, sensing-loop, spacer, and ASO inducer)

Base pairing between the ASO and the iSBH-sgRNA nucleates on the sensing-loop. As such, the sensing-loop sequence as to be chosen to guarantee that the segment will assume an open conformation once the iSBH folds. For example, Nupack predicts the iSBH sequence above to fold as follows:

```
5′-AGGACAGTACAGCGAGATACGAACATCCCTAACAGTAAGTCGGAGTACTGTCCT-3′
5′-((((((((((..(((..(((..............)))..)))..))))))))))-3′
```

While sensing-loop sequences can be manually derived, I recommend using the iSBHfold webtool (http://apps.molbiol.ox.ac.uk/iSBHfold/) in order to guarantee minimal interaction between the sensing-loop and the rest of the iSBH sequence. For a given spacer sequence, iSBHfold determines the corresponding $SBH^{(0B)}$ and outputs a list of potential sensing-loop sequences compatible with the hairpin. These candidate sequences are derived using a genetic algorithm to ensure adequate folding of the hairpin. Once the ASO-iSBH sequence has been chosen, the sequence of the matching ASO triggers can simply be obtained by taking the reverse complement of the segment comprising the three-last back-fold nucleotides, the sensing-loop, and the three first nucleotide of the spacer (as shown above). *See* Fig. 5 for step-by-step explanations on how to design an ASO-responsive iSBH-sgRNA using the iSBHfold webtool.

6. In the previous paragraph I describe how to design an iSBH-sgRNA–ASO inducer pair. Expression of this iSBH-sgRNA in the cell links the presence of the matching ASO trigger with the activation/repression of the gene targeted by the guide. Additionally, Ferry et al. have shown that the transcriptional output of multiple target genes can be concomitantly or independently controlled by expressing several iSBH-sgRNAs in the same cell: expression of two iSBH-sgRNAs sharing the same sensing-loop makes it possible to simultaneously activate two distinct gene targets with the delivery of a single ASO. Conversely, the same two genes can be controlled independently by expressing two iSBH-sgRNAs that specifically respond to two distinct ASO inducers.

Orthogonal iSBH-sgRNAs are designed similarly to single iSBH-sgRNA–ASO inducer pair (see previous section), with the extra steps that one as to carefully choose the two sensing-loop sequences to be significantly different. Any significant overlap between these sequences might reduce orthogonality between the iSBH-sgRNA–ASO inducer pairs. While the degree of similarity tolerated might vary based on the ASO

Step 1: Set the number of ASO to 1 (1), enter the name of your gene target (2), the sequence of the your spacer (3), and select ASO[1] as ASO inducer for this given guide (4).

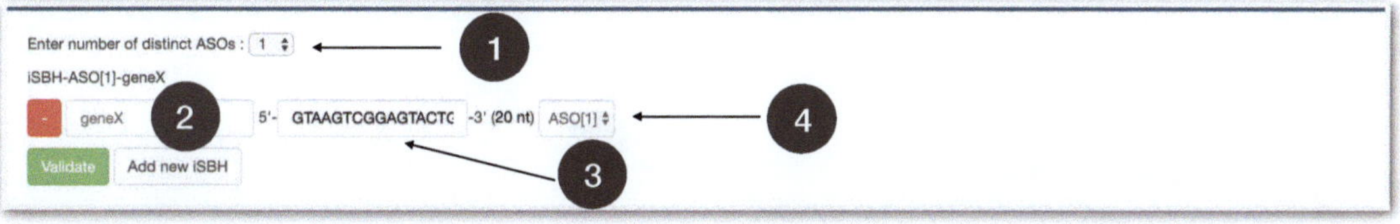

Step 2: Choose a candidate sensing-loop sequences for ASO[1] (5). The selected sensing-loop (bold font) will be used to derive the final iSBH sequence, along with its predicted RNA secondary structure (6). Use the "generate ZIP file" button (not shown) to export the iSBH and ASO sequences as well as primers required to clone the iSBH-sgRNA.

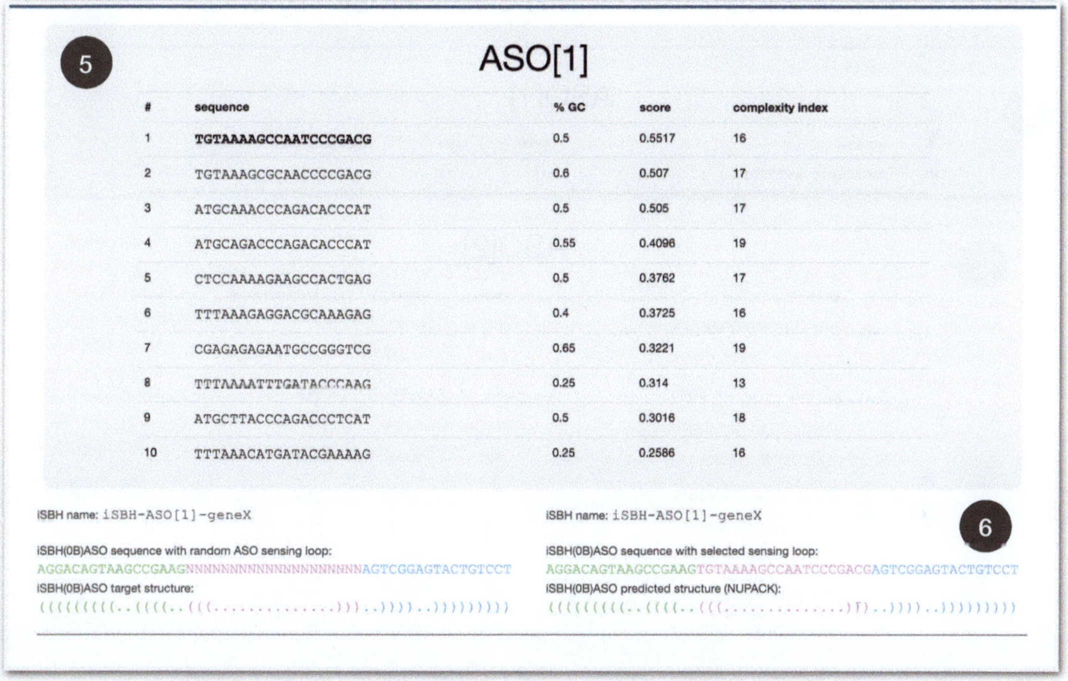

Fig. 5 iSBHfold: design ASO-responsive iSBH

sequences, I recommend avoiding sequence homologies greater than 20%. *See* Fig. 6 for a step-by-step explanation on how to design two orthogonal ASO-responsive iSBH-sgRNAs using the iSBHfold webtool.

Concomitant activation is achieved by expressing two iSBH-sgRNAs responding to the same inducer. Consequently, the sensing-loop and collar on both iSBHs have to be identical to ensure that the ASO inducer fully hybridizes with both hairpins. Accordingly, the iSBHfold algorithm was created so that it evolves the entire segment of the iSBH intended to base-

Step 1: Set the number of ASO to 2 or more (1), use "Add new iSBH" button to create a second guide (2). For both guides, enter the name of your gene target (3), the sequence of the your spacer (4), and select a distinct ASO inducer (5).

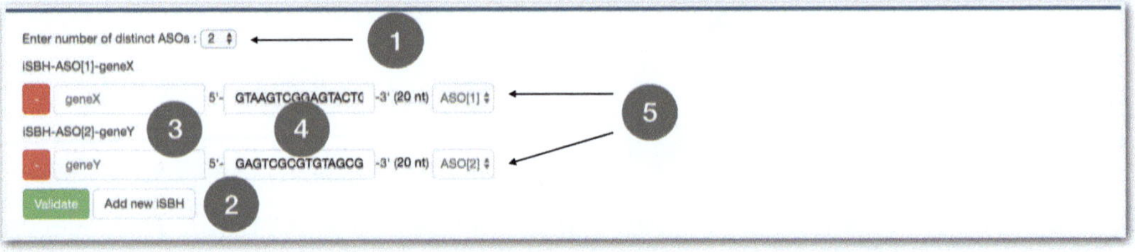

Step 2: Choose candidate sensing-loop sequences for ASO[1] (6) and ASO[2] (7) such that: (i) both iSBHs fold as desired; (ii) sensing-loop sequences are as different as possible. Use the "generate ZIP file" button (not shown) to export the iSBH and ASO sequences as well as primers required to clone the iSBH-sgRNAs.

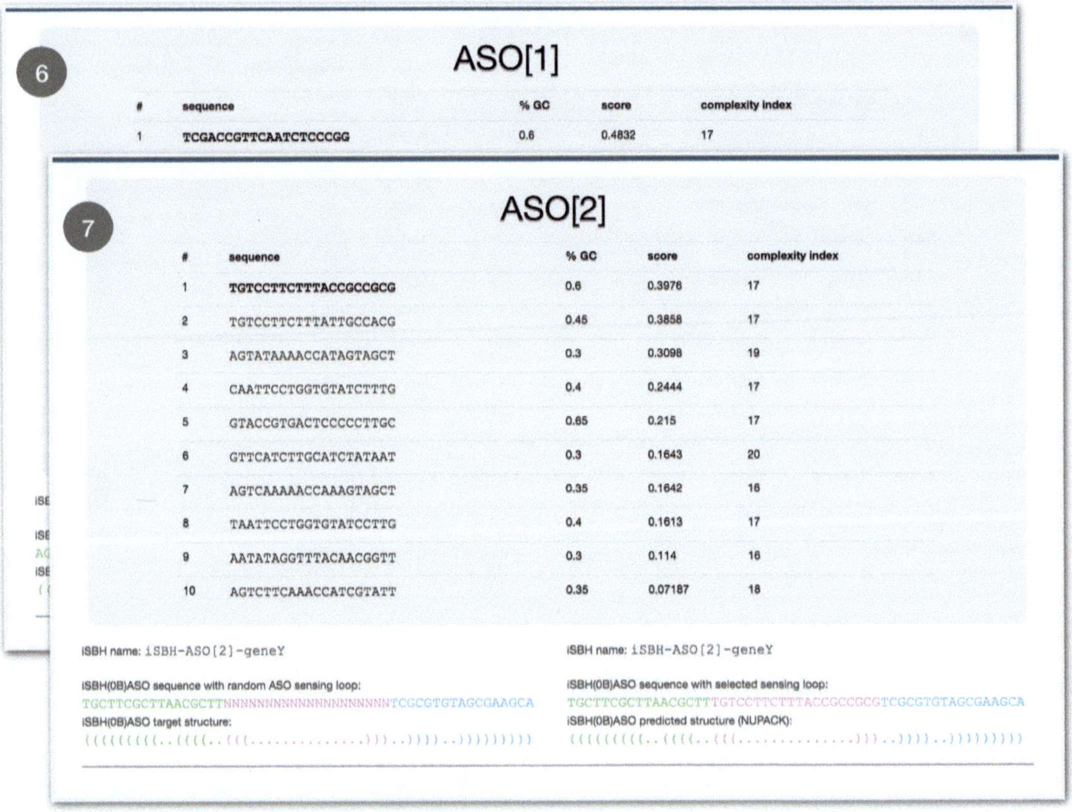

Fig. 6 iSBHfold: design of orthogonal ASO-iSBHs

pair with the ASO (sensing-loop plus 3 flanking nucleotides 5′ and 3′-ends). Note that doing so will most likely lead to an alteration of the original 5′-end of the spacer. This is caveated by the fact that (a) RNase-H is expected to degrade the entire RNA–DNA duplex and as such will shorten the spacer to 17 nt,

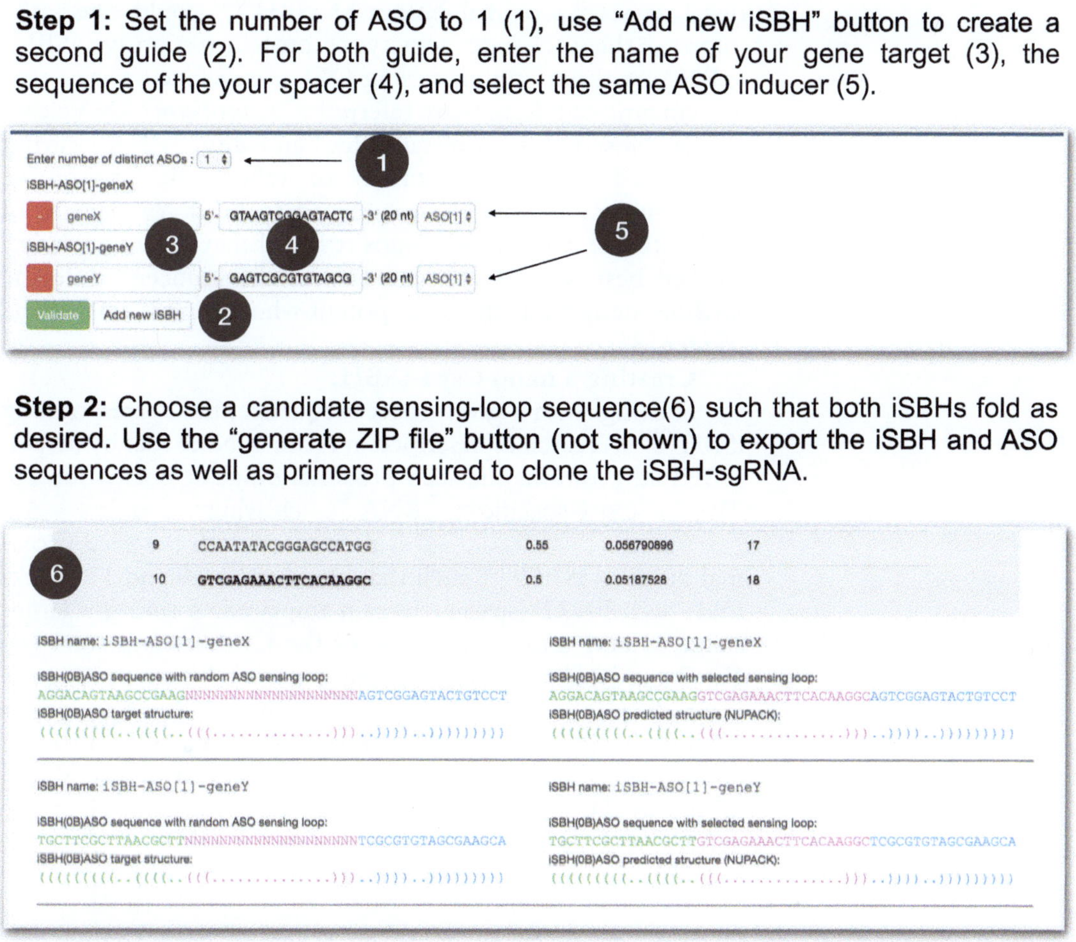

Fig. 7 iSBHfold: design of inducer-matching ASO-iSBHs

(b) Ferry et al. and others have shown that native guides with up to 10 nt-long spacers can successfully/efficiently direct dCas9 on target to mediate activation or repression [3, 22, 23]. Thus, the cropped guide should be functional without causing more off-target effects. *See* Fig. 7 for a step-by-step explanation on how to use iSBHfold webtool to design two ASO-responsive iSBH-sgRNAs responsive to a single ASO. Note that this time around iSBHfold considers both iSBH$^{(0B)}$ sequences when evolving the list of potential sensing-loops.

7. Ferry et al. have shown that iSBH-sgRNAs can be designed to specifically activate in the presence of specific endoribonucleases from the Cas6 family. Protein-responsive iSBH-sgRNA is created by replacing iSBH$^{(0B)}$ default "GAAA" linker segment with short RNA motifs that act as substrate for Csy4 or Cas6A. While this strategy yields functional guides, the authors found that fusing the RNA cleaving unit to either the seed

distal or seed proximal bulges of iSBH$^{(0B)}$ could drastically improve the ON-state activity of protein-responsive iSBH-sgRNAs without compromising OFF-state silencing. The two aforementioned designs are referred to as ***medium*** (RNA cleaving unit fused to seed distal bulge) and ***nano*** (RNA cleaving unit fused to seed proximal) to reflect the progressive shortening of the iSBH stem. Head to head comparison of *full*, ***medium***, and ***nano*** designs revealed that the ***nano*** design worked best when Csy4 was used as the inducer, while the ***medium*** design was the most potent when using Cas6A as the trigger [3].

Creating a nano Csy4-iSBH.

Starting from a given spacer sequence, the corresponding iSBH$^{(0B)}$ is created as described in **Note 4**. Note that by default iSBHfold positions the seed proximal bulge at nucleotide 10. The Csy4-specific RNA cleaving unit 5′-ACTGCCGTATAGGCAGC-3′ is then fused with the seed proximal bulge of iSBH$^{(0B)}$ such that nucleotides 10 and 11 (backfold) and 30–32 (spacer) remain unpaired, forming a larger bulge (3 and 4 nt) at the base of the Csy4 sensing stem (*see* Fig. 2 and RNA secondary structure below). When creating the stem, iSBHfold tests all possible combination for nucleotides 10 and 11 to guarantee that the bulge remains in an open conformation. Additionally, an extra C is added on the 5′-end of the iSBH sequence to base-pair with the first G of the sgRNA scaffold and increase stem stability.

```
AGGACAGTAAAACTGCCGTATAGGCAGCGAGTACTGTCCT
((((((((...(((((.....)))))....))))))))
|        |         |         |         |  .
1        10        20        30        40
```

(back-fold, cleaving unit, spacer)

Creating a medium Cas6A-iSBH

Starting from a given spacer sequence, the corresponding iSBH$^{(0B)}$ is created as described in **Note 4**. Note that by default iSBHfold positions the seed proximal bulge at nucleotide 10. The Cas6A-specific RNA cleaving unit 5′-CAAGGGATTGAGCCCCGTAAGGGGATTG -3′ is fused to iSBH$^{(0B)}$ immediately after the seed proximal bulge closes (*see* Fig. 3). As for Csy4-responsive design, iSBHfold will add an extra C on the 5′-end of the iSBH sequence to base-pair with the first G of the sgRNA scaffold and increase stem stability.

```
AGGACAGTAAGCCAAGGGATTGAGCCCCGTAAGGGGATTGGAGTACTGTCCT
(((((((((..((((.........((((....)))).))))..)))))))))
|        |         |         |         |         |
1        10        20        30        40        50
```

(back-fold, cleaving unit, spacer)

Figure 8 provides a step-by-step explanation on how to use iSBHfold to create both Csy4- and Cas6A-responsive hairpins.

Orthogonal and multiplexed gene activation with protein-iSBH

Orthogonal and multiplexed gene activation with protein-responsive iSBH-sgRNA does not require any additional design principle. Follow the information given above to create both iSBH-sgRNAs (Fig. 8).

8. Such vector should have a cassette comprising a U6 promoter, a placeholder for cloning of the sgRNA spacer, the sgRNA scaffold sequence and a U6 terminator sequence as shown below. The method described in this chapter has been notably validated for both the px330 and SAM sgRNA-2×MS2 vectors, whose sequences are given below.

   Px330 sgRNA expression cassette:

```
GAGGGCCTATTTCCCATGATTCCTTCATATTTGCATATACGATACAAGGCTGTTAGAGA
GATAATTGGAATTAATTTGACTGTAAACACAAAGATATTAGTACAAAATACGTGACGTA
GAAAGTAATAATTTCTTGGGTAGTTTGCAGTTTTAAAATTATGTTTTAAAATGGACTAT
CATATGCTTACCGTAACTTGAAAGTATTTCGATTTCTTGGCTTTATATATCTTGTGGAA
AGGACGAAACACCGGGTCTTCGAGAAGACCTGTTTTAGAGCTAGAAATAGCAAGTTAAA
ATAAGGCTAGTCCGTTATCAACTTGAAAAAGTGGCACCGAGTCGGTGCTTTTTT
```

(U6 promoter, spacer placeholder, sgRNA scaffold, U6 terminator)

.

SAM sgRNA expression cassette:

```
GAGGGCCTATTTCCCATGATTCCTTCATATTTGCATATACGATACAAGGCTGTTAGAGA
GATAATTGGAATTAATTTGACTGTAAACACAAAGATATTAGTACAAAATACGTGACGTA
GAAAGTAATAATTTCTTGGGTAGTTTGCAGTTTTAAAATTATGTTTTAAAATGGACTAT
CATATGCTTACCGTAACTTGAAAGTATTTCGATTTCTTGGCTTTATATATCTTGTGGAA
AGGACGAAACACCGGGTCTTCGAGAAGACCTGTTTTAGAGCTAGGCCAACATGAGGATC
ACCCATGTCTGCAGGGCCTAGCAAGTTAAAATAAGGCTAGTCCGTTATCAACTTGGCCA
ACATGAGGATCACCCATGTCTGCAGGGCCAAGTGGCACCGAGTCGGTGCTTTTTTT
```

(U6 promoter, spacer placeholder, sgRNA scaffold, U6 terminator)

.

Step 1: Use "Add new iSBH" button to create a second guide (1). For both guide, enter the name of your gene target (2), the sequence of the your spacer (3), and select the the protein inducer (4).

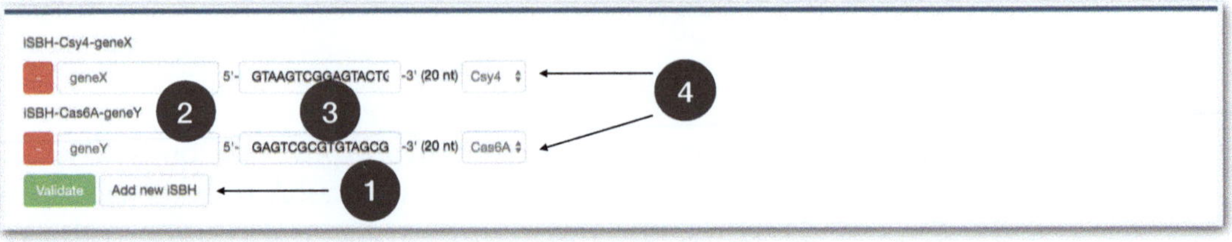

Step 2: iSBHfold outputs the iSBH sequences (5), primer sequences for cloning (6), and SVG rendering of the hairpins (7) for you to export.

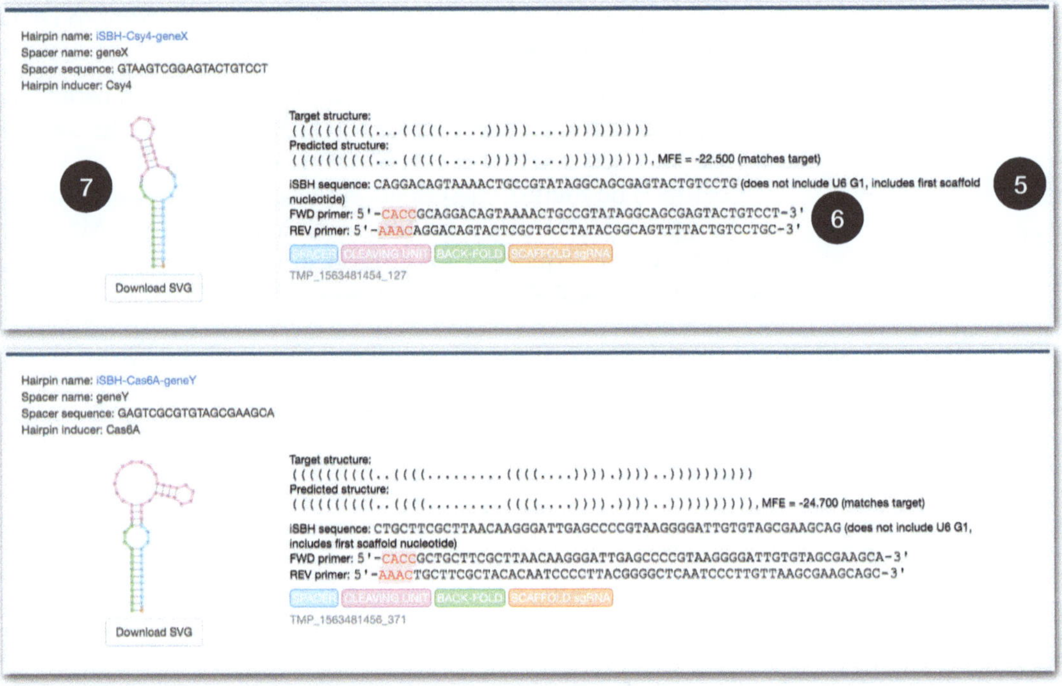

Fig. 8 iSBHfold: design of Csy4 and Cas6A-responsive iSBHs

9. Both Px330 and SAM sgRNA expression cassettes have a spacer placeholder segment directly following the U6 promoter. This segment contains two recognition sites for the type II restriction enzyme BbsI. After BbsI digestion, spacer or iSBH sequences can be cloned into the digested backbone as two annealed oligonucleotides using the steps below.
10. U6 transcription requires the first nucleotide downstream to promoter sequence to be a G. Accordingly, the sgRNA sequence should start with a G on its 5′-end. In the case

where G is not part of the original spacer or iSBH sequence, add it to the 5′-end of the sequence as shown in the example below:

```
5'-CACCCTAAGTCGGAGTACTGTCCT-3'
       ||||||||||||||||||||          .
    3'-GATTCAGCCTCATGACAGGACAAA-5'


5'-CACCGCTAAGTCGGAGTACTGTCCT-3'
       |||||||||||||||||||||          .
    3'-CGATTCAGCCTCATGACAGGACAAA-5'
```

11. For example, if the spacer sequence to clone is 5′-GTAAGTCG-GAGTACTGTCCT-3′, order forward primer 5′-**CACC**GTAAGTCG-GAGTACTGTCCT-3′ and reverse primer 5′-**AAAC**AGGACAGTACTCCGACTTAC-3′, where the bold sequences match the overhang post-BbsI digestion on the sgRNA expression pDNA backbone.
12. Using the running example given in **Note 10**, this step should create the following double-stranded DNA inserts:

```
5'CACCGTAAGTCGGAGTACTGTCCT-3'
      ||||||||||||||||||||          .
   3'-CATTCAGCCTCATGACAGGACAAA-5'
```

13. Alternatively, the digested backbone can be stored for up to 1 month at −20 °C and used for subsequent spacer or iSBH cloning. After BbsI digestion, the linearized vector should have the following overhangs:

```
5'-...GAAA        GTTTTAGA...-3'
      ||||            ||||        .
3'-...CTTTGTGG        ATCT...-5'
```

14. The ligated vector should be as follows:

```
5'-...GAAACACCGTAAGTCGGAGTACTGTCCTGTTTTAGA...-3'
      ||||||||||||||||||||||||||||||||||||        .
3'-...CTTTGTGGCATTCAGCCTCATGACAGGACAAAATCT...-5'
```

(insert is shown in black and insert in green).

15. When the iSBH sequence exceed 50 nt long, use two sets of oligonucleotides rather than one as long DNA oligonucleotides can form secondary structures that might prevent the forward and reverse primers to anneal properly. For example, the ASO-responsive iSBH below is 54 nt long:

5′-AGGACAGTACAGCGAGATACGAACATCCCTAACAG-TAAGTCGGAGTACTGTCCT-3′.

Its sequence can be split in half and cloned using the following two inserts:

Insert 1

```
CACCGAGGACAGTACAGCGAGATACGAACAT
    |||||||||||||||||||||||||||             .
    CTCCTGTCATGTCGCTCTATGCTTGTAGGGA
```

Insert 2

```
CCCTAACAGTAAGTCGGAGTACTGTCCT
    ||||||||||||||||||||||||        .
    TTGTCATTCAGCCTCATGACAGGACAAA
```

Note that the split has to be chosen such that the overhang sequence used to ligate the two oligo duplexes differs form the backbone overhangs. Prepare two separate oligonucleotide mixes (*oligo duplex 1 and 2*) and set up the following ligation reaction:

Reagent	**Volume, μl**
Nuclease-free H_2O	5.5
Annealed and phosphorylated oligo duplex 1 (10 μl reaction)	0.5
Annealed and phosphorylated oligo duplex 2 (10 μl reaction)	0.5
Gel-purified pDNA backbone (30 μl elution)	2
10× T4 ligase buffer	1
T4 ligase	0.5

16. I provide below the human codon-optimized amino acid sequence of Csy4 and Cas6A as used in [3].

Human codon-optimized *Pseudomonas aeruginosa Csy4* (Kozak sequence, 5′ NLS, Csy4).

GCCACCATGGCCCCAAAGAAGAAGCGGAAGGTCGGTATCCACGGAGTCCCAGCAGCCGA
AAACCTGTACTTCCAATCCAATGCAGCTAGCGACCACTATCTGGACATCAGACTGAGGC
CCGATCCTGAGTTCCCTCCCGCCCAGCTGATGAGCGTGCTGTTTGGCAAGCTGCATCAG
GCTCTGGTCGCCCAAGGCGGAGACAGAATCGGCGTGTCCTTCCCCGACCTGGACGAGTC
CCGGAGTCGCCTGGGCGAGCGGCTGAGAATCCACGCCAGCGCAGACGATCTGCGCGCCC
TGCTGGCCCGGCCTTGGCTGGAGGGCCTGCGGGATCATCTGCAGTTTGGCGAGCCCGCC
GTGGTGCCACACCCAACACCCTACCGCCAGGTGAGCCGCGTGCAGGCCAAGTCAAATCC
CGAGAGACTGCGGCGGAGGCTGATGAGGCGACATGATCTGAGCGAGGAGGAGGCCAGAA
AGAGAATCCCCGACACAGTGGCCAGAGCCCTGGATCTGCCATTTGTGACCCTGCGGAGC
CAGAGCACTGGCCAGCATTTCAGACTGTTCATCAGACACGGGCCCCTGCAGGTGACAGC
CGAGGAGGGCGGATTTACATGCTATGGCCTGTCTAAAGGCGGCTTCGTGCCCTGGTTCT
GA

Human codon-optimized *Streptococcus thermophilus* Cas6A (Kozak sequence, 5′ NLS, Cas6A).

GCCACCATGGCCCCAAAGAAGAAGCGGAAGGTCGGTATCCACGGAGTCCCAGC
AGCCGAAAACCTGTACTTCCAATCCAATGCAGCTAGCGTGCTTGCCGCGCTTGT
GCTCGTGTTGGAAGGGGAAGGCCTCCCGGAGCCGTTGGGTCTTAGGGGTTTCT
TTTACGGCCTTTTGCGCGAGGTAGCGCCAGAGGTACACGATCAAGGAGAAAATC
CTTTTGCTCTGGGATTCGGCGGGAGGGAAGGCGCAGCTTGGGCTAGGGTTAGT
CTTCTCGTAGAAGGGTTGTATGCGCGACTCGCACCTAGACTCTATGCGCTTGAA
GGTGAGGAAGTTCGCCTTGGACCTCCGTTCCGGGTACGCGCAGTCTTGCAAGA
GGGCCATCCTTGGGCTGGAGTTTCAACGTATCCACGCTTGTTCCAGGACCTCC
CTCCAGAGATCTGGCCCTCCGATTTGCAAGTCCAACCTTCTTCCGCCGGAAAGG
CGTTCACTACCCCGTGCCGGAACCTCGCCTGGTTCTGGAGTCCCTCCTGCGGC
GGTTGCAAGCATTCGGTCCACTTAAAGCCCCGGAAGGAGTGAGAGAAGCTCTTC
TGGAGAGAACCACGGTGCGATCATTGGAGGGACGCACACTTCCGGCGAGGACG
GAAGTTGACACAGCTGGATTTGTAGGCAGGGTCGTGTACCACCTGCCTAGAGCG
ACAGAGGAAGAAGCACTCTGGTTGTCTGCGTTGGGCCGATTCGCGTTCTATTCA
GGAGTAGGTGCAAAGACCAGCTTGGGATATGGTAGAGCACGAGCAGAATCTGCG
TAG

For constitutive expression in mammalian cells I recommend cloning the recombinant protein in PGK1p-Csy4-pA, in between HindIII and NheI restriction sites.

17. I recommend using a hemocytometer to measure cell concentration in the parent flask, seed several wells with different concentrations, and find out what is the optimal seeding concentration to reach the desired confluency 24h later. Determining the optimal seeding concentration will speed up subsequent experiments and help make the results more consistent.
18. For 12-well plates, I recommend keeping the total amount of pDNA under 2 μg per well to reduce cellular toxicity.

19. Additionally, one can check using a cell culture light microscope that the cells are still attached to the bottom of the well and proliferating. Failure to satisfy these two criteria might be indicative that the transfection mix is being toxic. In this case, reduce the amount of pDNA and PEI delivered per well.
20. If using the SAM system replace transfection mix by:

Reagent	Volume	Note
Opti-MEM	100 μl	
iSBH-sgRNA-2×MS2 pDNA	500 ng	Guide expression pDNA, ~5 kb. iSBH has to be designed to target *geneX* and respond to the selected ASO
SAM dCas9-VP64 pDNA	500 ng	
SAM MCP-p65-HSF1	500 ng	
PEI	2:3 ratio	

21. If using multiple iSBH-sgRNAs, scale down the amount of each iSBH-sgRNA pDNA delivered by the total number of guides. The same applies to ASO inducers when delivering several simultaneously.
22. If using the SAM system replace transfection mix by:

Reagent	Volume	Note
Opti-MEM	100 μl	
iSBH-sgRNA-2×MS2 pDNA	500 ng	Guide expression pDNA, ~5 kb. iSBH has to be designed to target *geneX* and respond to the selected ASO
SAM dCas9-VP64 pDNA	500 ng	
SAM MCP-p65-HSF1	500 ng	
Csy4 or Cas6A expressing vector (~4.2 kb)	250 ng	
PEI	2:3 ratio	

23. The amount of pDNA expressing the endoribonuclease might be scaled up or down depending on the size of the construct as well as the strength of the promoter driving Csy4/Cas6A expression.
24. If using multiple iSBH-sgRNAs, scale down the amount of each iSBH-sgRNA pDNA delivered by the total number of guides. The same applies to endoribonuclease-expressing pDNAs when delivering several protein inducers simultaneously.
25. I recommend using nonspecific primers (random hexamers, provided in the QuantiTect Reverse Transcription Kit) to obtain total RNA from the cells, including both the gene of interest and housekeeping genes whose levels are necessary to control for differences in cell number between conditions.
26. For each condition, I advise setting up a "non-RT" control: Follow the same reverse transcription protocol but omit adding the reverse transcriptase. This control is to ensure that the signal later picked up by the qPCR comes from reverse-transcribed RNA rather and genomic DNA.
27. Select primer pair so as to generate ~100 to ~150 nt amplicons. I recommend aiming for primers with 50% GC content and a ~60 °C melting temperature.
28. Make sure that: (a) The PCR reaction only amplifies one species. This can be checked by looking at the melting curve or running the product of the qPCR on a 1% gel to verify that there is only a single band. (b) Ct values obtain for each dilution reflect the dilution factors (run standard curves). I recommend testing several primer pairs for each target gene and selecting the pair with the best efficiency.
29. In addition to measuring *geneX* mRNA, run qPCR for dCas9-VP64 and housekeeping gene *GAPDH* levels (*see* Table 1 for primer sequences) as they will be required to compute the ON-OFF fold change between conditions.
30. When trying to assess the ON/OFF characteristics of a new iSBH-sgRNA, I advise setting up and testing all the following conditions:

Table 1
Primer pair sequences for qPCR analysis

Target mRNA	Forward primer 5′–3′	Reverse primer 5′–3′
GAPDH	AACAGCGACACCCACTCCTC	CATACCAGGAAATGAGCTTGACAA
dCas9-VP64 (SAM)	AACCTATGCCCACCTGTTCG	ATCCAGGATTGTCTTGCCGG

(1) Native sgRNA and (2) scramble native sgRNA conditions. If your iSBH-sgRNA targets *geneX* through the spacer sequence sp-*geneX*, test the native sgRNA with the same spacer sequence to validate that the activated iSBH-sgRNA will be able to drive CRISPR-TR of *geneX* in the ON-state. Condition (2), scramble native sgRNA, is used as a negative control and provides background *geneX* levels for comparison with condition (1).

Transfection mix for condition (1):

- pDNA expressing the native sgRNA with spacer against *geneX*.
- CRISPR-TR effector (e.g., dCas9-VP64)

Transfection mix for condition (2):

- pDNA expressing a native sgRNA with scramble spacer.
- CRISPR-TR effector (e.g., dCas9-VP64)

iSBH-sgRNA minus (3) and plus inducer (4) conditions. Condition (3) is used to confirm that the iSBH-sgRNA does not drive CRISPR-TR in the absence of the matching inducer. Condition (4) corresponds to the ON-state of the system. The presence of the matching inducer should trigger activation of the iSBH-sgRNA followed by activation or repression of the downstream gene target. Note that in the OFF-state condition, a decoy inducer or decoy pDNA is used to control for the absence of the matching inducer.

Transfection mix for condition (3):

- pDNA expressing the iSBH-sgRNA.
- CRISPR-TR effector (e.g., dCas9-VP64).
- pDNA expressing the endoribonuclease inducer or ASO matching the iSBH RNA cleaving unit.

Transfection mix for condition (4):

- pDNA expressing the iSBH-sgRNA.
- CRISPR-TR effector (e.g., dCas9-VP64).
- Decoy pDNA or ASO.

When testing an iSBH-sgRNA, two issues might arise: (1) The system might be leaky, in that CRISPR-TR activity above background is observed in the absence of the matching inducer (OFF-state); (2) The system might fail to drive CRISPR-TR activity in the presence of the inducer. Below I provide ideas on how to possibly address some of these issues.

Leakiness.

It is possible that CRISPR-TR activity might be observed in the absence of the iSBH-sgRNA inducer. Such problem might notably arise when the back-fold segment of the hairpin fails to stay base-paired with the spacer. In this case, I recommend taking steps to increase the stability of the hairpin.

Increase stability around the spacer seed: sgRNAs can drive CRISPR-TR with as little as ten spacer nucleotides (spacer seed). In order to increase the probability of the spacer seed sequence to interact with the back-fold, and consequently be unavailable to interact with the DNA target, I recommend adding an extra +C on the 5′-end of the iSBH sequence that will base-pair with the first G of the sgRNA scaffold.

Increasing stability around the hairpin collar: Most of the iSBH instability comes from the RNA-cleaving unit part of the design. As such, melting of the back-fold/spacer duplex is most likely to nucleate from that part. Accordingly, and in the case of OFF-state leakiness, I recommend increasing the stability of the hairpin "collar." This can be achieved by (1) moving down the seed distal bulge of the hairpin closer to the seed. Additionally, I know from our work on ASO-iSBH that the collar sequence can be altered without affecting the ability of the guide RNA to drive CRISPR-TR in the ON-state. Therefore, one can also increase collar stability by increasing the GC content of this region.

Increasing stability in the middle of the stem: In general, I advise adding two 2 nt bulges along the hairpin in order to reduce its stability and improve the system's ON-state performances. Nevertheless, in the case where increasing stem stability in the seed proximal and collar region of the hairpin does not solve the leakiness, one could play with the number and position of the bulges along the stem.

Weak ON-state activation.

In the case where no CRISPR-TR activity is observed in the presence of the inducer, I recommend the following steps:

- Ensure that the spacer sequence of the iSBH is functional. This can easily be tested using condition (1) described above. If the native sgRNA with the same spacer as the one used to design the iSBH fails to drive CRISPR-TR, one cannot expect the iSBH-sgRNA to be functional.
- Confirm that the inducer is functional in your cell culture. All inducers introduced in this chapter have in common that they slice a particular RNA sequence. Accordingly, one can test the potency of the inducer by cloning their respective RNA-cleaving unit in the 3′-untranslated region of a constitutively expressed reporter gene.
- Attempt reducing the hairpin stability. I discuss above increasing stem stability to prevent leakiness. Likewise, a poor on-state activation could be solved by reducing the overall stem stability. Try reducing stability in the collar first before destabilizing the seed proximal region.

References

1. Hsu PD, Lander ES, Zhang F (2014) Development and applications of CRISPR-Cas9 for genome engineering. Cell 157(6):1262–1278
2. Chavez A et al (2016) Comparison of Cas9 activators in multiple species. Nat Methods 13 (7):563–567
3. Ferry QRV, Lyutova R, Fulga TA (2017) Rational design of inducible CRISPR guide RNAs for de novo assembly of transcriptional programs. Nat Commun 8:14633
4. Jain PK et al (2016) Development of light-activated CRISPR using guide RNAs with photocleavable protectors. Angew Chem Int Ed Engl 55(40):12440–12444
5. Lee YJ et al (2016) Programmable control of bacterial gene expression with the combined CRISPR and antisense RNA system. Nucl Acids Res 44(5):gkw056-2473
6. Liu Y et al (2016) Directing cellular information flow via CRISPR signal conductors. Nat Methods 66(4):1173–1179
7. Tang W, Hu JH, Liu DR (2017) Aptazyme-embedded guide RNAs enable ligand-responsive genome editing and transcriptional activation. Nat Commun 8:15939
8. Baeumler TA, Ahmed AA, Fulga TA (2017) Engineering synthetic signaling pathways with programmable dCas9-based chimeric receptors. Cell Rep 20(11):2639–2653
9. Aricescu AR, Lu W, Jones EY (2006) A time- and cost-efficient system for high-level protein production in mammalian cells. Acta Crystallogr D Biol Crystallogr 62(Pt 10):1243–1250
10. Untergasser A et al (2012) Primer3—new capabilities and interfaces. Nucleic Acids Res 40 (15):e115–e115
11. Bennett CF, Swayze EE (2010) RNA targeting therapeutics: molecular mechanisms of antisense oligonucleotides as a therapeutic platform. Annu Rev Pharmacol Toxicol 50 (1):259–293
12. Zheng C, Baum BJ (2008) Evaluation of promoters for use in tissue-specific gene delivery. Methods Mol Biol (Clifton, NJ) 434(Chapter 13):205–219
13. Gossen M et al (1995) Transcriptional activation by tetracyclines in mammalian cells. Science (New York, NY) 268(5218):1766–1769
14. Gilbert LA et al (2013) CRISPR-mediated modular RNA-guided regulation of transcription in eukaryotes. Cell 154(2):442–451
15. Mali P et al (2013) CAS9 transcriptional activators for target specificity screening and paired nickases for cooperative genome engineering. Nat Biotechnol 31(9):833–838
16. Konermann S et al (2014) Genome-scale transcriptional activation by an engineered CRISPR-Cas9 complex. Nature 517 (7536):583–588
17. Chavez A et al (2015) Highly efficient Cas9-mediated transcriptional programming. Nat Methods 12(4):326–328
18. Gilbert LA et al (2014) Genome-scale CRISPR-mediated control of gene repression and activation. Cell 159(3):647–661
19. Shechner DM et al (2015) Multiplexable, locus-specific targeting of long RNAs with CRISPR-Display. Nat Methods 12 (7):664–670
20. Zalatan JG et al (2014) Engineering complex synthetic transcriptional programs with CRISPR RNA scaffolds. Cell 160 (1-2):339–350
21. Zadeh JN et al (2011) NUPACK: analysis and design of nucleic acid systems. J Comput Chem 32(1):170–173
22. Dahlman JE et al (2015) Orthogonal gene knockout and activation with a catalytically active Cas9 nuclease. Nat Biotechnol 33 (11):1159–1161
23. Kiani S et al (2015) Cas9 gRNA engineering for genome editing, activation and repression. Nat Methods 12(11):1051–1054

Chapter 10

Gene Manipulation Using Fusion Guide RNAs for Cas9 and Cas12a

Ha Rim Shin, Jiyeon Kweon, and Yongsub Kim

Abstract

The clustered regularly interspaced short palindromic repeats (CRISPR)-associated (Cas) protein has emerged as a genome engineering tool for various organisms. Known as the CRISPR-Cas system, Cas endonucleases such as Cas9 and Cas12a (also known as Cpf1) and guide RNA (gRNA) complexes recognize and cleave the target DNA, allowing for targeted gene manipulation. Along with the Cas protein engineering, gRNA engineering has broadened the applications of the CRISPR-Cas system. Recently, we have developed fusion guide RNAs (fgRNAs) for orthogonal gene manipulation using Cas9 and Cas12a. Here, we describe the methods for designing and generating fgRNAs-expression constructs to achieve multiplex genome editing and gene manipulation in human cells.

Key words CRISPR-Cas9, CRISPR-Cas12a, Genome engineering, Transcriptional activation, fusion guide RNA

1 Introduction

Genome engineering in living organisms using the bacteria-derived clustered regularly interspaced short palindromic repeats (CRISPR)-associated (Cas) system holds promise as a revolutionary tool for biomedical research [1]. The CRISPR-Cas system uses Cas endonucleases such as Cas9 and Cas12a (also known as Cpf1) and guide RNA (gRNA) complexes to cleave chromosomal DNA in a targeted manner resulting in DNA double-strand breaks (DSBs) [2]. These DSBs are then repaired by cellular DNA repair mechanisms such as nonhomologous end joining (NHEJ) or homology-directed repair (HDR) resulting in gene disruption, addition and correction [3].

Cas9, a single effector protein found in type II CRISPR-Cas system, is widely used for genome editing. When Cas9 proteins recognize a protospacer adjacent motif (PAM), G-rich sequence at the 3′ end of target sequences and generate blunt DSBs at the PAM proximal region [4–6]. However, Cas12a, found in type V-A

Tudor A. Fulga et al. (eds.), *CRISPR Guide RNA Design: Methods and Protocols*, Methods in Molecular Biology, vol. 2162, https://doi.org/10.1007/978-1-0716-0687-2_10, © Springer Science+Business Media, LLC, part of Springer Nature 2021

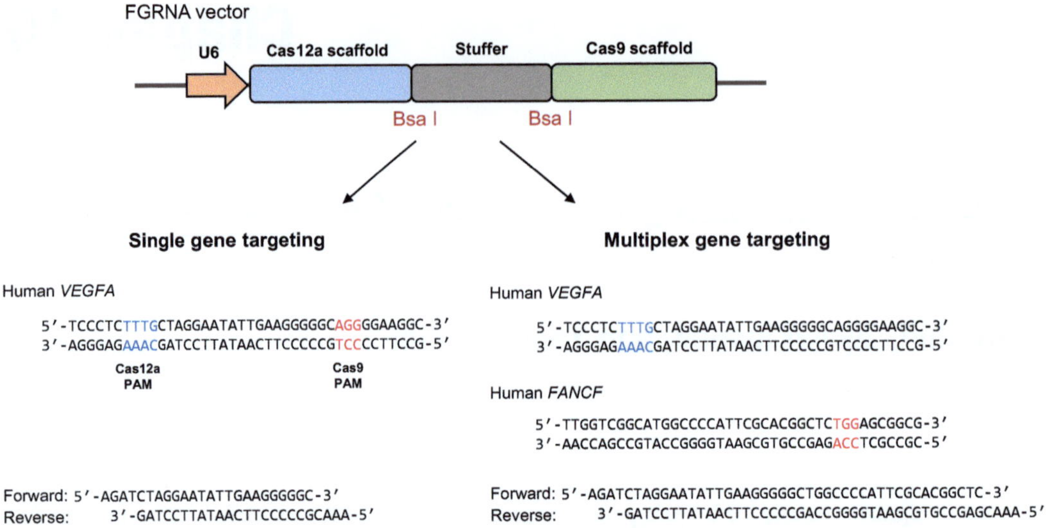

Fig. 1 Schematic of fgRNA design for both single and multiplexed gene targeting. Fusion gRNA (fgRNA) contains both the 5′-scaffold of Cas12a and the 3′-scaffold of Cas9. The fgRNA enable single or multiplex gene manipulation by recruiting both Cas9 and Cas12 protein at the endogenous target loci in mammalian cells

CRISPR-Cas system, provides an alternative to type II Cas9 by recognizing T-rich PAM sequences at the 5′ end of target sequences and generates cohesive DSBs at PAM distal region[7]. Cas9 and Cas12a can be used together to expand targetable regions, broadening the field of genome editing.

Recently, we developed synthetic fusion guide RNAs (fgRNAs) which can interact with both Cas9 and Cas12a proteins and demonstrated that the fgRNA can induce targeted mutagenesis at target sites of both Cas9 and Cas12a [8]. By extending the length of fgRNAs, fgRNAs can induce targeted mutation in two independent endogenous sites with both Cas9 and Cas12a proteins. Furthermore, fgRNAs can be used for orthogonal gene manipulation using both Cas12a and Cas9-derived transcriptional activators (Fig. 1). In this chapter, we describe how to design and construct fgRNAs for gene manipulation.

2 Materials

2.1 DNA Constructs

1. p3s-Cas9HC (Addgene, #43945).
2. pcDNA3.1-hLbCpf1 (Addgene, #69988).
3. pcDNA3.1-hAsCpf1 (Addgene, #69982).
4. lenti dCAS-VP64_Blast (Addgene, #61425).
5. lenti MS2-P65-HSF1_Hygro (Addgene, #61426).

6. pFGRNA-LbCpf1-SpCas9 (Addgene, #119673).
7. pFGRNA-AsCpf1-SpCas9 (Addgene, #119674).
8. pFGRNA-LbCpf1-spCas9-MS2 (Addgene, #119675).

2.2 Cloning Components

1. Oligonucleotides for constructing fgRNAs.
2. Restriction enzymes and buffers (New England Biolabs).
3. Quick ligase and buffer (New England Biolabs).
4. Gibson Assembly Master Mix (New England Biolabs).
5. T4 DNA ligase Buffer (New England Biolabs).
6. Competent cell of Bacterial DH5α strain.
7. LB agar plates containing ampicillin (100 μg/mL).
8. LB liquid medium containing ampicillin (50 μg/mL).
9. Agarose and electrophoresis equipment.
10. Heat block.
11. NanoDrop spectrophotometer (Thermo Fisher).
12. QIAquick Gel Extraction Kit (QIAGEN).
13. QIAGEN Plasmid Mini Kit (QIAGEN).

2.3 Transfection Components

1. HEK293T/17 (ATCC, CRL-11268) and HeLa (ATCC, CCL-2) cell lines.
2. Dulbecco's Modified Eagle's Medium (DMEM), 10% FBS, and 1% penicillin–streptomycin.
3. Opti-MEM (Gibco).
4. 24- and 96-well plates (Corning).
5. Lipofectamine 2000 (Life Technologies).
6. 4D-Nucleofector equipment (Lonza).

2.4 Measurement of the Genome Manipulation Efficiency

1. DNeasy Blood & Tissue Kit (QIAGEN).
2. RNeasy Mini Kit (QIAGEN).
3. QIAquick PCR Purification Kit (QIAGEN).
4. Phusion High-Fidelity DNA Polymerase (New England Biolabs).
5. T7 endonuclease I enzyme (T7E1, New England Biolabs).
6. Illumina MiniSeq System or appropriate NGS machine.
7. Illumina MiniSeq Mid Output Kit (300 Cycles).
8. AccuPower CycleScript RT PreMix (Bioneer).
9. iQ SYBR Green Supermix (Bio-Rad).
10. Target-specific PCR primers.
11. Bio-Rad CFX96 or appropriate quantitative real-time PCR instrument.

3 Methods

3.1 fgRNA Design

1. In order to generate fgRNA targeting a single locus in the genome, the target sequence of fgRNAs must contain PAM sequences of both SpCas9 and LbCas12a, such as 5′-TTTN (N_{X20}) NGG-3′. For cloning the target sequence into the fgRNA-expressing plasmid, the forward and reverse oligonucleotides are both 24-mer in length, with the forward designed using 5′-AGAT-(N_{X20})-3′ sequences and the reverse using 5′-AAAC-(N_{X20})-3′ sequences.
2. For multiplex genome editing or orthogonal gene manipulation using fgRNAs, select each target sequence of LbCas9 and SpCas9 such as 5′-TTTN (N_{X20})-3′ and 5′-(N_{X20}) NGG-3′, respectively. To generate a multitargeting fgRNA construct, combine the target sequence of LbCas12a and SpCas9 in order and design oligonucleotides with the 5′-AGAT-(N_{X40})-3′ for forward oligonucleotides and 5′-AAAC-(N_{X40})-3′ for reverse oligonucleotides.
3. To reduce unwanted mutations in the target genome, we recommend using online tools for gRNA design and off-target analysis such as Cas-Designer or Cas-OFFinder (http://www.rgenome.net) [9, 10].

3.2 fgRNA Construct Cloning

1. Digest 1 μg of appropriate pFGRNA vectors (*see* **Note 1**) with BsaI restriction enzyme for 2 h at 37 °C.
2. Load the reaction mixture into wells of a 1% agarose gel and separate the digested vector DNA fragment.
3. Extract vector DNA fragment using the QIAquick Gel Extraction Kit.
4. Measure DNA concentration with a NanoDrop spectrophotometer.

Prepare an insert DNA by melting-reannealing the target sequence–specific forward and reverse oligonucleotides (each 10 μM) using a PCR machine Forward oligonucleotides (100 μM)	5 μl
Reverse oligonucleotides (100 μM)	5 μl
T4 DNA ligase buffer	5 μl
Distilled water	35 μl
Total	50 μl

5. Dilute the annealed oligonucleotides duplex to 1/1000.
6. Ligate the vector DNA fragment and insert oligonucleotides under the following conditions for 5 min at room temperature.

Vector DNA fragment	50 ng
Insert oligonucleotides duplex	1 μl
Quick ligase	0.5 μl
2× Quick ligase buffer	5 μl
Distilled water	Up to 10 μl
Total	10 μl

7. Transform the ligate mixture with 100 μl competent cells.
8. Spread the transformant on the LB agar plate containing ampicillin (50 μg/ml).
9. Incubate the plate overnight at 37 °C.
10. Pick up several single colonies and extract the plasmid DNA using QIAGEN Plasmid Mini Kit.
11. Confirm the pFGRNA constructs by sequencing analysis.

3.3 Mammalian Cell Culture and Transfection

1. Maintain HEK293T/17 or HeLa cells in DMEM supplemented with 10% FBS and 1% penicillin–streptomycin. Incubate the cells in a 37 °C incubator with 5% CO_2.
2. One day before transfection, seed 2×10^5 HEK293T/17 cells or 8×10^4 HeLa cells per well of a 24-well plate.
3. On the day of transfection, the plasmid DNA mixture (*see* **Note 2**) is transfected using Lipofectamine 2000 according to the manufacturer's protocol.
4. Incubate the transfected cells for 72 h in a 37 °C incubator.

3.4 Measurement of the Genome Manipulation Efficiency

To measure the mutation frequency, genomic DNA is extracted using the DNeasy Blood & Tissue Kit (QIAGEN) according to the manufacturer's protocol after 72 h posttransfection. The mutation frequency can be measured by T7 endonuclease I (T7E1) assay or next generation sequencing (NGS) analysis [11, 12]. Total RNA is extracted using an RNeasy Mini Kit (QIAGEN) according to the manufacturer's protocols and the gene expression levels are analyzed by quantitative real-time PCR (qRT-PCR).

3.4.1 T7 Endonuclease I (T7E1) Assay

1. Design PCR primer pairs with PCR product size to be less than 1 kb using the web-based program Primer3 (http://bioinfo.ut.ee/primer3-0.4.0/) [13]. Nested PCR is recommended to amplify the target region more efficiently.
2. Perform the PCR amplification using Phusion High-Fidelity DNA Polymerase (New England Biolabs) according to the manufacturer's protocol. For nested PCR, 20 cycles for first PCR and 30 cycles for second PCR are recommended.

3. Melt and reanneal second PCR amplicon using PCR machine to make the heteroduplex form of PCR products. The melting–reannealing cycles are described below.

Cycle	Temperature (°C)	Time
1	95	3 min
2	95	0.01 s
	−1 °C/cycle	
3	Goto 2, 39×	
4	55	0.02 s
	−1 °C/cycle	
5	Go to 4, 39×	
6	16	∞

4. Mix the following reagents and incubate them for 20 min at 37 °C.

NEBuffer 2.1	2 μl
T7E1 enzyme	0.5 μl
PCR product	10 μl
Distilled water	7.5 μl
Total	20 μl

5. Load the T7E1 reaction mixture on 2% agarose gel and perform gel electrophoresis. Determine the mutation frequency using the band intensity of cleaved DNA fragments.

3.4.2 Next-Generation Sequencing (NGS) Analysis

1. Design appropriate PCR primer pairs with PCR product size to be less than 300 bp using the web-based program Primer3 (http://bioinfo.ut.ee/primer3-0.4.0/) for NGS analysis. As shown in Fig. 2, the PCR procedure for constructing NGS library consists of three steps. The amplicon size of second PCR products should be adjusted according to the read length of the NGS reagent kit. Primers for second PCR should be contained partial Illumina sequencing adapters as follows: Deep F: 5′-ACACTCTTTCCCTACACGACGCTCTTCCG ATCT-3′ and Deep R: 5′-GTGACTGGAGTTCAGACG TGTGCTCTTCCGATCT-3′. It is recommended to use high-fidelity polymerase in each step of the PCR to minimize errors in PCR.

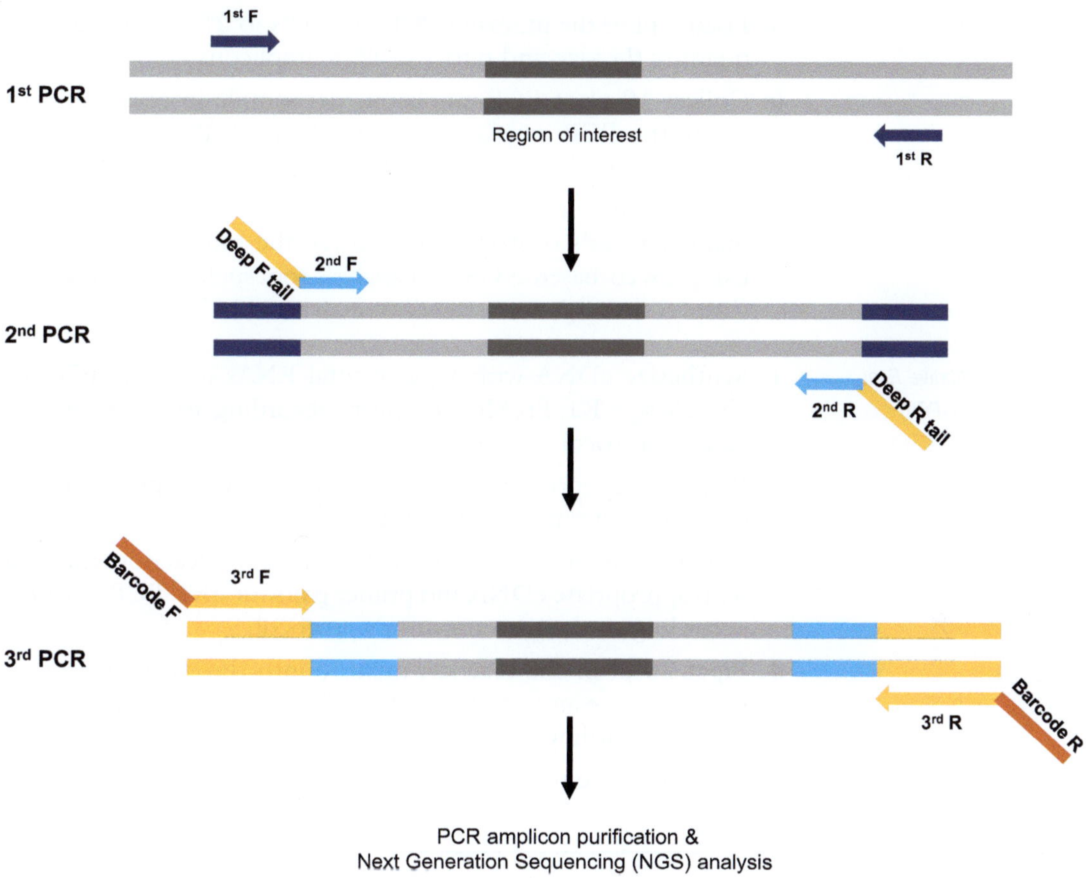

Fig. 2 PCR procedure for constructing an NGS library. Target sites of fgRNAs are amplified with target specific primers and amplicons are again amplified using TruSeq HT Dual index-containing primers to generate next-generation sequencing libraries. The libraries are sequenced using Illumina MiSeq or MiniSeq with paired-end sequencing systems

2. Perform the first PCR using Phusion High-Fidelity DNA Polymerase (New England Biolabs) according to the manufacturer's protocol. Fewer than 20 cycles are recommended.
3. 1 μl of the tenfold diluted first PCR product is used as the template DNA of second PCR. Perform second PCR using Phusion High-Fidelity DNA Polymerase (New England Biolabs) according to the manufacturer's protocol. Fewer than 30 cycles are recommended.
4. To make the NGS library, the indexing PCR is conducted using NGS index sequences-containing primer pairs (forward: 5′-AATGATACGGCGACCACCGAGATCTACACtatagcct ACACTCTTTCCCTACACGAC-3′, reverse: 5′-CAAGCA GAAGACGGCATACGAGATcgagtaatGTGACTGGAGTTC AGACGTGT-3′). 1 μl of the tenfold diluted second PCR product is used as the template DNA of the indexing PCR.

5. Load 5 μl of the indexing PCR products to 2% agarose gel and quantify the size and purity of PCR amplicons.
6. Collect 10 μl of PCR amplicons per sample in one tube and purify the PCR amplicons using QIAquick PCR Purification Kit (QIAGEN) according to the manufacturer's protocol.
7. Perform NGS using Illumina MiniSeq system according to the manufacturer's protocol and analyze the mutation frequency using a web-based NGS data analysis tool such as Cas-Analyzer (http://www.rgenome.net/cas-analyzer) [14].

3.4.3 Quantitative Real-Time PCR (qRT-PCR) Analysis

1. Synthesize cDNA with 1 μg of total RNAs using AccuPower CycleScript RT PreMix (Bioneer) according to the manufacturer's protocol.
2. Prepare appropriate primers for quantifying the expression level of target genes using qRT-PCR.
3. Using iQ SYBR Green Supermix, prepare reaction mixture with appropriate cDNA and primer pairs for qRT-PCR reaction according to the manufacturer's protocol.
4. Pipet 20 μl of the reaction mixture into the wells of 96-well reaction plate and run the plate in Bio-Rad CFX96 using parameters according to the manufacturer's protocol.
5. Analyze of the results of qRT-PCR using appropriate software.

4 Notes

1. There are three types of pFGRNA, pFGRNA-LbCpf1-SpCas9, pFGRNA-AsCpf1-SpCas9, and pFGRNA-LbCpf1-spCas9-MS2. The appropriate pFGRNA vector should be selected according to the experimental design.
2. A 1:1:2 ratio of Cas9, Cas12a, and fgRNA plasmid DNA is used to single or multiple genome engineering. For orthogonal genome manipulation, a 1:1:1:2 ratio of dCas9-VP64, MS2-p65-HSF1, C12a, and fgRNA plasmid DNA is mixed.

Acknowledgments

This work was supported by the National Research Foundation of Korea (grants 2017M3A9B4062419 and 2018R1A5A2020732 to Y.K.), the Korean Health Technology R&D Project, Ministry of Health and Welfare, Republic of Korea (grant HI17C0337 to Y.K.),

References

1. Knott GJ, Doudna JA (2018) CRISPR-Cas guides the future of genetic engineering. Science 361(6405):866–869. https://doi.org/10.1126/science.aat5011
2. Wright AV, Nunez JK, Doudna JA (2016) Biology and applications of CRISPR systems: harnessing nature's toolbox for genome engineering. Cell 164(1-2):29–44. https://doi.org/10.1016/j.cell.2015.12.035
3. Kim H, Kim JS (2014) A guide to genome engineering with programmable nucleases. Nat Rev Genet 15(5):321–334. https://doi.org/10.1038/nrg3686
4. Cho SW, Kim S, Kim JM, Kim JS (2013) Targeted genome engineering in human cells with the Cas9 RNA-guided endonuclease. Nat Biotechnol 31(3):230–232. https://doi.org/10.1038/nbt.2507
5. Cong L, Ran FA, Cox D, Lin S, Barretto R, Habib N, Hsu PD, Wu X, Jiang W, Marraffini LA, Zhang F (2013) Multiplex genome engineering using CRISPR/Cas systems. Science 339(6121):819–823. https://doi.org/10.1126/science.1231143
6. Mali P, Yang L, Esvelt KM, Aach J, Guell M, DiCarlo JE, Norville JE, Church GM (2013) RNA-guided human genome engineering via Cas9. Science 339(6121):823–826. https://doi.org/10.1126/science.1232033
7. Zetsche B, Gootenberg JS, Abudayyeh OO, Slaymaker IM, Makarova KS, Essletzbichler P, Volz SE, Joung J, van der Oost J, Regev A, Koonin EV, Zhang F (2015) Cpf1 is a single RNA-guided endonuclease of a class 2 CRISPR-Cas system. Cell 163(3):759–771. https://doi.org/10.1016/j.cell.2015.09.038
8. Kweon J, Jang AH, Kim DE, Yang JW, Yoon M, Rim Shin H, Kim JS, Kim Y (2017) Fusion guide RNAs for orthogonal gene manipulation with Cas9 and Cpf1. Nat Commun 8(1):1723. https://doi.org/10.1038/s41467-017-01650-w
9. Park J, Bae S, Kim JS (2015) Cas-Designer: a web-based tool for choice of CRISPR-Cas9 target sites. Bioinformatics 31(24):4014–4016. https://doi.org/10.1093/bioinformatics/btv537
10. Bae S, Park J, Kim JS (2014) Cas-OFFinder: a fast and versatile algorithm that searches for potential off-target sites of Cas9 RNA-guided endonucleases. Bioinformatics 30(10):1473–1475. https://doi.org/10.1093/bioinformatics/btu048
11. Kim Y, Kweon J, Kim A, Chon JK, Yoo JY, Kim HJ, Kim S, Lee C, Jeong E, Chung E, Kim D, Lee MS, Go EM, Song HJ, Kim H, Cho N, Bang D, Kim S, Kim JS (2013) A library of TAL effector nucleases spanning the human genome. Nat Biotechnol 31(3):251–258. https://doi.org/10.1038/nbt.2517
12. Cho SW, Kim S, Kim Y, Kweon J, Kim HS, Bae S, Kim JS (2014) Analysis of off-target effects of CRISPR/Cas-derived RNA-guided endonucleases and nickases. Genome Res 24(1):132–141. https://doi.org/10.1101/gr.162339.113
13. Untergasser A, Cutcutache I, Koressaar T, Ye J, Faircloth BC, Remm M, Rozen SG (2012) Primer3—new capabilities and interfaces. Nucleic Acids Res 40(15):e115. https://doi.org/10.1093/nar/gks596
14. Park J, Lim K, Kim JS, Bae S (2017) Cas-analyzer: an online tool for assessing genome editing results using NGS data. Bioinformatics 33(2):286–288. https://doi.org/10.1093/bioinformatics/btw561

Part IV

Characterization of CRISPR Efficacy and Specificity

Chapter 11

Methods for Measuring CRISPR/Cas9 DNA Cleavage in Cells

Christopher R. Cromwell, Juan Jovel, and Basil P. Hubbard

Abstract

The CRISPR/Cas9 system has transformed how gene knockout and knock-in studies are performed in the lab, and it is poised to revolutionize medicine. However, one of the present limitations of this technology is its imperfect specificity. While CRISPR/Cas9 can be programmed to cut a specific DNA target sequence with relative precision, off-target sequence cleavage can occur in large genomes. Importantly, several techniques have recently been developed to measure CRISPR/Cas9 on- and off-target DNA cleavage in cells. Here, we present detailed protocols for evaluating the specificity of CRISPR/Cas9 and related systems in cells using both targeted-approaches, in which off-target sites are known a priori, and unbiased approaches which are able to identify off-target cleavage events throughout an entire genome. Together, these techniques can be used to assess the reliability of experimental models generated using CRISPR/Cas9 as well as the safety of therapeutics employing this technology.

Key words Genome engineering, Gene editing, CRISPR/Cas9, DNA cleavage specificity, Off-target effects, T7 endonuclease I assay, Targeted high-throughput sequencing in cells, GUIDE-seq

1 Introduction

The clustered regularly interspaced short palindromic repeats (CRISPR)/Cas9 system, originally characterized as a mediator of the prokaryotic adaptive immune response [1], has emerged as an effective tool for performing gene editing in a wide variety of organisms [2]. Cas9 directs sequence-specific DNA cleavage using two noncoding RNA elements: a CRISPR RNA (crRNA), which contains a 20-nt RNA sequence complementary to the target DNA sequence, and a transactivating crRNA (tracrRNA), which acts as a bridge between the crRNA and Cas9 enzyme [3]. These two molecules can be hybridized to form a guide RNA (gRNA) [3] or covalently fused to produce a single guide RNA (sgRNA) [4]. Target recognition by Cas9 proceeds through binding of a protospacer adjacent motif (PAM) sequence (5′-NGG-′3 in *S. pyogenes*), followed by hybridization of the gRNA to its complementary sequence. Complete hybridization between the guide segment and target DNA induces conformational changes in the nuclease

Tudor A. Fulga et al. (eds.), *CRISPR Guide RNA Design: Methods and Protocols*, Methods in Molecular Biology, vol. 2162, https://doi.org/10.1007/978-1-0716-0687-2_11, © Springer Science+Business Media, LLC, part of Springer Nature 2021

domains of Cas9 that result in double strand DNA cleavage [5, 6]. In cells, these DNA breaks can be exploited for gene inactivation via insertion or deletion of information (indel) during nonhomologous end joining repair [7] or gene editing via homology-directed repair in the presence of a repair template [8].

Specificity is one of the primary concerns when performing gene editing [9]. While CRISPR/Cas9 displays superior specificity comparable to previous state-of-the-art technologies [10], it can induce off-target DNA cleavage appreciably in large genomes [11]. DNA cleavage specificity is highly dependent on the crRNA sequence and correlates with the folding stability of the target–crRNA duplex [12]. While some crRNAs result in few off-target events, others are highly promiscuous [13]. Several strategies have been proposed for decreasing off-target DNA cleavage by Cas9 [4]. First, protein engineering and directed evolution have been used to generate a number of Cas9 variants with improved inherent specificity [14, 15]. These include eSpCas9 [14] and SpCas9-HF1 [15] which display reduced nonspecific affinity for the nontarget and target DNA strands, respectively. Second, computational algorithms have been designed to help select optimal crRNAs based on uniqueness from other sequences in the genome, self-complementarity, and RNA stability [16]. Finally, several studies have shown that incorporation of alternative nucleic acids including DNA [17], bridged and locked nucleic acids (BNA/LNA) [18], and 2′ OMe phosophonoacetate (PACE) nucleic acids [19] into crRNAs dramatically reduces off-target DNA cleavage by Cas9. Despite these advances, off-target DNA cleavage remains a concern, especially in therapeutic contexts where single off-target cellular events could result in detrimental consequences [16].

A number of experimental approaches have been developed to evaluate CRISPR/Cas9 off-target effects in cells. Broadly, these may be divided into two categories: targeted approaches [20], in which cleavage at predicted or known off-targeted loci is measured, and unbiased approaches [21] which identify off-target DNA cleavage events throughout a genome. Examples of targeted strategies include the Surveyor assay [22] and the T7 endonuclease I assay [20], which employ locus-specific PCR amplification to examine DNA cleavage in a semiquantitative manner. A more quantitative assessment of cleavage at preselected loci can be obtained by performing high-throughput sequencing following PCR amplification of the putative regions [18]. Methods for identifying off-target DNA double strand breaks (DSBs) in an unbiased manner include classical techniques such as GUIDE-seq [21], which relies on integration of double stranded oligonucleotide tags into the regions where DSBs occur, Digenome-seq [23], which involves in vitro digestion of a genome by Cas9, and BLISS [24], a semiquantitative approach that involves in situ DNA digestion by Cas9 following cell/tissue fixation. Several second-generation unbiased techniques have recently been described. These include CIRCLE-seq [25], SITE-seq [26], and DISCOVER-seq [27], which identifies DNA double-strand breaks in cells by exploiting binding of endogenous DNA repair factors such as Mre11.

Here, we present methods for evaluating the specificity of CRISPR/Cas9 in cells using established techniques. We describe a protocol for measuring predicted off-target cleavage events in cells using a T7 endonuclease I assay [20] and a high-throughput sequencing assay [18]. We outline how GUIDE-Seq [21] can be used to identify cellular off-target cleavage events in an unbiased manner. Together, these approaches can be used to comprehensively characterize the specificity of CRISPR/Cas9 and related systems, including those using modified Cas proteins and guide RNAs, in a wide variety of contexts.

2 Materials

2.1 T7 Endonuclease I Assay

1. Sets of primers designed to amplify a genomic region of ~400–1200 bp encompassing either the Cas9 on-target site or an off-target site being evaluated. Primers should be designed such that the cut site is positioned asymmetrically within the amplified region (e.g., closer to one end) to facilitate resolution of digested fragments following treatment with T7 endonuclease I.
2. Q5 High Fidelity DNA Polymerase (NEB) or similar.
3. dNTP mix (10 mM).
4. 0.2 mL PCR tubes.
5. PCR thermocycler.
6. QIAquick PCR Purification Kit (Qiagen).
7. NanoDrop.
8. T7 Endonuclease I (NEB).
9. MinElute PCR Purification Kit (Qiagen).
10. 2× Loading Buffer: 90% formamide, 10% glycerol.
11. 50× TAE buffer: 2 M Tris (pH 7.6), 1 M glacial acetic acid, 50 mM EDTA (pH 8.0).
12. Agarose.
13. 10 mg/mL ethidium bromide solution.
14. 1 kb DNA ladder.
15. UV imager.
16. Horizontal gel electrophoresis unit and power supply; gel casting tray and combs.
17. ImageJ or similar software for densitometry analysis.

2.2 Targeted Cellular Cleavage High-Throughput Sequencing Assay

1. Sets of primers containing Illumina overhang adapter sequences (underlined in primer below) designed to amplify a genomic region of ~200–280 bp encompassing either the Cas9 on-target site or an off-target site being evaluated. Example primer sequences:

   5′-TCGTCGGCAGCGTCAGATGTGTATAAGAGACAG-{fwd gene sequence}-′3.

   5′-GTCTCGTGGGCTCGGAGATGTGTATAAGAGACAG-{rev gene sequence}-′3.
2. Q5 Hot Start High-Fidelity 2× Master Mix (NEB) or similar.
3. 0.2 mL PCR tubes.
4. PCR thermocycler.
5. GeneRead Size Selection Kit (Qiagen).
6. Nextera XT Index Kit (Illumina).
7. Qubit fluorometer.
8. QIAxcel (Qiagen).
9. BioAnalyzer (Agilent).

2.3 GUIDE-Seq Analysis

1. Cas9-expressing cell line either stably transduced or transiently transfected.
2. Oligonucleotides for integration and library preparation (Table 1).
3. TE buffer.
4. 0.2 mL PCR tubes.
5. PCR thermocycler.
6. TrypLE Express (Gibco) or similar cell dissociation reagent.
7. Hemocytometer or similar instrument for counting cells.
8. 1.7 mL microcentrifuge tubes.
9. Tabletop centrifuge.
10. 12-Well cell culture plate
11. PBS, pH 7.4 (1×).
12. Amaxa SE Cell Line 4D-Nucleofector X Kit S (32 RCT) (different kits may be required depending on the cell line being nucleofected).
13. Lonza 4D Nucleofector Core and X-Unit.
14. crRNA.
15. tracrRNA.
16. Cell culture incubator.
17. DNeasy Blood & Tissue Kit (Qiagen).
18. Qubit fluorometer.

Table 1
Oligonucleotide sequences used for GUIDE-seq tag integration and library preparation

ODN_F	/5Phos/G*T*T TAA TTG AGT TGT CAT ATG TTA ATA ACG GT*A *T
ODN_R	/5Phos/A*T*A CCG TTA TTA ACA TAT GAC AAC TCA ATT AA*A *C
MiSeq Common Oligo	/5Phos/GAT CGG AAG AGC *C*A
A01	AAT GAT ACG GCG ACC ACC GAG ATC TAC ACT AGA TCG CNN WNN WNN ACA CTC TTT CCC TAC ACG ACG CTC TTC CGA TC*T
A02	AAT GAT ACG GCG ACC ACC GAG ATC TAC ACC TCT CTA TNN WNN WNN ACA CTC TTT CCC TAC ACG ACG CTC TTC CGA TC*T
A03	AAT GAT ACG GCG ACC ACC GAG ATC TAC ACT ATC CTC TNN WNN WNN ACA CTC TTT CCC TAC ACG ACG CTC TTC CGA TC*T
A04	AAT GAT ACG GCG ACC ACC GAG ATC TAC ACA GAG TAG ANN WNN WNN ACA CTC TTT CCC TAC ACG ACG CTC TTC CGA TC*T
GSP1(+)	GGA TCT CGA CGC TCT CCC TAT ACC GTT ATT AAC ATA TGA CA
GSP1(−)	GGA TCT CGA CGC TCT CCC TGT TTA ATT GAG TTG TCA TAT GTT AAT AAC
GSP2(+)	CCT CTC TAT GGG CAG TCG GTG ATA CAT ATG ACA ACT CAA TTA AAC
GSP2(−)	CCT CTC TAT GGG CAG TCG GTG ATT TGA GTT GTC ATA TGT TAA TAA CGG TA
P5_1	AAT GAT ACG GCG ACC ACC GAG ATC TA
P5_2	AAT GAT ACG GCG ACC ACC GAG ATC TAC AC
P701	CAA GCA GAA GAC GGC ATA CGA GAT TCG CCT TAG TGA CTG GAG TCC TCT CTA TGG GCA GTC GGT GA
P702	CAA GCA GAA GAC GGC ATA CGA GAT CTA GTA CGG TGA CTG GAG TCC TCT CTA TGG GCA GTC GGT GA
P703	CAA GCA GAA GAC GGC ATA CGA GAT TTC TGC CTG TGA CTG GAG TCC TCT CTA TGG GCA GTC GGT GA
P704	CAA GCA GAA GAC GGC ATA CGA GAT GCT CAG GAG TGA CTG GAG TCC TCT CTA TGG GCA GTC GGT GA
Index1	ATC ACC GAC TGC CCA TAG AGA GGA CTC CAG TCA C
Read2	GTG ACT GGA GTC CTC TCT ATG GGC AGT CGG TGA T

/5Phos/ indicates the inclusion of a 5′ phosphate modification, while an asterisk indicates a phosphorothioate modification. *N* indicates a position with a mixture of A, C, G, or T, while W indicates a mixture of A or T. Sequences underlined in the A01–A04 and P701–P704 primers indicate unique indices which are used to facilitate sample identification following sequencing

19. *Nde*I endonuclease.
20. T7 Endonuclease I (NEB).
21. Agarose.
22. 50× TAE buffer: 2 M Tris (pH 7.6), 1 M glacial acetic acid, 50 mM EDTA (pH 8.0).

23. 10 mg/mL ethidium bromide solution.
24. UV imager.
25. ImageJ or similar software for densitometry analysis.
26. Covaris S2.
27. Covaris MicroTube AFA Slit SnapCap.
28. AMPure XP beads.
29. BioAnalyzer (Agilent).
30. Bioanalyzer DNA 2100 chip.
31. dNTP mix (10 mM).
32. T4 DNA Ligase (NEB).
33. End-Repair Mix (Low Concentration) (Enzymatics Inc.).
34. Taq DNA Polymerase (Invitrogen).
35. TMAC (5 M).
36. Platinum Taq DNA Polymerase (Invitrogen).
37. NEBNext Library Quant Kit for Illumina (NEB).

3 Methods

3.1 T7 Endonuclease I Assay

This protocol was adapted from Guschin et al. [22] and Cromwell et al. [18]. An outline of the assay is shown in Fig. 1.

1. Design and order primers to PCR amplify a region of ~400–1200 bp surrounding the on- and/or off-target site (s) of interest using a tool such as Primer3Plus (http://www.bioinformatics.nl/primer3plus) (*see* **Note 1** for additional considerations). The primer must be very specific and only a single PCR product should be amplified during the PCR reaction (*see* **Note 2**).
2. Dissolve lyophilized primers in ultrapure water to a final concentration of 100 μM and dilute to 10 μM in ultrapure water to generate stock solutions.
3. Prepare 50 μL PCR reactions containing 10 μL of 5× Q5 Reaction Buffer, 2.5 μL of 10 μM forward and reverse primer, 1 μL of dNTP mix (200 μM final), and 100 ng of genomic DNA (*see* **Note 3**). Bring the volume to 49.5 μL with ultrapure water and add 0.5 μL of Q5 Hot Start High-Fidelity DNA Polymerase.
4. Perform PCR using verified cycling conditions to amplify the genomic region(s).
5. Purify the amplified DNA using the QIAquick PCR Purification Kit. Quantify the purified DNA by NanoDrop.
6. Prepare reactions for T7 endonuclease I digestion on ice. Mix 2 μL of 10× NEBuffer 2 and 200 ng of purified PCR product

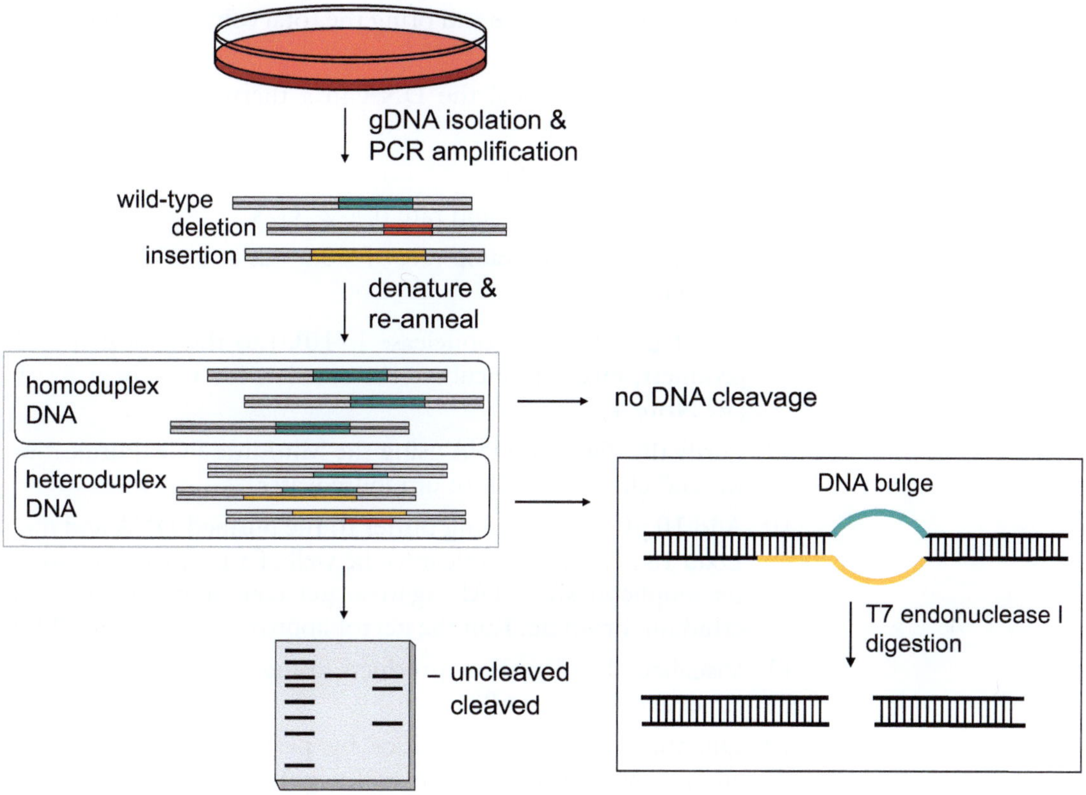

Fig. 1 Diagram of the T7 endonuclease I assay. Genomic DNA from Cas9-treated cells is used as a template to PCR amplify the locus of interest. PCR products are denatured and reannealed to form homo- and heteroduplex structures. Heteroduplexes containing a mismatch are recognized and digested by T7 endonuclease I. The resulting digestion products are resolved via agarose gel electrophoresis and quantified using densitometry

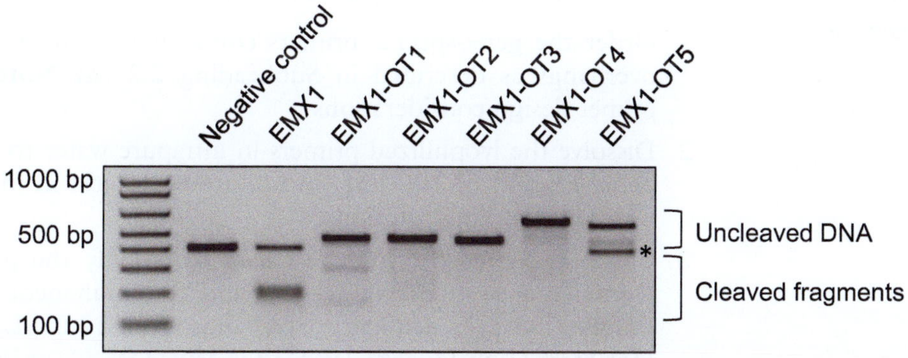

Fig. 2 Example T7 endonuclease I assay results. Agarose gel showing the relative cellular on- and off-target (OT) cleavage efficiencies for a previously described EMX1-targeting crRNA [18]. Cas9-expressing U2OS cells were nucleofected with 100 pmol of EMX1-gRNA. Cells were grown for 48 h prior to genomic DNA isolation. EMX1 on- and off-target loci were individually amplified by PCR using validated conditions. A mock transfection lacking gRNA was used as a control. The band marked by an asterisk indicates a Cas9-independent T7 endonuclease I cleavage product

in a 0.2 mL PCR tube and bring the total volume to 19 μL with ultrapure water.

7. Denature and anneal the DNA in a thermocycler using the following program:
   (a) 95 °C for 5 min.
   (b) 95–85 °C at a ramp rate of −2 °C/s.
   (c) 85–25 °C at a ramp rate of −0.1 °C/s.
   (d) Hold at 4 °C.
8. Add 1 μL of T7 endonuclease I (10U) to the annealed PCR products, mix, and incubate the reaction for 15 min at 37 °C (*see* **Note 4**).
9. Purify the digested DNA using the MinElute PCR Purification Kit and elute in 10 μL of ultrapure water.
10. Add 10 μL of 2× loading buffer to the purified DNA and mix. Load 18 μL of the sample into the well of a 1–1.5% (depending on amplicon size) TAE agarose gel containing 0.5 μg/mL ethidium bromide. Run the gel for approximately 1 h at 100 V.
11. Visualize the digestion products using a UV imager (e.g., Amersham Imager 600).
12. Quantify the band intensities by performing densitometry analysis with software such as ImageJ. Determine the insertion/deletion (indel) percentage using the formula: % indel $= 100 \times (1 - (1 - \text{fraction cleaved})^{0.5})$. Example results from the assay are provided in Fig. 2.

3.2 Targeted Cellular Cleavage High-Throughput Sequencing Assay

This protocol was adapted from the Illumina 16S Metagenomic Sequencing Library Preparation Guide and Cromwell et al. [18]. An outline of the assay is shown in Fig. 3.

1. Order the gene-specific primers containing Illumina adaptor overhangs as described in Subheading 2.2 (*see* **Note 5** for primer design considerations).
2. Dissolve the lyophilized primers in ultrapure water to a stock concentration of 100 μM, and dilute to 10 μM in ultrapure water for working solutions.
3. The purpose of the initial PCR is to amplify the genomic sequence around the target site and to simultaneously add Illumina adaptor sequence overhangs. Mix 12.5 μL of Q5 Hot Start High-Fidelity 2× Master Mix, 1.25 μL of both the forward and reverse primers (10 μM), and 100 ng of genomic DNA (*see* **Note 6**) in a 0.2 mL PCR tube or 96-well plate and bring the volume to 25 μL with ultrapure water.
4. Purify the PCR product by removing free primers and primer-dimer species using the GeneRead Size Selection Kit according to the manufacturer's instructions. Elute in 20 μL of ultrapure water.

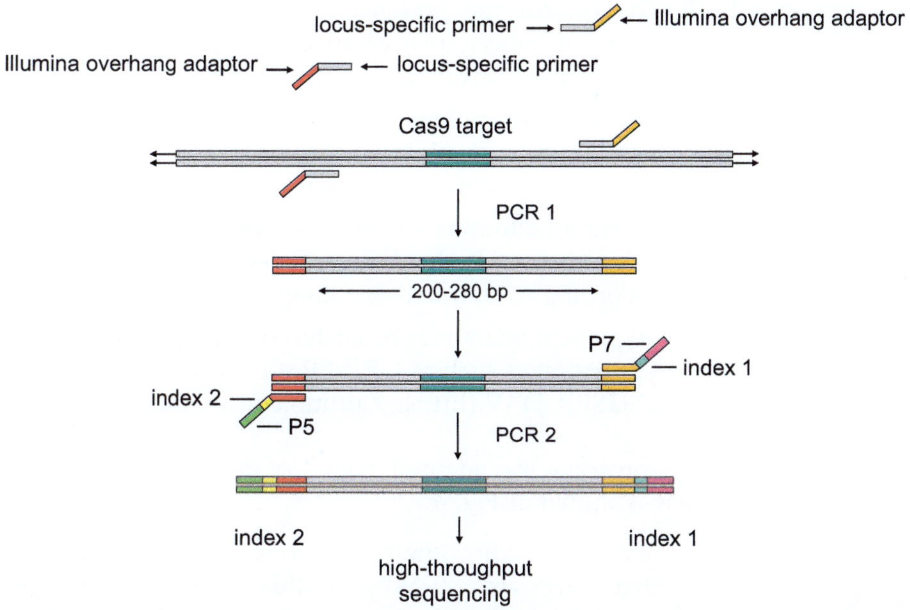

Fig. 3 Diagram of the targeted cellular cleavage high-throughput sequencing assay. Forward and reverse primers containing adaptor sequence overhangs are designed to amplify the region surrounding the locus of interest. Using Cas9-cleaved DNA as a template, the first PCR amplifies the target region of interest and adds Illumina adaptor sequences to the product. The second PCR reaction (PCR 2) adds unique barcode sequences to the products of PCR 1 to enable pooled high-throughput sequencing

5. Unique indices are added to the Illumina sequencing adaptors via PCR using the Nextera XT Index Kit to tag each of the various samples being analyzed. Each sample (on or off-target site) should be labeled with a unique pair of indices. To begin, mix 25 μL of Q5 Hot Start High-Fidelity 2× Master Mix, 5 μL of Nextera XT Index Primer 1, 5 μL of Nextera XT Index Primer 2, 13 μL of ultrapure water, and 2 μL of DNA from PCR 1 (**step 4**) in a 0.2 mL PCR tube or a 96-well plate.
6. Move the plate to a thermocycler and run the following program:
   (a) 95 °C for 3 min.
   (b) 95 °C for 30 s.
   (c) 55 °C for 30 s.
   (d) 72 °C for 30 s.
   (e) Go to step b and repeat for seven additional cycles.
   (f) 72 °C for 5 min.
   (g) Hold at 4 °C.
7. Purify the PCR product using the GeneRead Size Selection Kit, elute in 17 μL of ultrapure water, and quantify using a Qubit fluorometer.

8. To confirm the presence of a single PCR product of the intended length, analyze an aliquot of each final library using the QIAxcel Advanced System.
9. Dilute each sample to a final concentration of 4 nM in 10 mM Tris, pH 8.5 (*see* **Note 7**). At this stage, multiple samples may be pooled, maintaining a fixed concentration of 4 nM.
10. Using an Illumina MiSeq Reagent V2—300 cycle (2 × 150 paired end) kit, load the sample onto an Illumina MiSeq according to standard operating guidelines.
11. Sequencing data may be analyzed using publicly available analysis packages such as CRISPResso (http://crispresso.rocks) or CRISPR-DAV (https://github.com/pinetree1/crispr-dav).

3.3 GUIDE-Seq Analysis

This protocol was adapted from Tsai et al. [21]. A diagram of the assay is shown in Fig. 4.

1. Seed Cas9-expressing cells 2–3 days prior to transfection such that they are 80–90% confluent at the time of gRNA transfection.
2. Prior to transfection, prepare the dsODN tag by annealing the ODN_F (Table 1) and ODN_R oligonucleotides. Briefly, mix 5 μL of ODN_F (1 mM) and 5 μL of ODN_R (1 mM) with 40 μL of TE buffer in a 0.2 mL PCR tube. Move the mixture to a thermocycler and heat at 95 °C for 5 min and then slowly cool to 25 °C over the course of 1 h. The 100 μM annealed dsODN may be stored at −20 °C.
3. On the day of transfection, detach cells using a nonenzymatic dissociation reagent (*see* **Note 8**), and resuspend in an antibiotic-free growth media. Count cells and pipet a total of 2×10^5 cells into a microcentrifuge tube. Pellet the cells by centrifuging at 90 × *g* for 10 min. While the cells are centrifuging, add 1 mL of growth media to the wells of a 12-well plate. Warm the plate and media to 37 °C in a cell culture incubator. After centrifugation, carefully remove the supernatant, wash the cell pellet with 500 μL of PBS, and repellet the cells by centrifuging at 90 × *g* for 10 min. Aspirate off the PBS solution.
4. Resuspend the cells in 20 μL of SE Solution and pipet up and down to achieve a single-cell suspension. Add 1 μL of gRNA (100 μM) and 1 μL of dsODN (100 μM) (*see* **Note 9**) and mix by pipetting. Transfer the entire cell suspension to the well of a 20 μL, 16-well Nucleocuvette™ Strip. Include an additional nucleofection reaction without gRNA as a negative control.
5. Place the 16-well strip(s) into the Lonza 4D Nucleofector X-Unit and nucleofect using the appropriate program.

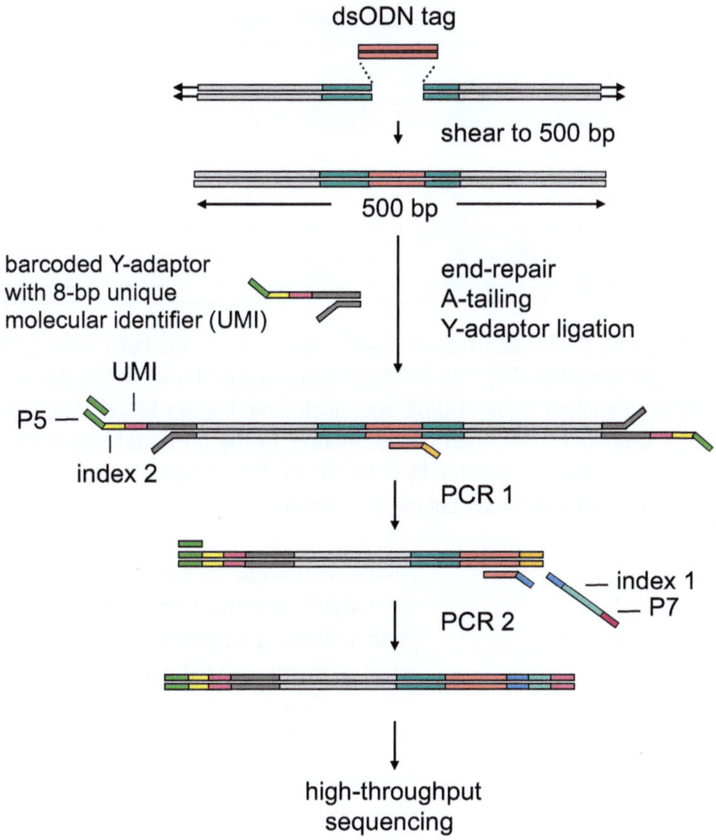

Fig. 4 Diagram illustrating how to perform GUIDE-seq. Genomic DNA isolated from Cas9-treated cells is sheared to an average length of 500 bp, end-repaired, A-tailed, and ligated with a barcoded Y-adaptor. Through the series of PCR reactions illustrated in the diagram, sheared DNA containing the dsODN is amplified and prepared for high throughput sequencing via addition of the indicated adapters/indices

6. Using a P200 tip, add 80 μL of prewarmed media to each well and gently pipet up and down to resuspend the cells. Transfer the suspensions to the wells of the 12-well plate containing prewarmed media. Add an additional 100 μL of prewarmed media to the 16-well strip to wash out any remaining cells and add to the 12-well plate. Allow cells to grow for 48 h in a cell culture incubator.
7. Isolate genomic DNA using the DNeasy Blood & Tissue Kit and quantify by Qubit. Isolated gDNA may be stored at −20 °C.
8. Prior to downstream library preparation, confirm the presence of Cas9 genome-editing events as well as sufficient dsODN integration. Since the dsODN contains an *Nde*I restriction site, quantification of integration can be estimated through a

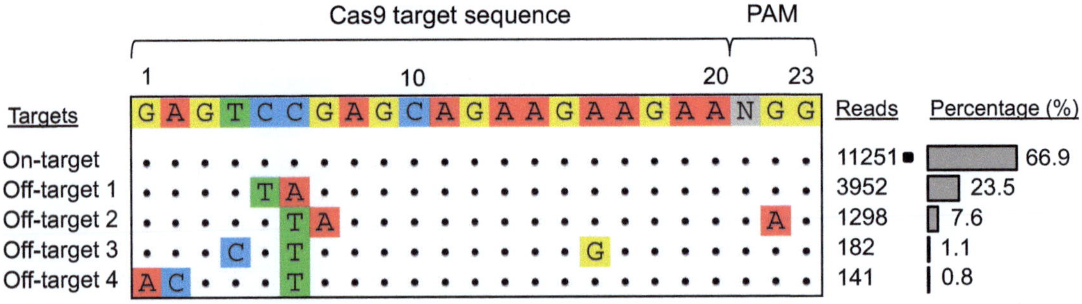

Fig. 5 Example GUIDE-seq results. Cas9-expressing U2OS cells were treated with a previously described EMX1-targeting crRNA [18]. The on-target sequence is shown on the top line. Beneath are the highest scoring sequences identified using GUIDE-seq, including the on-target sequence and the four most prevalent off-target sequences. Matched bases, relative to the on-target sequence, are indicated by a period while mismatched bases are indicated by highlighting. The number of reads and the percentage relative to the total number of reads for each sequence is indicated

restriction digest. To perform this analysis, PCR amplify the on-target genomic region of interest as described above and digest 500 ng of product with 20 U of *Nde*I endonuclease in a total volume of 50 μL in CutSmart Buffer (1×) for 3 h at 37 °C. Resolve the digestion products using agarose gel electrophoresis and quantify by densitometry.

9. Prepare the barcoded Y-adaptor by annealing the MiSeq Common Oligo with the barcoded sample adaptor (A01–A04) (Table 1) (*see* **Note 10**). As each treatment requires the use of a differentially barcoded Y-adaptor, prepare the adaptors using the oligos A01 through A04 as needed (*see* **Note 11**). Add 5 μL of adaptor A0# (100 μM) and 5 μL of MiSeq Common Adaptor (100 μM) to 40 μL of TE buffer (final concentration 10 μM) in a PCR tube. Move the reaction to a thermocycler and run the following program:
   (a) 95 °C for 10 s.
   (b) 60 °C for 30 s.
   (c) Return to step b, decreasing by 1 °C per cycle for 55 cycles.
   (d) Hold at 4 °C.

   The annealed Y-adaptor may be stored at −20 °C.

10. Mix 400 ng of isolated gDNA with TE buffer to a final volume of 130 μL. Using a Covaris S2, shear the gDNA to an average length of approximately 500 bp. Clean up the sheared DNA using 130 μL of AMPure XP beads (1× ratio) and elute in 15 μL of TE buffer. Validate using a Bioanalyzer DNA 2100 chip.
11. Prepare the sample for adaptor ligation by end-repairing/A-tailing the sheared DNA. Mix 0.5 μL of ultrapure water, 1 μL

of dNTP mix (5 mM), 2.5 μL of 10× T4 DNA Ligase Buffer, 2 μL of End-Repair Mix (Low Concentration), 2 μL of 10× Taq Polymerase Buffer (Mg^{2+} free), 0.5 μL of Taq DNA Polymerase (non-hot start), and 14 μL of sheared DNA in a 0.2 mL PCR tube. Run the end-repair thermocycler program below:

(a) 12 °C for 15 min.

(b) 37 °C for 15 min.

(c) 72 °C for 15 min.

(d) Hold at 4 °C.

12. Add 1 μL of the annealed Y-adaptor (10 μM) from **step 9** and 2 μL of T4 DNA Ligase to the reaction. Return the tube to the thermocycler and run the adaptor ligation program:

(a) 16 °C for 30 min.

(b) 22 °C for 30 min.

(c) Hold at 4 °C.

13. Clean up the adaptor ligated DNA using 22.95 μL of AMPure XP beads (0.9× ratio), according to the manufacturer's instructions. Elute in 22 μL of TE buffer.

14. During PCR 1, the Y-adaptor-ligated DNA is amplified using primers that anneal to the adaptor sequence and the dsODN tag. In order to capture sequences flanking both sides of the tag, the DNA from **step 13** is divided into two fractions. Separate fractions are used as template DNA for independent PCR reactions that amplify either the sense strand (using the GSP1+ primer—Table 1), or the antisense strand (using the GSP1− primer) of the tag. Add 11.9 μL of ultrapure water, 3 μL of 10× Taq Polymerase Buffer (Mg^{2+} free), 0.6 μL of dNTP mix (10 mM), 1.2 μL of $MgCl_2$ (50 mM), 0.3 μL of Platinum Taq DNA Polymerase, 1 μL of either the GSP1+ or GSP1− primer (10 μM), 1.5 μL of TMAC (0.5 M), 0.5 μL of the P5_1 primer (10 μM), and 10 μL of the adaptor-ligated gDNA from **step 13** to two separate PCR tubes.

15. Move the reaction to a thermocycler and run the following PCR program:

(a) 95 °C for 5 min.

(b) 95 °C for 30 s.

(c) 70 °C for 2 min (−1 °C/cycle).

(d) 72 °C for 30 s.

(e) Return to step b and repeat for 14 additional cycles.

(f) 95 °C for 30 s.

(g) 55 °C for 1 min.

(h) 72 °C for 30 s.

(i) Return to step f and repeat for nine additional cycles.

(j) 72 °C for 5 min.

(k) Hold at 4 °C.

16. Clean up the two PCR reactions using 36 μL of AMPure XP beads (1.2× ratio), according to the manufacturer's instructions. Elute each in 15 μL of TE Buffer.

17. In PCR 2, the products from PCR 1 are further amplified, and sequencing adaptors and unique barcodes for sample identification are added. Using the corresponding PCR 1 product as template, set up two separate PCR reactions using either the GSP2+ or the GSP2− primers. Add 5.4 μL of ultrapure water, 3 μL of 10× Taq Polymerase Buffer (Mg^{2+} free), 0.6 μL of dNTP mix (10 mM), 1.2 μL of $MgCl_2$ (50 mM), 0.3 μL of Platinum Taq DNA Polymerase, 1 μL of either the GSP2+ or the GSP2− primer (10 μM), 1.5 μL of TMAC (0.5 M), 0.5 μL of P5_2 primer (10 μM), 1.5 μL of P70# primer (10 μM), and 15 μL of the corresponding PCR products from **step 16** to each reaction. The P70# primer (P701–P704) contains a unique barcode which must be varied between different treatment samples (*see* **Note 11**). However, it is unnecessary to independently barcode the sense and antisense reactions. Move the samples to a thermocycler and run the program outlined in **step 15.**

18. Clean up the PCR 2 reactions using 21 μL of AMPure XP beads (0.7× ratio) and elute in 30 μL of TE Buffer.

19. Quantify the PCR 2 products using the NEBNext Library Quant Kit for Illumina. To determine the average size of each library, analyze an aliquot on a Bioanalyzer DNA 2100 chip. Using the molarity from the qPCR run, as well as the average size, determine the size-adjusted concentration of each library.

20. Dilute each PCR 2 product to a final concentration of 4 nM using 10 mM Tris (pH 8.5). Add 5 μL of each diluted PCR 2 product to a single tube for pooled sequencing. If multiple treatments are being compared, combine all samples at this step (all at 4 nM).

21. Load the sample into an Illumina MiSeq Reagent Kit V2—300 cycle (2 × 150 paired end) kit. Add 3 μL of 100 μM custom sequencing primer Index1 to the MiSeq reagent cartridge position 13 (Index Primer Mix), and 3 μL of 100 μM custom sequencing primer Read2 to MiSeq reagent cartridge position 14 (Read 2 Primer Mix). Using the paired end Nextera sequencing protocol, sequence the sample using the following number of cycles: 151 | 8 | 16 | 151.

22. Analyze the sequencing data using a publicly available package such as http://www.jounglab.org/guideseq/ or https://

bioconductor.org/packages/release/bioc/html/GUIDEseq.html. Example results for a previously described EMX1-targeting crRNA [18] are outlined in Fig. 5.

4 Notes

1. If an experimentally determined list of off-target sites is not available for the crRNA being tested, an in silico prediction tool such as Cas-OFFinder (http://www.rgenome.net/cas-offinder/) [28] may be used.
2. Prior to T7 endonuclease I digestion, it is imperative to ensure that only a single PCR product has been amplified. The presence of secondary products may lead to unreliable results.
3. To ensure that T7 endonuclease I digestion is the result of Cas9-mediated indels, and not the result of an impure PCR product and/or polymorphisms between alleles, genomic DNA from a mock-transfected population should be included as a negative control. In some cases, a single PCR amplification product may still be strongly digested by the T7 endonuclease I (as seen in Fig. 2: EMX1-OT5). This digestion pattern can result from inherent polymorphisms between alleles. Using primers outside of the polymorphic region can circumvent this issue.
4. Following the denaturation/annealing step, it is common to see condensation on the insides of the 0.2 mL PCR tubes. Prior to the addition of T7 endonuclease I, briefly spin down the tubes in a pulse spinner to collect the liquid.
5. Primers for PCR should be designed such that the resulting product is ~200–280 bp in length (excluding Illumina adaptor sequences) with a T_m between 60–65 °C. To ensure adequate coverage depth, the target site should be located within 100 bp of either the 5′ or 3′ end of the insert. Each primer set should be tested to ensure that only a single amplification product is generated using genomic DNA isolated from the cell line of interest prior to sequencing.
6. In addition to Cas9-treated cells, genomic DNA isolated from a mock-transfected sample should be included for all on- and off-target sites being investigated.
7. Prior to loading on the MiSeq, the concentration and size of the final pooled library should be quantified using a Qubit and Bioanalyzer, respectively.
8. A nonenzymatic dissociation reagent is used since trypsin enzyme preparations may contain RNase activity that can degrade the gRNA.

9. The amount of gRNA and dsODN being used should be experimentally determined for each cell line of interest.
10. In addition to Illumina indices, the adaptors also contain an 8-bp unique molecular identifier in the format NNWNNNWNN (N = A, C, T or G; W = A or T).
11. If additional indices are required (e.g., for more than 4 samples), the barcodes in the sequences of the P701–P704 and A01–A04 oligonucleotides (*see* Table 1) may be replaced using alternative barcodes selected from the list present in the Illumina Nextera XT Index Kit.

Acknowledgments

This work was supported by NSERC Discovery/Discovery Accelerator Supplement (RGPIN-2016-0638) and CIHR grants (PS-408552) to BPH. We thank Sudip Subedi and the members of the Applied Genomics Core at the University of Alberta for their technical assistance with next-generation sequencing.

References

1. Horvath P, Barrangou R (2010) CRISPR/Cas, the immune system of bacteria and archaea. Science 327(5962):167–170. https://doi.org/10.1126/science.1179555
2. Doudna JA, Charpentier E (2014) Genome editing. The new frontier of genome engineering with CRISPR-Cas9. Science 346 (6213):1258096. https://doi.org/10.1126/science.1258096
3. Cong L, Ran FA, Cox D, Lin S, Barretto R, Habib N, Hsu PD, Wu X, Jiang W, Marraffini LA, Zhang F (2013) Multiplex genome engineering using CRISPR/Cas systems. Science 339(6121):819–823. https://doi.org/10.1126/science.1231143
4. Hsu PD, Scott DA, Weinstein JA, Ran FA, Konermann S, Agarwala V, Li Y, Fine EJ, Wu X, Shalem O, Cradick TJ, Marraffini LA, Bao G, Zhang F (2013) DNA targeting specificity of RNA-guided Cas9 nucleases. Nat Biotechnol 31(9):827–832. https://doi.org/10.1038/nbt.2647
5. Jinek M, Chylinski K, Fonfara I, Hauer M, Doudna JA, Charpentier E (2012) A programmable dual-RNA-guided DNA endonuclease in adaptive bacterial immunity. Science 337 (6096):816–821. https://doi.org/10.1126/science.1225829
6. Jiang F, Zhou K, Ma L, Gressel S, Doudna JA (2015) STRUCTURAL BIOLOGY. A Cas9-guide RNA complex preorganized for target DNA recognition. Science 348 (6242):1477–1481. https://doi.org/10.1126/science.aab1452
7. Ablain J, Zon LI (2016) Tissue-specific gene targeting using CRISPR/Cas9. Methods Cell Biol 135:189–202. https://doi.org/10.1016/bs.mcb.2016.03.004
8. Cheng R, Peng J, Yan Y, Cao P, Wang J, Qiu C, Tang L, Liu D, Tang L, Jin J, Huang X, He F, Zhang P (2014) Efficient gene editing in adult mouse livers via adenoviral delivery of CRISPR/Cas9. FEBS Lett 588 (21):3954–3958. https://doi.org/10.1016/j.febslet.2014.09.008
9. Hubbard BP, Badran AH, Zuris JA, Guilinger JP, Davis KM, Chen L, Tsai SQ, Sander JD, Joung JK, Liu DR (2015) Continuous directed evolution of DNA-binding proteins to improve TALEN specificity. Nat Methods 12 (10):939–942. https://doi.org/10.1038/nmeth.3515
10. Wei C, Liu J, Yu Z, Zhang B, Gao G, Jiao R (2013) TALEN or Cas9 - rapid, efficient and specific choices for genome modifications. J Genet Genomics 40(6):281–289. https://doi.org/10.1016/j.jgg.2013.03.013
11. Tycko J, Myer VE, Hsu PD (2016) Methods for optimizing CRISPR-Cas9 genome editing

specificity. Mol Cell 63(3):355–370. https://doi.org/10.1016/j.molcel.2016.07.004

12. Xu X, Duan D, Chen SJ (2017) CRISPR-Cas9 cleavage efficiency correlates strongly with target-sgRNA folding stability: from physical mechanism to off-target assessment. Sci Rep 7 (1):143. https://doi.org/10.1038/s41598-017-00180-1
13. O'Geen H, Yu AS, Segal DJ (2015) How specific is CRISPR/Cas9 really? Curr Opin Chem Biol 29:72–78. https://doi.org/10.1016/j.cbpa.2015.10.001
14. Slaymaker IM, Gao L, Zetsche B, Scott DA, Yan WX, Zhang F (2016) Rationally engineered Cas9 nucleases with improved specificity. Science 351(6268):84–88. https://doi.org/10.1126/science.aad5227
15. Kleinstiver BP, Pattanayak V, Prew MS, Tsai SQ, Nguyen NT, Zheng Z, Joung JK (2016) High-fidelity CRISPR-Cas9 nucleases with no detectable genome-wide off-target effects. Nature 529(7587):490–495. https://doi.org/10.1038/nature16526
16. Labun K, Montague TG, Gagnon JA, Thyme SB, Valen E (2016) CHOPCHOP v2: a web tool for the next generation of CRISPR genome engineering. Nucleic Acids Res 44 (W1):W272–W276. https://doi.org/10.1093/nar/gkw398
17. Yin H, Song CQ, Suresh S, Kwan SY, Wu Q, Walsh S, Ding J, Bogorad RL, Zhu LJ, Wolfe SA, Koteliansky V, Xue W, Langer R, Anderson DG (2018) Partial DNA-guided Cas9 enables genome editing with reduced off-target activity. Nat Chem Biol 14(3):311–316. https://doi.org/10.1038/nchembio.2559
18. Cromwell CR, Sung K, Park J, Krysler AR, Jovel J, Kim SK, Hubbard BP (2018) Incorporation of bridged nucleic acids into CRISPR RNAs improves Cas9 endonuclease specificity. Nat Commun 9(1):1448. https://doi.org/10.1038/s41467-018-03927-0
19. Ryan DE, Taussig D, Steinfeld I, Phadnis SM, Lunstad BD, Singh M, Vuong X, Okochi KD, McCaffrey R, Olesiak M, Roy S, Yung CW, Curry B, Sampson JR, Bruhn L, Dellinger DJ (2018) Improving CRISPR-Cas specificity with chemical modifications in single-guide RNAs. Nucleic Acids Res 46(2):792–803. https://doi.org/10.1093/nar/gkx1199
20. Vouillot L, Thelie A, Pollet N (2015) Comparison of T7E1 and surveyor mismatch cleavage assays to detect mutations triggered by engineered nucleases. G3 (Bethesda) 5 (3):407–415. https://doi.org/10.1534/g3.114.015834
21. Tsai SQ, Zheng Z, Nguyen NT, Liebers M, Topkar VV, Thapar V, Wyvekens N, Khayter C, Iafrate AJ, Le LP, Aryee MJ, Joung JK (2015) GUIDE-seq enables genome-wide profiling of off-target cleavage by CRISPR-Cas nucleases. Nat Biotechnol 33 (2):187–197. https://doi.org/10.1038/nbt.3117
22. Guschin DY, Waite AJ, Katibah GE, Miller JC, Holmes MC, Rebar EJ (2010) A rapid and general assay for monitoring endogenous gene modification. Methods Mol Biol 649:247–256. https://doi.org/10.1007/978-1-60761-753-2_15
23. Kim D, Bae S, Park J, Kim E, Kim S, Yu HR, Hwang J, Kim JI, Kim JS (2015) Digenome-seq: genome-wide profiling of CRISPR-Cas9 off-target effects in human cells. Nat Methods 12(3):231–243. https://doi.org/10.1038/nmeth.3284
24. Yan WX, Mirzazadeh R, Garnerone S, Scott D, Schneider MW, Kallas T, Custodio J, Wernersson E, Li Y, Gao L, Federova Y, Zetsche B, Zhang F, Bienko M, Crosetto N (2017) BLISS is a versatile and quantitative method for genome-wide profiling of DNA double-strand breaks. Nat Commun 8:15058. https://doi.org/10.1038/ncomms15058
25. Tsai SQ, Nguyen NT, Malagon-Lopez J, Topkar VV, Aryee MJ, Joung JK (2017) CIRCLE-seq: a highly sensitive in vitro screen for genome-wide CRISPR-Cas9 nuclease off-targets. Nat Methods. https://doi.org/10.1038/nmeth.4278
26. Cameron P, Fuller CK, Donohoue PD, Jones BN, Thompson MS, Carter MM, Gradia S, Vidal B, Garner E, Slorach EM, Lau E, Banh LM, Lied AM, Edwards LS, Settle AH, Capurso D, Llaca V, Deschamps S, Cigan M, Young JK, May AP (2017) Mapping the genomic landscape of CRISPR-Cas9 cleavage. Nat Methods. https://doi.org/10.1038/nmeth.4284
27. Wienert B, Wyman SK, Richardson CD, Yeh CD, Akcakaya P, Porritt MJ, Morlock M, Vu JT, Kazane KR, Watry HL, Judge LM, Conklin BR, Maresca M, Corn JE (2019) Unbiased detection of CRISPR off-targets in vivo using DISCOVER-Seq. Science 364 (6437):286–289. https://doi.org/10.1126/science.aav9023
28. Bae S, Park J, Kim JS (2014) Cas-OFFinder: a fast and versatile algorithm that searches for potential off-target sites of Cas9 RNA-guided endonucleases. Bioinformatics 30 (10):1473–1475. https://doi.org/10.1093/bioinformatics/btu048

Chapter 12

In Vitro Assays for Comparing the Specificity of First- and Next-Generation CRISPR/Cas9 Systems

Christopher R. Cromwell and Basil P. Hubbard

Abstract

CRISPR/Cas9 has revolutionized the ability to edit cellular DNA and is poised to transform the treatment of genetic diseases. One of the major concerns regarding its therapeutic use is the potential for off-target DNA cleavage, which could have detrimental consequences in vivo. To circumvent this, a number of strategies have been employed to develop next-generation CRISPR/Cas9 systems with improved specificity. These include the development of new protein variants of Cas9, as well as chemically modified guide RNA molecules. Here, we provide detailed protocols for two in vitro methods that enable the specificity of first- and next-generation CRISPR/Cas9 systems to be compared, and we demonstrate their applicability to evaluating chemically modified guide RNAs. One of these assays allows the specificity of different guide RNA/Cas9 complexes to be compared on a set of known off-target DNA sequences, while the second provides a broad specificity profile based on cleavage of a massive library of potential off-target DNA sequences. Collectively, these assays may be used to evaluate the specificity of different CRISPR/Cas9 systems on any DNA target sequence in a time- and cost-effective manner.

Key words Genome engineering, Gene editing, CRISPR/Cas9, DNA cleavage specificity, Next-generation CRISPR/Cas9, Cas9 variants, Chemically modified guide RNAs, In vitro DNA cleavage assay, In vitro high-throughput specificity profiling assay

1 Introduction

Clustered regularly interspersed short palindromic repeat (CRISPR)-Cas systems allow bacteria to mount adaptive immune responses toward bacteriophage and other non-native DNA [1]. To date, over 400 CRISPR-Cas homologs, composed of diverse enzymes with different DNA binding mechanisms, have been identified in bacteria [2]. In addition to their endogenous biological role, several of these have been successfully repurposed for gene editing [3, 4]. For example, CRISPR/Cas9, the most widely used system, has been used to knockout and knock-in genes in a wide variety of organisms ranging from yeast to mice [5]. Moreover, proof-of-principle studies using CRISPR/Cas9 in mice have displayed promising results in relation to the treatment of monogenic

Tudor A. Fulga et al. (eds.), *CRISPR Guide RNA Design: Methods and Protocols*, Methods in Molecular Biology, vol. 2162, https://doi.org/10.1007/978-1-0716-0687-2_12, © Springer Science+Business Media, LLC, part of Springer Nature 2021

diseases such as sickle cell anemia [6] and hemophilia [7]. Cas9 is a DNA endonuclease which utilizes two RNA elements, a CRISPR RNA (crRNA) that contains a 20-nt sequence complementary to the target DNA and a transactivating crRNA (tracrRNA) to direct sequence-specific DNA cleavage [8]. When hybridized together, these two RNA elements constitute a guide RNA (gRNA), or a single guide RNA (sgRNA) when covalently fused [8]. In addition to complementarity between the gRNA and the target DNA, Cas9 also requires the presence of an upstream protospacer adjacent motif (PAM) sequence (5′-NGG-3′ in *Streptococcus pyogenes*) in order to bind DNA [6, 8, 9]. Once cellular DNA has been cut, endogenous repair pathways may be exploited to perform gene knockout (via nonhomologous end joining repair) [10], or gene editing (via homology-directed repair) [11].

One of the major concerns when using CRISPR/Cas9 to edit genes is its imperfect specificity [12]. While Cas9 and related endonucleases display specificity profiles comparable to other genome engineering technologies such as zinc finger nucleases and transcription activator like effector nucleases [13, 14], they do induce off-target DNA cleavage in cells at relevant levels [14, 15]. For Cas9, DNA cleavage specificity is highly dependent on the crRNA sequence [16]. For example, while certain crRNAs result in few to no detectable (<0.1% of cells) off-target cleavage events, others are highly promiscuous and cut many sequences with relatively high efficiency [17]. In addition to optimizing guide RNA design computationally, a number of protein engineering strategies have been applied to improving Cas9 specificity [17]. These include rationally designed Cas9 variants with reduced non-specific DNA affinity such as eSpCas9, which bears three mutations within the nt-groove of Cas9 that weaken its interaction with the nontarget DNA strand [18], and SpCas9-HF1, which bears four mutations that weaken its interaction with the target DNA strand [19]. Additional protein engineering and directed evolution experiments have yielded further next-generation Cas9 systems including evoCas9 [20], HypaCas9 [21], and Opti-SpCas9 [22], which boast improved specificity without a decrease in activity or targeting range.

A variety of guide RNA modifications have also been shown to improve Cas9 DNA cleavage specificity. First, reducing the number of nucleotides in the guide RNA spacer sequence from 20 to 17–18 bp improves specificity, but decreases on-target activity [23]. Second, a number of studies have found that incorporating chemically modified or synthetic nucleotides into guide RNAs can boost specificity. For example, inclusion of DNA [24, 25], locked or bridged nucleic acid (LNA/BNA) [26], and 2′OMe phosphonoacetate (PACE) nucleic acid [27] into guide RNAs decreases off-target DNA cleavage. It has been reported that numerous other chemical modifications may be tolerated within Cas9 gRNAs, although their effect on specificity remains unexplored [28]. Finally, inclusion of certain secondary structures into guide RNAs can also improve targeting specificity [29].

The large selection of next-generation CRISPR/Cas9 components now available, and the possibility of combining various modified Cas9 proteins and guide RNAs, can make it difficult for researchers to identify the best system to use for a particular target sequence. In vitro methods are well-suited tools for aiding in this decision making, as they can evaluate the efficiency and specificity of different systems in a time- and cost-effective manner. Here, we provide protocols for two in vitro assays that allow the specificity of first- and next-generation CRISPR/Cas9 systems to be compared. Also, we demonstrate how these techniques can be used to compare the specificity of unmodified, and BNA- or 2′OMe PACE-modified guide RNAs. The first technique evaluates the activity of each guide RNA–Cas9 complex against its on-target sequence and a series of known off-target DNA sequences, while the second generates a global specificity profile for each complex based on cleavage a library of $>10^{12}$ potential off-target sequences related to the on-target sequence. These assays can help identify the optimal next-generation CRISPR/Cas9 system to cut any DNA target sequence with precision.

2 Materials

2.1 In Vitro Cas9 DNA Cleavage Assay

1. Complementary single-stranded DNA oligonucleotides containing the target sequence of interest. The sequence of the first oligo should be 5′-GCCGAAGCTTCTNNNNNNN NNNNNNNNNNNNNNNNNCTTCTAGAGGCC-3′, where N's are replaced by a 20 bp target sequence, followed by a 3 bp PAM sequence (e.g., 5′-NGG-3′ in *S. pyogenes*). The second oligonucleotide should be the reverse complement. *Hin*dIII and *Xba*I restriction consensus sequences used for cloning into pUC19 are underlined.
2. 0.2 mL PCR tubes.
3. Duplex buffer: 30 mM HEPES (pH 7.5), 100 mM potassium acetate.
4. PCR thermocycler.
5. *Xba*I endonuclease.
6. *Hin*dIII endonuclease.
7. Alkaline phosphatase, calf intestinal (CIP).
8. MinElute PCR Purification Kit (Qiagen).
9. Quick Ligation Kit (NEB).
10. DH5α *E. coli* or similar.
11. Water bath or heating block set to 42 °C.

12. Luria-Bertani (LB) liquid medium.
13. Luria-Bertani (LB) agar.
14. Carbenicillin.
15. Bacterial incubator set to 37 °C.
16. Bacterial shaking incubator set to 37 °C.
17. 15 mL culture tubes.
18. QIAprep Spin Miniprep Kit (Qiagen).
19. NanoDrop or equivalent.
20. pUC19_fwd primer: 5′-CAGCGAGTCAGTGAGCGA-3′.
21. pUC19_rev primer: 5′-GCGACACGGAAATGTTGAATACT CAT-3′.
22. Q5 High-Fidelity DNA Polymerase (NEB) or similar.
23. dNTP mix (10 mM).
24. QIAquick PCR Purification Kit (Qiagen).
25. crRNA.
26. tracrRNA.
27. Cas9 endonuclease (or next-generation variant).
28. 10× Cas9 reaction buffer: 200 mM HEPES, 1 M NaCl, 50 mM $MgCl_2$, 1 mM EDTA.
29. 2× Loading buffer: 90% formamide, 10% glycerol.
30. 50× TAE buffer: 2 M Tris (pH 7.6), 1 M glacial acetic acid, 50 mM EDTA (pH 8.0).
31. Agarose.
32. 10 mg/mL ethidium bromide solution.
33. 1 kb DNA ladder.
34. Horizontal gel electrophoresis unit and power supply; gel casting tray and combs.
35. UV imager.
36. ImageJ or similar software for densitometry analysis.

2.2 In Vitro High-Throughput Specificity Profiling Assay

1. A library of partially randomized oligonucleotides with the sequence 5′-_TCTTCT_NNNNC*C*NT*T*C*T*T*C*T*T*C*T*G*C*T*C*G*G*A*C*T*C*NNNN**ACC TGCCGAG**_TCTTCT_-′3, where N refers to a fully randomized position, and an asterisk indicates a partially randomized position where 79% of the mixture corresponds to the base preceding the asterisk and 7% corresponds to each of the other three bases (ordered as a hand mix). The reverse complement of the desired target sequence should be used in place of the underlined bases (the example above corresponds to a previously characterized EMX1 target sequence [26] and includes a

Table 1
Oligonucleotide sequences used for high-throughput library selection amplification and sequencing. Sequences in bold may be varied to barcode multiple reactions for pooled sequencing (*see* Note 3)

Adaptor1	AAT GAT ACG GCG ACC ACC GAG ATC TAC ACT CTT TCC CTA CAC GAC GCT CTT CCG ATC T**AC TGT**
Adaptor2	**ACA GT**A GAT CGG AAG AGC GTC GTG TAG GGA AAG AGT GTA GAT CTC GGT GG
Preselection Adaptor1	GAC GGC ATA CGA GAT
Preselection Adaptor2	*TCT T*AT CTC GTA TGC CGT CTT CTG CTT G
Postselection Fwd	AAT GAT ACG GCG ACC ACC GA
EMX1 Postselection Rev	CAA GCA GAA GAC GGC ATA CGA GAT ACC TGC CGA G*TC TTC T*
Preselection Fwd	CAA GCA GAA GAC GGC ATA CGA GAT
EMX1 Preselection Rev	AAT GAT ACG GCG ACC ACC GAG ATC TAC ACT CTT TCC CTA CAC GAC GCT CTT CCG ATC TNN NNA CCT ACC TGC CGA G*TC TTC T*

The bolded region in Adaptor2 is the reverse complement of the barcode in Adaptor1. Italicized sequences must be varied if multiple target libraries are present in the same sequencing run, as they must match the target barcode of the library. N indicates a position with a mixture of A, C, G, or T

5′-NGG-3′ PAM). Italicized bases denote a customizable 6 bp target barcode used to identify the library following pooled next-generation sequencing. This target barcode should be varied for each unique target site to be profiled. Bolded bases denote a fixed sequence containing a *Bsp*MI restriction site. T his library of oligonucleotides should be ordered with a 5′ phosphate modification.

2. Oligonucleotides listed in Table 1.
3. 1 mM Tris (pH 8.0).
4. 0.2 mL PCR tubes.
5. CircLigase II ssDNA Ligase (Epicentre).
6. Betaine.
7. PCR thermocycler.
8. 15% TBE-Urea polyacrylamide gel.
9. 1× TBE buffer: 100 mM Tris–HCl (pH 8.0), 100 mM boric acid, 10 mM EDTA.
10. 10,000× SYBR Gold nucleic acid stain (Invitrogen).
11. UV imager.
12. Illustra TempliPhi Amplification Kit (GE).
13. Qubit Fluorometer.
14. Cas9 endonuclease (or next-generation variant).
15. crRNA.

16. tracrRNA.
17. 10× Cas9 reaction buffer: 200 mM HEPES, 1 M NaCl, 50 mM $MgCl_2$, 1 mM EDTA.
18. *Bsp*MI endonuclease.
19. QIAquick PCR Purification Kit (Qiagen).
20. Duplex buffer: 30 mM HEPES (pH 7.5), 100 mM potassium acetate.
21. T4 DNA Ligase.
22. 2× Phusion High-Fidelity DNA Polymerase Master Mix (NEB).
23. Agarose.
24. 50× TAE buffer: 2 M Tris (pH 7.6), 1 M glacial acetic acid, 50 mM EDTA (pH 8.0).
25. 10 mg/mL ethidium bromide solution.
26. 1 kb DNA ladder.
27. MinElute PCR Purification Kit (Qiagen).
28. UV imaging table.
29. Clean razor/scalpel.
30. MinElute Gel Extraction Kit (Qiagen).

3 Methods

3.1 *In Vitro* Cas9 DNA Cleavage Assay

This protocol was adapted from Jinek et al. [30] and Cromwell et al. [26]. An outline of the assay is shown in Fig. 1.

1. Resuspend lyophilized forward and reverse target site oligonucleotides to a final concentration of 100 μM in ultrapure water.
2. To generate the target site DNA duplex, mix 1.5 μL of the forward oligo and 1.5 μL of the reverse oligo in a 0.2 mL PCR tube, adjusting the final volume to 50 μL with duplex buffer. Using a thermocycler, heat the mixture to 95 °C for 5 min and then cool slowly to 25 °C over a 1 h period. If not proceeding immediately, the resulting double-stranded DNA may be stored at −20 °C.
3. In separate 25 μL reactions, digest 1 μL of the oligonucleotide duplex from **step 2** and 500 ng of pUC19 plasmid with 1 μL of *Xba*I (20 U) and 1 μL of *Hind*III (20 U) in 1× NEBuffer 2.1. Allow reactions to proceed for 4 hrs at 37 °C. Subsequently, add 1 μL of CIP (10 U) to the pUC19 reaction, and incubate at room temperature for an additional 30 min.

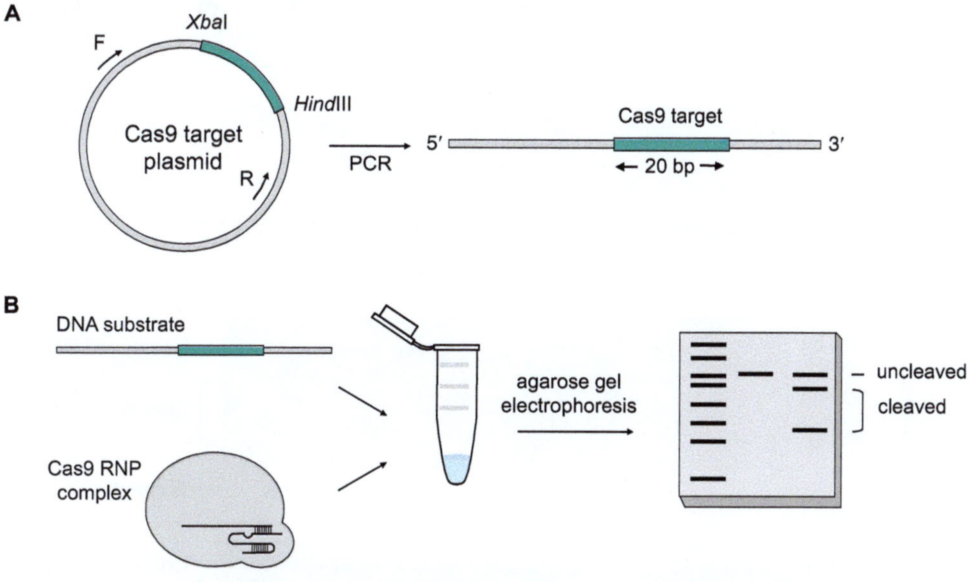

Fig. 1 Schematic of the Cas9 in vitro DNA cleavage assay. (**a**) The DNA substrate is prepared via PCR amplification of a pUC19-based plasmid containing the target site using forward (F) and reverse (R) primers. (**b**) The DNA substrate is digested by Cas9 complexed with a guide RNA. The cleavage products are purified and resolved using agarose gel electrophoresis

4. Purify the digested oligonucleotide duplex and pUC19 DNA pieces using the MinElute PCR Purification Kit. Elute each piece of DNA in 10 μL of ultrapure water. The products may be stored at −20 °C if not proceeding directly to ligation.
5. To ligate the oligonucleotide duplex into the pUC19 plasmid, mix 1 μL of linearized pUC19 (50 ng total) with 0.25 μL of digested oligonucleotide duplex (7 ng total) and adjust the volume to 9 μL with ultrapure water. Add 10 μL of 2× Quick Ligation Buffer and pipette up and down to mix. Add 1 μL of Quick T4 DNA Ligase and mix. Incubate the reaction for 5 min at 25 °C. If continuing directly to transformation, chill the reaction on ice. Alternatively, the reaction may be stored at −20 °C. A parallel reaction should be performed without insert DNA as a control.
6. Thaw a 50 μL aliquot of chemically competent DH5α bacteria on ice for 20 min. Add 1 μL of the crude ligation mixture to the bacteria and mix by stirring with a pipette tip. Incubate the bacteria on ice for an additional 20 min. Next, transfer the bacteria to a 42 °C incubator for 30 s, and then return to ice for another 5 min. Add 200 μL of prewarmed liquid LB media without antibiotics and incubate for 1 h at 37 °C. Streak 100 μL of the bacteria on an LB agar plate containing 50 μg/mL carbenicillin, invert the plate, and incubate at 37 °C overnight.

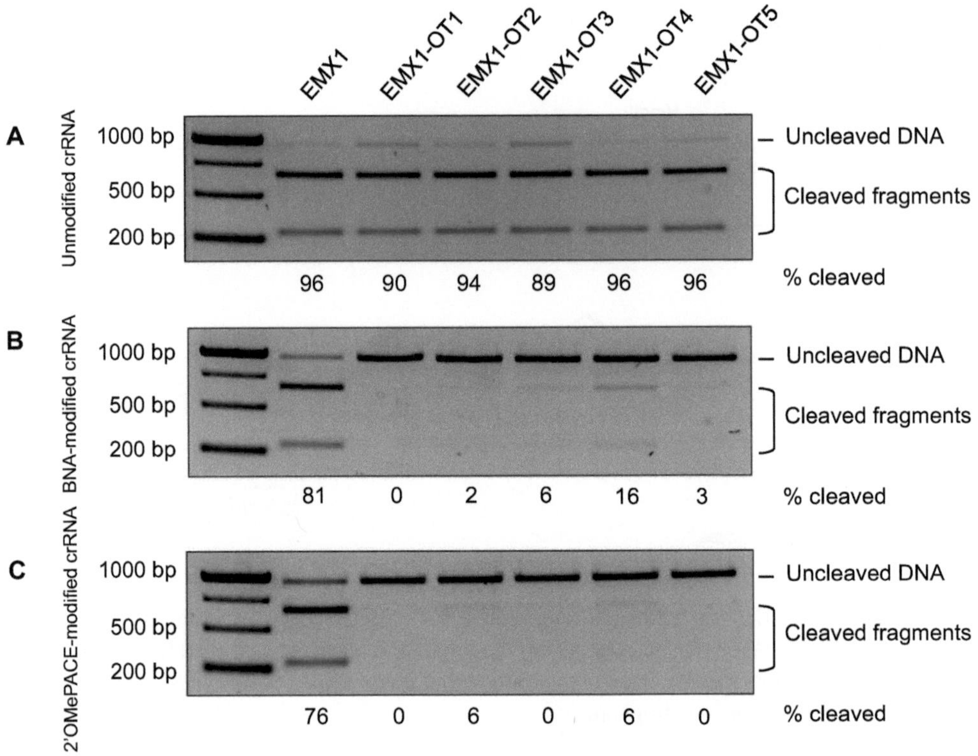

Fig. 2 Example in vitro DNA cleavage assay results. Image of an agarose gel displaying the results of an in vitro Cas9 DNA cleavage assay using either an (**a**) unmodified (5′-GAGTCCGAGCAGAAGAAGAA-3′), (**b**) BNA-modified (5′-GAGTCCGAGCAGAAGAAGAA-3′), or (**c**) 2′OMe-PACE-modified (5′-GAGTCCGAGCAGAAGAAG AA-3′) EMX1 targeting gRNA using the on-target and previously identified off-target (OT) DNA sequences [26]. The location of the gRNA modifications is underlined. The top band in the figure is the full-length DNA substrate, while the bottom two bands are cleavage products. Quantification (included underneath the image) was performed using densitometry

7. Inspect the agar plates the following morning and ensure that colonies did not grow on the control plate. Plates may be stored at 4 °C in a fridge short-term, or used immediately to pick colonies.
8. Using sterile pipette tips, pick 3–5 bacterial colonies and transfer each tip to a 15-mL culture tube containing 3–5 mL of liquid LB supplemented with 50 μg/mL carbenicillin. Grow the cultures for 12–16 h at 37 °C in a bacterial shaker (200 rpm).
9. Isolate and purify plasmids using a QIAprep Spin Miniprep Kit according to the manufacturer's instructions. Purified DNA may be quantified using a NanoDrop.
10. Using the pUC19_fwd primer, perform Sanger sequencing on all purified plasmids and select a sequence-validated clone for subsequent procedures.

11. Prepare the double-stranded DNA (dsDNA) substrate by amplifying the appropriate region from the pUC19 target construct by PCR. To begin, mix 10 μL of 5× Q5 Reaction Buffer, 1 μL of dNTP mix (10 mM), 2.5 μL of pUC19_fwd primer (10 μM), 2.5 μL of pUC19_rev primer (10 μM), and 10 ng of pUC19 target construct with ultrapure water to final volume of 49.5 μL on ice. Then, add 0.5 μL of Q5 High-Fidelity DNA Polymerase and mix by pipetting.
12. Move the reaction to a thermocycler and run the program below:
    (a) 98 °C for 30 s.
    (b) 98 °C for 10 s.
    (c) 68 °C for 20 s.
    (d) 72 °C for 10 s.
    (e) Repeat from step b for 34 additional cycles.
    (f) 72 °C for 2 min.
    (g) Hold at 4 °C.
13. Purify the PCR product using the QIAquick PCR Purification Kit according to the manufacturer's instructions, eluting in 30 μL of ultrapure water. Quantify the DNA using a NanoDrop or equivalent.
14. Prepare the gRNA by mixing 1.25 μL of crRNA (100 μM) and 1.25 μL of tracrRNA (100 μM) in 22.5 μL of duplex buffer (final concentration 5 μM). Place the tube in a thermocycler, heat for 5 min at 95 °C and then cool to 25 °C over the course of 1 h. The annealed gRNA may be stored at −20 °C if necessary.
15. Assemble Cas9 ribonucleoprotein (RNP) complexes by mixing 20 μL of water, 3 μL of 10× Cas9 reaction buffer, 3 μL of gRNA (1 μM) and 1.5 μL of Cas9 (1 μM) (final Cas9 RNP concentration will be 50 nM). Incubate the mixture for 10 min at 25 °C (*see* **Note 1**).
16. Start the reaction by adding 2.5 μL of substrate DNA (38 ng/μL) (final concentration of DNA substrate is 5 nM) to the precomplexed Cas9 RNPs. Incubate for 1 h at 37 °C (*see* **Note 2**).
17. Purify the Cas9-digested DNA using the MinElute PCR Purification Kit according to the manufacturer's instructions, and elute in 10 μL of ultrapure water. Mix the entire volume of purified Cas9-digested DNA with 10 μL of 2× loading buffer, and load samples into the wells of a 1% TAE agarose gel containing 0.5 μg/mL ethidium bromide. Run the gel for 1 h at 100 V.

18. Visualize digestion products using a UV imager (e.g., Amersham Imager 600).
19. Quantify the band intensities using densitometry. The fraction cleaved may be calculated using the formula: % cleaved = 100 × [intensity of (cut 1 + cut 2)/intensity of (cut 1 + cut 2 + uncut)]. Example results for this assay are displayed in Fig. 2.

3.2 In Vitro High-Throughput Specificity Profiling Assay

This protocol was adapted from Pattanayak et al. [31] and Cromwell et al. [26]. An outline of the assay is shown in Fig. 3.

1. Order the partially randomized oligonucleotide library described in Subheading 2.2, as well as the required oligonucleotides listed in Table 1. Prior to performing the selection experiments, determine the number of samples that will be included in the sequencing run, as this will determine the number of barcoded oligonucleotides to order. Two types of barcodes are used for sample identification following pooled high-throughput sequencing: a target barcode and a reaction barcode. The target barcode is used to identify samples containing different starting libraries and is part of the partially randomized library oligo. The reaction barcode is used to identify unique reaction conditions (e.g., with different Cas9 variants or modified guide RNAs) against the same target library and is added during adaptor ligation. Each sample to be sequenced must contain a unique combination of the two barcodes. *See* **Note 3** for a list of previously utilized barcode sequences.
2. Upon arrival, resuspend the oligonucleotides in ultrapure water to a stock concentration of 100 μM. Subsequently, dilute the partially randomized library oligo to a working concentration of 10 μM in 1 mM Tris, pH 8.0.
3. Set up the circularization reaction on ice. Combine 1 μL of library oligonucleotide (10 pmol total), 2 μL of CircLigase II 10× Reaction Buffer, 1 μL of 50 mM $MnCl_2$, 15 μL of ultrapure water, and 1 μL of CircLigase II ssDNA Ligase (100 U) in a 0.2 mL PCR tube. For difficult ligation substrates, betaine may be included in the reaction mixture at a final concentration of 1 M. Incubate the reaction for 16 h at 60 °C, and then heat inactivate for 10 min at 85 °C. A positive control reaction should be performed as described in the CircLigase II ssDNA Ligase kit.
4. To confirm library circularization, load 2.5 pmol of uncircularized library oligo and 1.25 pmol (2.5 μL) of the crude CircLigase reaction in separate wells on a 15% TBE-Urea polyacrylamide gel and run for 1.5 h at 100 V. Stain the gel

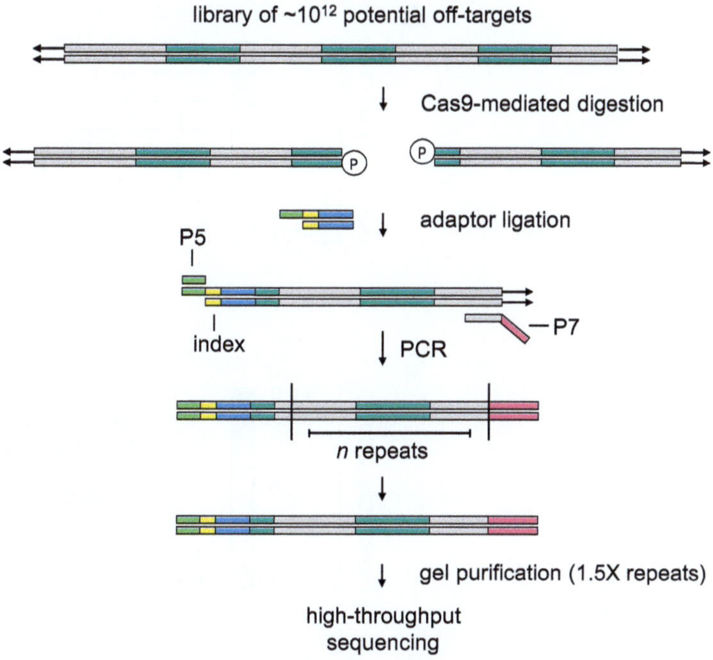

Fig. 3 Diagram of the in vitro high-throughput specificity profiling assay. Concatemeric preselection DNA libraries containing ~10^{12} potential off-target sites derived from the on-target sequence (green) [26] are digested with Cas9-gRNA complex. Digested library members contain a free 5′ phosphate group (white circles) rendering them substrates for adaptor ligation. The adaptor-ligated postselection library members are amplified by PCR (these contain multiple target site sequences due to concatemerization). Postselection library members containing 1.5 target-site repeats are isolated by gel purification and subjected to high-throughput sequencing

using 50 mL of 0.5× TBE containing 1× SYBR Gold for 1 hr at room temperature. Rinse briefly with water before imaging on a UV imager. The circularized oligo will run more slowly than the linear control.

5. Amplify the circularized library using the Illustra TempliPhi Amplification Kit (*see* **Note 4**). Briefly, combine 0.5 μL of the crude CircLigase reaction with 5 μL of TempliPhi sample buffer in a 0.2 mL PCR tube on ice. Incubate for 3 min at 95 °C, then cool to 4 °C at 0.5 °C/s in a thermocycler. Add 5 μL of TempliPhi reaction buffer and 0.2 μL of TempliPhi enzyme mix to the cooled sample. Incubate the reaction for 16 h at 30 °C and subsequently heat inactivate for 10 min at 65 °C. If more product is required, these quantities can be scaled up as necessary.
6. Since the amplified library is digested without purification, it should be quantified using a Qubit fluorometer, or an alternative double-stranded DNA sensitive assay. Dilute the

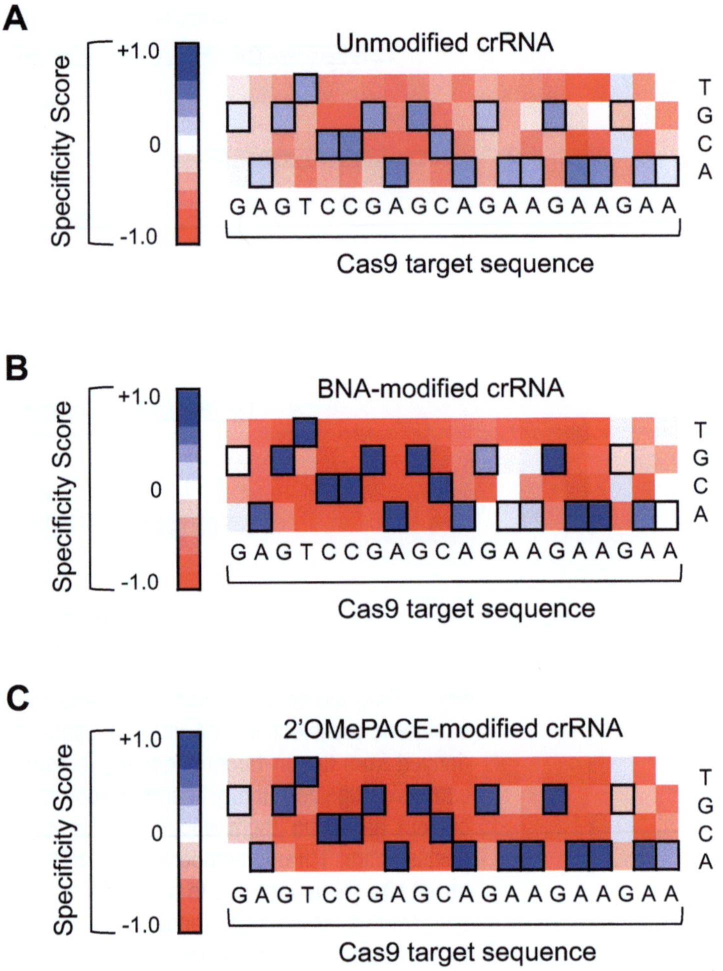

Fig. 4 Example heat maps displaying the results of an in vitro high-throughput specificity profiling assay using two modified crRNAs. Heat maps illustrating DNA cleavage specificity scores for an (**a**) unmodified, (**b**) BNA-modified or (c) 2′OMePACE-modified EMX1-targeting gRNA against ~10^{12} potential off-target sequences. The sequences of the modified gRNAs are indicated in the legend for Fig. 2. A specificity score of 1.0 (indicated by dark blue) represents a 100% enrichment for a particular base at a particular position, while a specificity score of −1.0 (indicated by dark red) indicates a 100% enrichment against a particular base at a particular position. Black boxes indicate the intended target nucleotide. 200 nM of DNA substrate and 1000 nM of Cas9: gRNA (EMX1) complex were used in this experiment

preselection library to a working concentration of 2 μM in ultrapure water.

7. Prior to digestion of the preselection library, assemble Cas9 RNP complexes. Mix 1000 nM of Cas9 and 1000 nM of gRNA with 5 μL of 10× Cas9 reaction buffer and adjust to a total

volume of 45 μL with ultrapure water. Incubate for 10 min at 25 °C to allow RNP complexes to form.

8. Start the digestion reaction by adding 5 μL of diluted preselection library (final concentration 200 nM) (*see* **Note 5**). Incubate the reaction for 1 h at 37 °C.
9. In a separate reaction, digest 200 nM of preselection library with 2 U of *Bsp*MI endonuclease in a total volume of 50 μL 1× NEBuffer 3.1 for 1 h at 37 °C. Since all library members contain a *Bsp*MI restriction site, this nonspecific digestion captures the composition of the preselection library.
10. Purify both the Cas9-digested and restriction-digested libraries using the QIAquick PCR Purification Kit according to the manufacturer's instructions, and elute in 30 μL of ultrapure water.
11. Prepare the barcoded adaptors for ligation by mixing 5 μL of Adaptor1 (100 μM) (Table 1) and 5 μL of Adaptor2 (100 μM) for Cas9-digested samples (*see* **Note 6**), or 5 μL of Preselection Adaptor1 (100 μM) and 5 μL of Preselection Adaptor2 (100 μM) for restriction-digested samples (*see* **Note 7**) with 40 μL of duplex buffer. Place the tubes in a thermocycler, heat for 5 min at 95 °C, and then cool to 25 °C over the course of 1 h.
12. Ligate the double-stranded barcoded adaptor sequences to postselection library members by mixing 1 μL of annealed adaptor from **step 11**, 5 μL of 10× T4 DNA Ligase Buffer, 2.5 μL of T4 DNA Ligase, 11.5 μL of ultrapure water, and 30 μL of postselection library DNA from **step 10**. Incubate the ligation reaction overnight at 25 °C. Repeat this protocol for the restriction-digested preselection library, replacing the annealed Adaptors 1 and 2 with the annealed Preselection Adaptors 1 and 2.
13. Purify the adaptor-ligated libraries using the QIAquick PCR Purification Kit and elute in 50 μL of ultrapure water.
14. Perform a small-scale test PCR of the postselection library to identify the amplification saturation point by mixing 19 μL of ultrapure water, 2.5 μL of Postselection Fwd primer (10 μM), 2.5 μL of EMX1 Postselection Rev. primer (10 μM), 1 μL of adaptor-ligated library from **step 13** and 25 μL of 2× Phusion High-Fidelity PCR Master Mix in a 0.2 mL PCR tube. For the restriction-digested library, prepare the reaction as above but replace the primers with the Preselection Fwd primer and EMX1 Preselection Rev. primers. Move the reaction to a thermocycler and run the following program, removing an aliquot every 5 cycles:
    (a) 98 °C for 30 s.
    (b) 98 °C for 10 s.

(c) 60 °C for 30 s.

(d) 72 °C for 60 s.

(e) Repeat from step b for 34 additional cycles.

(f) 72 °C for 5 min.

(g) Hold at 4 °C.

15. Following completion of the PCR reactions, visualize the products on a 1.2% TAE agarose gel to determine the number of cycles required to achieve saturation. Subsequently, repeat the amplification using the full remaining volume of each postselection library, stopping at saturation. The number of cycles to run should be calculated as follows: Cycle number at saturation in test reaction − n, where 2^n equals the volume of template being added. For example, if the reaction is being scaled from 1 to 32 μL of template, 5 cycles should be subtracted from the saturation point determined in **step 14** (since $2^5 = 32$).

16. Purify the PCR amplified pre- and postselection libraries using the MinElute PCR Purification Kit and elute in 10 μL of ultrapure water.

17. Load the entire PCR product onto a 1.5% TAE agarose gel and run for 1.5 h at 100 V or until the dye has migrated roughly three-quarters down the gel. Visualize the gel on a UV imaging table. Due to the concatemeric nature of the library, PCR amplification results in a ladder of products corresponding to differing numbers of target site repeats. To normalize the size of the library for high-throughput sequencing, excise the band that corresponds to 1.5 repeats (which should be approximately 170 bp) with a clean razor blade. Extract the DNA fragment using the MinElute Gel Extraction Kit.

18. Quantify the gel-purified product using a Qubit fluorometer and dilute each sample to a concentration of 4 nM. Pool 5 μL of each diluted sample in a single microcentrifuge tube for high-throughput sequencing.

19. Load the sample onto an Illumina MiSeq according to standard operating guidelines for single-read sequencing (*see* **Note 8** for additional sequencing considerations).

20. Sequencing data may be analyzed using previously described, publicly available scripts [26]. An example sequencing read is outlined below:

*ACTGTgaangg*ACTT*AGAAGAAGAAGA*CTCGGCAGG-T**ACTTGAGTCCGAGCAGAAGAAGAAGGGGTC**-**G***AGAAGAAGAAGA*CTCGGCAGGT, where.

ACTGT represents the 5-bp reaction barcode for de-multiplexing multiple reaction conditions, *gaangg* corresponds to the half-site of the target which was cleaved by

Cas9 during digestion, and *AGAAGA AGAAGA* is the reverse complement of the target barcode. The target barcode is repeated twice due to the concatemeric nature of the library. CTCGGCAGGT is the constant sequence, the reverse complement of the bolded sequence in **step 1**. **ACTTGAGTCCGAGCAGAAGAAGAAGGGGTCG** is the sequence of the full, postselection library member, where the first and last four nucleotides are fully randomized. After binning, specificity scores may be calculated using the formulas: Positive specificity score = (frequency of base pair at position[postselection] − frequency of base pair at position[preselection])/(1 − frequency of base pair at position[preselection]) and Negative specificity score = (frequency of base pair at position[postselection] − frequency of base pair at position[preselection])/(frequency of base pair at position[preselection]). These specificity scores may subsequently be used to generate representative heat map diagrams [26] like those shown in Fig. 4.

4 Notes

1. A molar Cas9–gRNA ratio of 1:2 is used to ensure that all Cas9 molecules are bound to a gRNA.
2. These reaction conditions result in a final concentration of 50 nM of Cas9 RNP and 5 nM of DNA. A ten-fold molar excess of Cas9 RNP over DNA is typically enough to result in complete digestion of the target DNA. However, when studying the effects of gRNA modifications on Cas9 activity and/or specificity, it is important to perform reactions at subsaturating conditions. These can be determined by performing titrations of Cas9 RNP.
3. TCT TCT and AAC ACA have previously been used as target barcodes, while CTG AA, TGA CT, TGC AA, GCA TT, CAT GA, ATG CT, CTA GT, GCT AA, CAG TT, GTC AT, and ACG TA have been used as reaction barcodes [26]. In addition to these, new barcodes may be generated so long as they differ by at least two base pairs, to facilitate accurate and reliable sample identification.
4. The Φ29 DNA polymerase supplied in the enzyme mix is active at temperatures above 4 °C. Consequently, it must be stored at −70 °C and thawed only prior to use.
5. The concentration of Cas9 endonuclease and gRNA used to digest the preselection library may be varied, although concentrations should be selected to ensure sufficient cleavage for

robust detection but not complete digestion of the library. Use of excessive concentrations will result in lower apparent nuclease specificity. An in vitro cleavage assay may be performed to measure the amount of on-target digestion resulting from specific Cas9 and gRNA concentration parameters.

6. In addition to providing the necessary components for sequencing on an Illumina platform, adaptor ligation also adds unique indices for barcoding multiple reactions for pooled sequencing. Adaptor1 and Adaptor2 (listed in Table 1) are example adaptors sequences (barcodes are indicated in bold). The bolded sequence may be substituted for any sequence of choice, with Adaptor2 containing the reverse complement.
7. The first four bases of the preselection adaptor2 (italicized) must match the first four bases of the library oligonucleotide barcode used, as *Bsp*MI digestion will leave an overhang specific to the barcode sequence.
8. If the final library has high sequence uniformity, the sample should be supplemented with 25% PhiX Control Library to increase diversity and improve the quality of the run.

Acknowledgments

This work was supported by NSERC Discovery/Discovery Accelerator Supplement (RGPIN-2016-0638) and CIHR (PS-408552) grants to B.P.H. We thank Dr. Juan Jovel, Sudip Subedi, and the members of the Applied Genomics Core at the University of Alberta for their technical assistance with next-generation sequencing. We thank Bio-Synthesis Inc. (Texas) for providing the 2′OMePACE-modified guide RNA.

References

1. Horvath P, Barrangou R (2010) CRISPR/Cas, the immune system of bacteria and archaea. Science 327(5962):167–170. https://doi.org/10.1126/science.1179555
2. Makarova KS, Haft DH, Barrangou R, Brouns SJ, Charpentier E, Horvath P, Moineau S, Mojica FJ, Wolf YI, Yakunin AF, van der Oost J, Koonin EV (2011) Evolution and classification of the CRISPR-Cas systems. Nat Rev Microbiol 9(6):467–477. https://doi.org/10.1038/nrmicro2577
3. Jinek M, Jiang F, Taylor DW, Sternberg SH, Kaya E, Ma E, Anders C, Hauer M, Zhou K, Lin S, Kaplan M, Iavarone AT, Charpentier E, Nogales E, Doudna JA (2014) Structures of Cas9 endonucleases reveal RNA-mediated conformational activation. Science 343 (6176):1247997. https://doi.org/10.1126/science.1247997
4. Nakade S, Yamamoto T, Sakuma T (2017) Cas9, Cpf1 and C2c1/2/3—what's next? Bioengineered 8(3):265–273. https://doi.org/10.1080/21655979.2017.1282018
5. Barrangou R, Doudna JA (2016) Applications of CRISPR technologies in research and beyond. Nat Biotechnol 34(9):933–941. https://doi.org/10.1038/nbt.3659
6. Jiang F, Doudna JA (2017) CRISPR-Cas9 structures and mechanisms. Annu Rev Biophys 46:505–529. https://doi.org/10.1146/annurev-biophys-062215-010822
7. Huai C, Jia C, Sun R, Xu P, Min T, Wang Q, Zheng C, Chen H, Lu D (2017) CRISPR/

Cas9-mediated somatic and germline gene correction to restore hemostasis in hemophilia B mice. Hum Genet. https://doi.org/10.1007/s00439-017-1801-z
8. Cong L, Ran FA, Cox D, Lin S, Barretto R, Habib N, Hsu PD, Wu X, Jiang W, Marraffini LA, Zhang F (2013) Multiplex genome engineering using CRISPR/Cas systems. Science 339(6121):819–823. https://doi.org/10.1126/science.1231143
9. Jiang F, Zhou K, Ma L, Gressel S, Doudna JA (2015) STRUCTURAL BIOLOGY. A Cas9-guide RNA complex preorganized for target DNA recognition. Science 348 (6242):1477–1481. https://doi.org/10.1126/science.aab1452
10. Ablain J, Zon LI (2016) Tissue-specific gene targeting using CRISPR/Cas9. Methods Cell Biol 135:189–202. https://doi.org/10.1016/bs.mcb.2016.03.004
11. Cheng R, Peng J, Yan Y, Cao P, Wang J, Qiu C, Tang L, Liu D, Tang L, Jin J, Huang X, He F, Zhang P (2014) Efficient gene editing in adult mouse livers via adenoviral delivery of CRISPR/Cas9. FEBS Lett 588 (21):3954–3958. https://doi.org/10.1016/j.febslet.2014.09.008
12. Hubbard BP, Badran AH, Zuris JA, Guilinger JP, Davis KM, Chen L, Tsai SQ, Sander JD, Joung JK, Liu DR (2015) Continuous directed evolution of DNA-binding proteins to improve TALEN specificity. Nat Methods 12 (10):939–942. https://doi.org/10.1038/nmeth.3515
13. Wei C, Liu J, Yu Z, Zhang B, Gao G, Jiao R (2013) TALEN or Cas9 – rapid, efficient and specific choices for genome modifications. J Genet Genomics 40(6):281–289. https://doi.org/10.1016/j.jgg.2013.03.013
14. Kleinstiver BP, Tsai SQ, Prew MS, Nguyen NT, Welch MM, Lopez JM, McCaw ZR, Aryee MJ, Joung JK (2016) Genome-wide specificities of CRISPR-Cas Cpf1 nucleases in human cells. Nat Biotechnol 34(8):869–874. https://doi.org/10.1038/nbt.3620
15. Tycko J, Myer VE, Hsu PD (2016) Methods for optimizing CRISPR-Cas9 genome editing specificity. Mol Cell 63(3):355–370. https://doi.org/10.1016/j.molcel.2016.07.004
16. Xu X, Duan D, Chen SJ (2017) CRISPR-Cas9 cleavage efficiency correlates strongly with target-sgRNA folding stability: from physical mechanism to off-target assessment. Sci Rep 7 (1):143. https://doi.org/10.1038/s41598-017-00180-1
17. O'Geen H, Yu AS, Segal DJ (2015) How specific is CRISPR/Cas9 really? Curr Opin Chem Biol 29:72–78. https://doi.org/10.1016/j.cbpa.2015.10.001
18. Slaymaker IM, Gao L, Zetsche B, Scott DA, Yan WX, Zhang F (2016) Rationally engineered Cas9 nucleases with improved specificity. Science 351(6268):84–88. https://doi.org/10.1126/science.aad5227
19. Kleinstiver BP, Pattanayak V, Prew MS, Tsai SQ, Nguyen NT, Zheng Z, Joung JK (2016) High-fidelity CRISPR-Cas9 nucleases with no detectable genome-wide off-target effects. Nature 529(7587):490–495. https://doi.org/10.1038/nature16526
20. Casini A, Olivieri M, Petris G, Montagna C, Reginato G, Maule G, Lorenzin F, Prandi D, Romanel A, Demichelis F, Inga A, Cereseto A (2018) A highly specific SpCas9 variant is identified by in vivo screening in yeast. Nat Biotechnol. https://doi.org/10.1038/nbt.4066
21. Chen JS, Dagdas YS, Kleinstiver BP, Welch MM, Sousa AA, Harrington LB, Sternberg SH, Joung JK, Yildiz A, Doudna JA (2017) Enhanced proofreading governs CRISPR-Cas9 targeting accuracy. Nature 550 (7676):407–410. https://doi.org/10.1038/nature24268
22. Choi GCG, Zhou P, Yuen CTL, Chan BKC, Xu F, Bao S, Chu HY, Thean D, Tan K, Wong KH, Zheng Z, Wong ASL (2019) Combinatorial mutagenesis en masse optimizes the genome editing activities of SpCas9. Nat Methods. https://doi.org/10.1038/s41592-019-0473-0
23. Fu Y, Sander JD, Reyon D, Cascio VM, Joung JK (2014) Improving CRISPR-Cas nuclease specificity using truncated guide RNAs. Nat Biotechnol 32(3):279–284. https://doi.org/10.1038/nbt.2808
24. Rueda FO, Bista M, Newton MD, Goeppert AU, Cuomo ME, Gordon E, Kroner F, Read JA, Wrigley JD, Rueda D, Taylor BJM (2017) Mapping the sugar dependency for rational generation of a DNA-RNA hybrid-guided Cas9 endonuclease. Nat Commun 8(1):1610. https://doi.org/10.1038/s41467-017-01732-9
25. Yin H, Song CQ, Suresh S, Kwan SY, Wu Q, Walsh S, Ding J, Bogorad RL, Zhu LJ, Wolfe SA, Koteliansky V, Xue W, Langer R, Anderson DG (2018) Partial DNA-guided Cas9 enables genome editing with reduced off-target activity. Nat Chem Biol 14(3):311–316. https://doi.org/10.1038/nchembio.2559
26. Cromwell CR, Sung K, Park J, Krysler AR, Jovel J, Kim SK, Hubbard BP (2018) Incorporation of bridged nucleic acids into CRISPR RNAs improves Cas9 endonuclease specificity.

Nat Commun 9(1):1448. https://doi.org/10.1038/s41467-018-03927-0

27. Ryan DE, Taussig D, Steinfeld I, Phadnis SM, Lunstad BD, Singh M, Vuong X, Okochi KD, McCaffrey R, Olesiak M, Roy S, Yung CW, Curry B, Sampson JR, Bruhn L, Dellinger DJ (2018) Improving CRISPR-Cas specificity with chemical modifications in single-guide RNAs. Nucleic Acids Res 46(2):792–803. https://doi.org/10.1093/nar/gkx1199
28. O'Reilly D, Kartje ZJ, Ageely EA, Malek-Adamian E, Habibian M, Schofield A, Barkau CL, Rohilla KJ, DeRossett LB, Weigle AT, Damha MJ, Gagnon KT (2019) Extensive CRISPR RNA modification reveals chemical compatibility and structure-activity relationships for Cas9 biochemical activity. Nucleic Acids Res 47(2):546–558. https://doi.org/10.1093/nar/gky1214
29. Kocak DD, Josephs EA, Bhandarkar V, Adkar SS, Kwon JB, Gersbach CA (2019) Increasing the specificity of CRISPR systems with engineered RNA secondary structures. Nat Biotechnol 37(6):657–666. https://doi.org/10.1038/s41587-019-0095-1
30. Jinek M, Chylinski K, Fonfara I, Hauer M, Doudna JA, Charpentier E (2012) A programmable dual-RNA-guided DNA endonuclease in adaptive bacterial immunity. Science 337(6096):816–821. https://doi.org/10.1126/science.1225829
31. Pattanayak V, Lin S, Guilinger JP, Ma E, Doudna JA, Liu DR (2013) High-throughput profiling of off-target DNA cleavage reveals RNA-programmed Cas9 nuclease specificity. Nat Biotechnol 31(9):839–843. https://doi.org/10.1038/nbt.2673

Chapter 13

Profiling Genome-Wide Specificity of CRISPR-Cas9 Using Digenome-Seq

Daesik Kim and Jin-Soo Kim

Abstract

Digenome-seq is a highly sensitive method for analyzing the genome-wide specificity of CRISPR-Cas9 nuclease activity. In this procedure, genomic DNA is first subjected to digestion by CRISPR-Cas9 in vitro and then to whole genome sequencing, which results in unusual patterns of straight alignments at on-target and potential off-target sites. Analysis of these data with the Digenome-seq computer program allows for identification of the in vitro cleavage sites associated with the straight alignments. Here, we present a detailed Digenome-seq protocol for genome-wide profiling of in vitro CRISPR-Cas9 nuclease cleavage sites.

Key words CRISPR, Cas9, Genome editing, Off-target, Whole-genome sequencing

1 Introduction

Programmable nucleases, including zinc finger nucleases (ZFNs) [1–4], transcription activator-like effector nucleases (TALENs) [5–7], and CRISPR–Cas nucleases [8–11] enable targeted genome engineering in cultured cells and whole organisms [12]. CRISPR-Cas9, which includes the Cas9 protein and a guide RNA (gRNA), induces targeted insertion or deletion (indel) mutations via targeted DNA double-stranded breaks (DSBs) in the genome. However, CRISPR-Cas9 can also induce off-target mutations at sites that are similar in sequence to on-target sites [13–17]. To profile genome-wide CRISPR-Cas9 off-target effects in an unbiased manner, we developed CRISPR-Cas9 nuclease-digested whole genome sequencing (Digenome-seq) [18, 19]. In this method, genomic DNA (gDNA) isolated from cultured cells is incubated with CRISPR-Cas9 nuclease, which induces cleavage in vitro at the on-target and potential off-target sites. Subsequently, after DNA fragmentation and ligation with adaptors, the digested gDNA is subjected to whole genome sequencing (WGS). After mapping sequence reads to a reference genome, we could observe unusual

Tudor A. Fulga et al. (eds.), *CRISPR Guide RNA Design: Methods and Protocols*, Methods in Molecular Biology, vol. 2162, https://doi.org/10.1007/978-1-0716-0687-2_13,

patterns of straight alignments at the on- and off-target sites using Integrative Genomics Viewer (IGV) [20]. To identify CRISPR-Cas9-mediated in vitro cleavage sites on a genomic scale, we developed an algorithm that assigns high DNA cleavage scores to the genomic regions with straight alignments. In summary, Digenome-seq is a highly sensitive, unbiased in vitro profiling method for capturing nuclease-mediated DNA DSBs, which has been widely used in human cells [18, 19], mouse cells [21], and plants [22]. Here, we present a detailed Digenome-seq protocol.

2 Materials

2.1 sgRNA DNA Templates

1. sgRNA_F (Macrogen): GAA ATT AAT ACG ACT CAC TAT A NNNNNNNNNNNNNNNNNNNN GTTT TAG AGC TAG AAA TAG CAA GTT AAA ATA AGG CTA GTC CG (*see* **Note 1**).
2. sgRNA_R (Macrogen): AAA AAA AGC ACC GAC TCG GTG CCA CTT TTT CAA GTT GAT AAC GGA CTA GCC TTA TTT TAA C.
3. Phusion High-Fidelity DNA Polymerase.
4. 5× Phusion HF Buffer.
5. dNTPs.
6. TOPcloner™ TA-Blunt Kit (Enzynomics).
7. Exprep Plasmid SV (GeneAll).
8. EcoRI-HF.
9. NTPs (ATP, UTP, CTP, and GTP).
10. 100 mM $MgCl_2$.
11. T7 RNA polymerase.
12. 1× RNAPol Reaction Buffer (New England Biolabs): 400 mM Tris–HCl, 60 mM $MgCl_2$, 10 mM DTT, 20 mM spermidine.
13. RNase Inhibitor.
14. 1 M DTT
15. DNase I.
16. 10× DNase I Reaction Buffer: 100 mM Tris–HCl, 25 mM $MgCl_2$, and 5 mM $CaCl_2$.
17. PCR Product Purification Kit (MGmed).

2.2 Genomic DNA Extraction and In Vitro Digestion

1. DNeasy Blood and Tissue Kit (Qiagen).
2. Cas9 protein (Toolgen).
3. RNase A.
4. 10× NEB3.1 (New England Biolabs): 100 mM NaCl, 50 mM Tris–HCl, 10 mM $MgCl_2$, and 100 μg/ml BSA.

2.3 Quantitative Real-Time Polymerase Chain Reaction (qRT-PCR)

KAPA SYBR FAST qPCR Master Mix (2×) Kit (Kapa Biosystems).

2.4 Whole Genome Sequencing

1. Covaris systems M220 Focused-ultrasonicator.
2. TruSeq DNA PCR-Free Library Kit (Illumina).
3. Agilent Technologies 2100 Bioanalyzer (Agilent).
4. DNA 1000 Chip Kit (Agilent).
5. HiSeq X Ten (Illumina).

3 Methods

3.1 Preparation of the sgRNA

1. To obtain the DNA template for the sgRNA, resuspend two complementary oligonucleotides (sgRNA_F and sgRNA_R) at a final concentration of 1 μM in the reaction buffer as follows (Fig. 1):

Component	Volume (μl)	Final concentration
sgRNA_F (100 μM)	0.5	1 μM
sgRNA_R (100 μM)	0.5	1 μM
5× Phusion HF Buffer	10	1×
10 mM dNTPs	2.5	0.5 mM
Phusion High-Fidelity DNA Polymerase (2 U/μL)	0.5	
Nuclease-free water	36	
Total	50	

2. Anneal and extend oligonucleotides using the following program in a thermocycler:

Step	Temperature, °C	Time	Cycles
Denaturation	95	1 min	1
Denaturation Annealing Extension	95 54 72	15 s 15 s 15 s	25
Extension	72	3 min	1
Hold	12		1

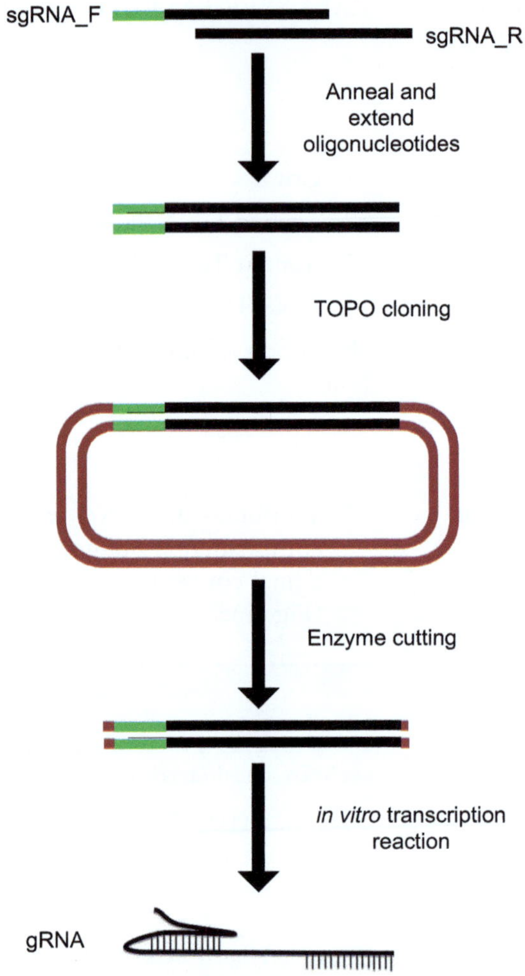

Fig. 1 In vitro transcription of gRNA via T7 polymerase. Two complementary oligonucleotides (sgRNA_F and sgRNA_R), which include the T7 promoter and the gRNA sequence, are annealed and extended by High-Fidelity DNA Polymerase. After TOPO cloning and restriction enzyme-mediated digestion to release the template fragment, the gRNA is transcribed by T7 polymerase from the DNA template, which contains the T7 promoter and the gRNA sequence

3. Purify the DNA template with an MG PCR Product Purification Kit, following the manufacturer's instructions.
4. To minimize oligonucleotide-mediated errors, clone the sgRNA DNA template in the pTOP TA-Blunt V2 vector (used for cloning blunt ended DNA templates) (Fig. 1) as follows (*see* **Notes 2** and **3**):

Component	Volume (μl)
6× TOPcloner™ buffer	1
pTOP TA-Blunt V2 (10 ng/μl)	0.5
DNA template (5 ng/μl)	1
D.W.	3.5
Total	6

5. Incubate at room temperature for 5 min and transform the ligated DNA into competent cells.
6. Isolate plasmid DNA from transformed colonies using a plasmid DNA mini-prep kit and confirm that the desired sequence has been inserted by Sanger sequencing.
7. Linearize plasmids containing the desired insert (the DNA template for the sgRNA) by restriction enzyme digestion as follows (Fig. 1):

Component	Volume (μl)
Plasmid DNA	10 μg (*X* μl)
EcoRI-HF	2
Cut smart	10
D.W.	(88-X)
Total	100

8. Incubate at 37 °C for 4 h and clean up the reaction with an MG PCR Product Purification Kit.
9. Mix the sgRNA template DNA with T7 RNA polymerase to generate the sgRNA as follows (Fig. 1) (*see* **Note 4**):

Component	Volume (μl)	Final concentration
sgRNA template DNA	1250 ng (*X* μl)	
ATP (100 mM)	2	4 mM
UTP (100 mM)	2	4 mM
CTP (100 mM)	2	4 mM
GTP (100 mM)	2	4 mM
100 mM $MgCl_2$	7	14 mM
T7 RNA polymerase	7.5	
RNase inhibitor	1.25	

(continued)

Component	Volume (μl)	Final concentration
1 M DTT	0.5	10 mM
10× RNAPol Reaction Buffer	5	1×
Nuclease-free water	(20.75-X)	
Total	50	

10. Incubate the in vitro transcription reaction at 37 °C for 4 h. Clean up the in vitro-transcribed sgRNA using an MG PCR Product Purification Kit, following the manufacturer's instructions.
11. To degrade the DNA template in the in vitro transcription reaction, incubate the in vitro-transcribed sgRNA with DNase I (final concentration: 2 U/50 μl) in 1× DNase I Reaction Buffer at 37 °C for 30 min. Clean up the mixture again with the MG PCR Product Purification Kit.

3.2 In Vitro Digestion of Genomic DNA

1. Purify genomic DNA from cells using a DNeasy Blood and Tissue Kit following the manufacturer's instructions (*see* **Note 5**).
2. To digest genomic DNA, set up the following reaction and incubate it at 37 °C for 8 h (Fig. 2) (*see* **Notes 6** and 7).

Component	Volume (μl)	Final concentration
Genomic DNA	10 μg (*X* μl)	
sgRNA	*Y*	300 nM
Cas9 protein	*Z*	100 nM
10× NEB3.1	50	
Nuclease-free water	(450-*X*-*Y*-*Z*)	
Total	500	

3. To remove the sgRNA, add RNase A (final concentration: 50 μg/ml) and incubate the mixture at 37 °C for 30 min. Purify the in vitro-digested DNA with a DNeasy Blood and Tissue Kit according to the manufacturer's instructions.
4. Measure the ratio of target site digestion compare to untreated gDNA with the comparative C_T method using qRT-PCR [23].

3.3 Cas9 Nuclease-Digested Whole-Genome (Digenome) Sequencing (Digenome-Seq)

1. To obtain a library of DNA fragments with an average size of 350 bp, fragment 1 μg of genomic DNA with a Covaris systems ultrasonicator (M220) using the following parameters:

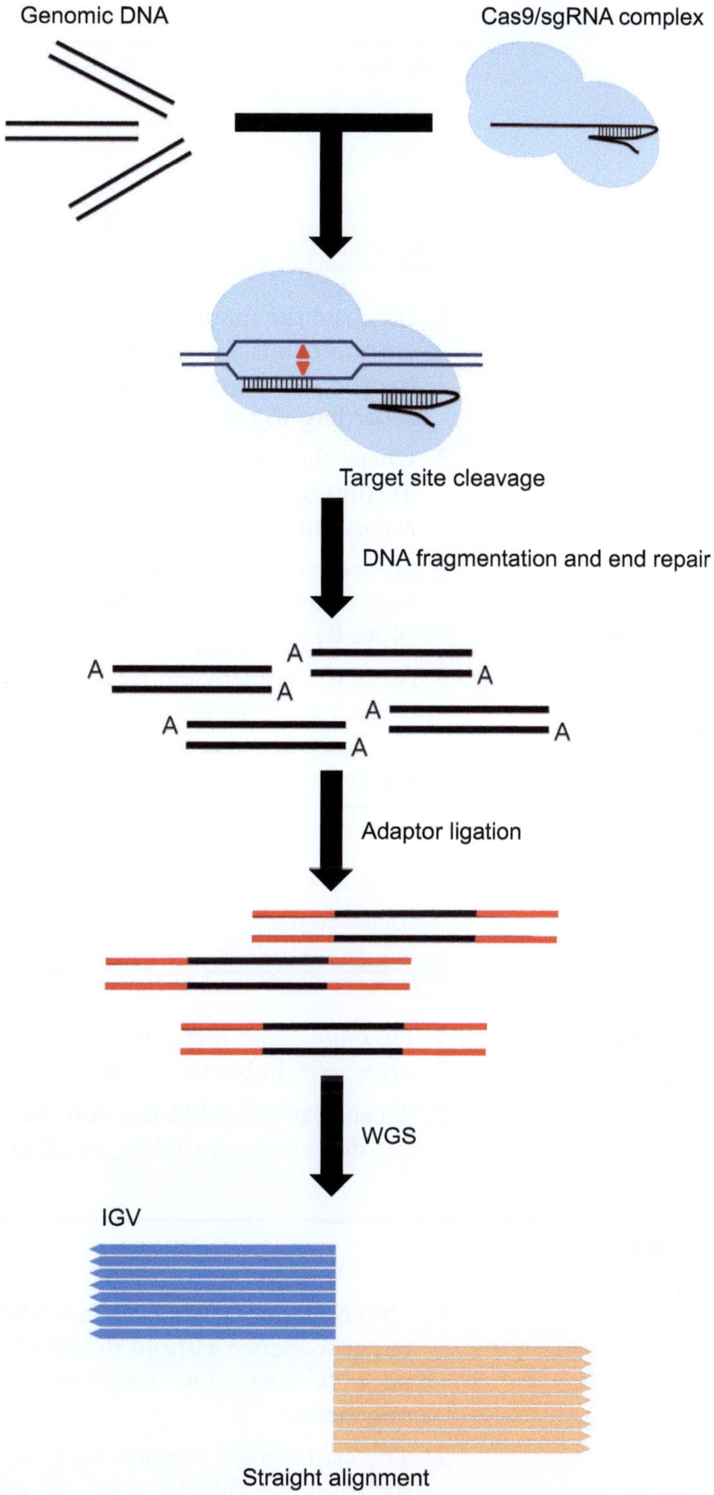

Fig. 2 Digenome-seq workflow for the identification of CRISPR-Cas9 nuclease off-target sites

Duty factor (%)	20
Peak/displayed power (W)	50
Cycles/burst	200
Duration (s)	65
Temperature (°C)	20

2. Incubate the fragmented DNA with End Repair Mix to generate blunt ends. Ligate the blunt-ended fragments with adaptors to produce libraries using a TruSeq DNA PCR-Free Library Kit according to the manufacturer's instructions (Fig. 2).
3. Check the library quality and size distribution with an Agilent Technologies 2100 Bioanalyzer using a DNA 1000 chip according to the manufacturer's instructions.
4. Subject libraries to WGS using an Illumina HiSeq X Ten Sequencer, following the manufacturer's instructions (*see* **Note 8**).
5. Align the sequencing data to the reference genome using Isaac Aligner with the following parameters:

Parameter	Value
--base-quality-cutoff	15
--keep-duplicates	1
--default-adapters	AGATCGGAAGAGC* *GCTCTTCCGATCT

3.4 Digenome-Seq Analysis

1. Download and install the Digenome-seq program, available at https://github.com/chizksh/digenome-toolkit2 (*see* **Note 9**).
2. To analyze the BAM file, run the script below:
   digenome-run BAM_file_PATH.

4 Notes

1. "NNNNNNNNNNNNNNNNNNNN" represents the Cas9 target sequence without the PAM sequence. The first N should be a "G" to allow for expression of the sgRNA with T7 polymerase.
2. The synthesized oligonucleotides can potentially carry errors such as nucleotide deletions or substitutions. To minimize such errors, clone the sgRNA DNA template and use it after restriction enzyme-mediated plasmid digestion to release the insert DNA.

3. Phusion DNA polymerase produces blunt-ended products. Therefore, the TOPO Cloning Kit for TA cloning should not be used for this experiment.
4. Additional DTT enhances in vitro transcription efficiency.
5. To ensure accurate quantitation of gDNA, RNA-free gDNA should be extracted.
6. Low-quality Cas9 protein induces genomic DNA degradation.
7. Digenome-seq is also available for profiling the off-target effects of base editors [24].
8. The WGS depth should be at least 30× for Digenome-seq analysis.
9. Alternatively, the Digenome-seq web tool is available at http://www.rgenome.net/digenome-js [25].

References

1. Bibikova M, Beumer K, Trautman JK, Carroll D (2003) Enhancing gene targeting with designed zinc finger nucleases. Science 300 (5620):764. https://doi.org/10.1126/science.1079512
2. Kim HJ, Lee HJ, Kim H, Cho SW, Kim JS (2009) Targeted genome editing in human cells with zinc finger nucleases constructed via modular assembly. Genome Res 19 (7):1279–1288. https://doi.org/10.1101/gr.089417.108
3. Porteus MH, Baltimore D (2003) Chimeric nucleases stimulate gene targeting in human cells. Science 300(5620):763. https://doi.org/10.1126/science.1078395
4. Urnov FD, Miller JC, Lee YL, Beausejour CM, Rock JM, Augustus S, Jamieson AC, Porteus MH, Gregory PD, Holmes MC (2005) Highly efficient endogenous human gene correction using designed zinc-finger nucleases. Nature 435(7042):646–651. https://doi.org/10.1038/nature03556
5. Kim Y, Kweon J, Kim A, Chon JK, Yoo JY, Kim HJ, Kim S, Lee C, Jeong E, Chung E, Kim D, Lee MS, Go EM, Song HJ, Kim H, Cho N, Bang D, Kim S, Kim JS (2013) A library of TAL effector nucleases spanning the human genome. Nat Biotechnol 31(3):251–258. https://doi.org/10.1038/nbt.2517
6. Kim YK, Wee G, Park J, Kim J, Baek D, Kim JS, Kim VN (2013) TALEN-based knockout library for human microRNAs. Nat Struct Mol Biol 20(12):1458–1464. https://doi.org/10.1038/nsmb.2701
7. Miller JC, Tan S, Qiao G, Barlow KA, Wang J, Xia DF, Meng X, Paschon DE, Leung E, Hinkley SJ, Dulay GP, Hua KL, Ankoudinova I, Cost GJ, Urnov FD, Zhang HS, Holmes MC, Zhang L, Gregory PD, Rebar EJ (2011) A TALE nuclease architecture for efficient genome editing. Nat Biotechnol 29(2):143–148. https://doi.org/10.1038/nbt.1755
8. Cho SW, Kim S, Kim JM, Kim JS (2013) Targeted genome engineering in human cells with the Cas9 RNA-guided endonuclease. Nat Biotechnol 31(3):230–232. https://doi.org/10.1038/nbt.2507
9. Cong L, Ran FA, Cox D, Lin S, Barretto R, Habib N, Hsu PD, Wu X, Jiang W, Marraffini LA, Zhang F (2013) Multiplex genome engineering using CRISPR/Cas systems. Science 339(6121):819–823. https://doi.org/10.1126/science.1231143
10. Jinek M, East A, Cheng A, Lin S, Ma E, Doudna J (2013) RNA-programmed genome editing in human cells. eLife 2:e00471. https://doi.org/10.7554/eLife.00471
11. Mali P, Yang L, Esvelt KM, Aach J, Guell M, DiCarlo JE, Norville JE, Church GM (2013) RNA-guided human genome engineering via Cas9. Science 339(6121):823–826. https://doi.org/10.1126/science.1232033
12. Kim H, Kim JS (2014) A guide to genome engineering with programmable nucleases. Nat Rev Genet 15(5):321–334. https://doi.org/10.1038/nrg3686
13. Cho SW, Kim S, Kim Y, Kweon J, Kim HS, Bae S, Kim JS (2014) Analysis of off-target effects of CRISPR/Cas-derived RNA-guided endonucleases and nickases. Genome Res 24

(1):132–141. https://doi.org/10.1101/gr.162339.113

14. Cradick TJ, Fine EJ, Antico CJ, Bao G (2013) CRISPR/Cas9 systems targeting beta-globin and CCR5 genes have substantial off-target activity. Nucleic Acids Res 41 (20):9584–9592. https://doi.org/10.1093/nar/gkt714
15. Fu Y, Foden JA, Khayter C, Maeder ML, Reyon D, Joung JK, Sander JD (2013) High-frequency off-target mutagenesis induced by CRISPR-Cas nucleases in human cells. Nat Biotechnol 31(9):822–826. https://doi.org/10.1038/nbt.2623
16. Hsu PD, Scott DA, Weinstein JA, Ran FA, Konermann S, Agarwala V, Li Y, Fine EJ, Wu X, Shalem O, Cradick TJ, Marraffini LA, Bao G, Zhang F (2013) DNA targeting specificity of RNA-guided Cas9 nucleases. Nat Biotechnol 31(9):827–832. https://doi.org/10.1038/nbt.2647
17. Pattanayak V, Lin S, Guilinger JP, Ma E, Doudna JA, Liu DR (2013) High-throughput profiling of off-target DNA cleavage reveals RNA-programmed Cas9 nuclease specificity. Nat Biotechnol 31(9):839–843. https://doi.org/10.1038/nbt.2673
18. Kim D, Bae S, Park J, Kim E, Kim S, Yu HR, Hwang J, Kim JI, Kim JS (2015) Digenome-seq: genome-wide profiling of CRISPR-Cas9 off-target effects in human cells. Nat Methods 12(3):231–243. https://doi.org/10.1038/nmeth.3284
19. Kim D, Kim S, Kim S, Park J, Kim JS (2016) Genome-wide target specificities of CRISPR-Cas9 nucleases revealed by multiplex digenome-seq. Genome Res 26(3):406–415. https://doi.org/10.1101/gr.199588.115
20. Robinson JT, Thorvaldsdottir H, Winckler W, Guttman M, Lander ES, Getz G, Mesirov JP (2011) Integrative genomics viewer. Nat Biotechnol 29(1):24–26. https://doi.org/10.1038/nbt.1754
21. Kim E, Koo T, Park SW, Kim D, Kim K, Cho HY, Song DW, Lee KJ, Jung MH, Kim S, Kim JH, Kim JH, Kim JS (2017) In vivo genome editing with a small Cas9 orthologue derived from *Campylobacter jejuni*. Nat Commun 8:14500. https://doi.org/10.1038/ncomms14500
22. Endo M, Mikami M, Endo A, Kaya H, Itoh T, Nishimasu H, Nureki O, Toki S (2018) Genome editing in plants by engineered CRISPR-Cas9 recognizing NG PAM. Nat Plants. https://doi.org/10.1038/s41477-018-0321-8
23. Schmittgen TD, Livak KJ (2008) Analyzing real-time PCR data by the comparative C (T) method. Nat Protoc 3(6):1101–1108
24. Kim D, Lim K, Kim ST, Yoon SH, Kim K, Ryu SM, Kim JS (2017) Genome-wide target specificities of CRISPR RNA-guided programmable deaminases. Nat Biotechnol 35 (5):475–480. https://doi.org/10.1038/nbt.3852
25. Park J, Childs L, Kim D, Hwang GH, Kim S, Kim ST, Kim JS, Bae S (2017) Digenome-seq web tool for profiling CRISPR specificity. Nat Methods 14(6):548–549. https://doi.org/10.1038/nmeth.4262

Chapter 14

Detection of CRISPR/Cas9-Generated Off-Target Effect by Integration-Defective Lentiviral Vector

Xiaoling Wang, Youjun Wu, and Jiing-Kuan Yee

Abstract

Clustered regularly interspaced short palindromic repeat (CRISPR) and other gene editing technologies such as zinc finger nucleases (ZFNs), transcription activator-like effector nucleases (TALENs) show great promises for research and therapeutic applications. One major concern is the off-target effects generated by these nucleases at unintended genomic sequences. In silico methods are usually used for off-target site prediction. However, based on currently available algorithms, the predicted cleavage activity at these potential off-target sites does not always reflect the true cleavage in vivo. Here we present an unbiased screening protocol using integration-defective lentiviral vector (IDLV) and deep sequencing to map the off-target sites generated by gene editing tools.

Key words CRISPR/Cas9, Genome editing, IDLV, Off-target activity

1 Introduction

CRISPR/Cas9 and other site-specific nucleases such as ZFNs and TALENs emerged as a revolutionary technology for genome engineering and potential therapeutic applications to the unmet needs in treating cancers and genetic diseases [1, 2]. Currently several clinical trials are underway, and the preliminary data showed great promises with these technologies [3, 4]. However, a lot of discussion has been focused on the utmost concern, the off-target effects of gene editing on unintended genomic sequences which can give rise to aberrant gene expression, disruption of normal gene functions, and other potential problems, such as genome instability, DNA rearrangement, and cancer [5–7]. These off-target sites can be recognized and cleaved by the nuclease used in gene editing. The resulting double-strand DNA break (DSB) is usually repaired by nonhomologous end-joining (NHEJ), leading to nucleotide substitution, insertion, or deletion of the DNA sequence [8]. To detect the off-target sites, in silico prediction is widely used to generate lists of potential off-target (POT) sites and deep

Tudor A. Fulga et al. (eds.), *CRISPR Guide RNA Design: Methods and Protocols*, Methods in Molecular Biology, vol. 2162, https://doi.org/10.1007/978-1-0716-0687-2_14,

sequencing is applied for verification [9]. However, improvement is still needed to enhance the predictability of these algorithms as the predicted cleavage activity at these POT sites does not always reflect the true off-target effect in vivo [8]. In transduced cells, linear double-stranded IDLV genomes can be incorporated preferentially into DSB via NHEJ [10]. These vector integration sites can be mapped via nonrestrictive linear amplification-mediated PCR (nrLAM-PCR), which makes it possible to directly measure the off-target activity of gene editing tools in the context of different chromatin status and DNA methylation [11]. Here, we present a step-by-step protocol for IDLV-mediated off-target detection in mammalian cells [12].

2 Materials

2.1 IDLV Preparation

2.1.1 Plasmids for IDLV Packaging

1. pHIV7/PGK-Puro (available upon request).
2. pcHELP [13].
3. pCMV-G [14] (Addgene #8454).

2.1.2 IDLV Packaging

1. HEK293T cells (ATCC CRL-3216) cultured in Dulbecco's Modified Eagle Medium (DMEM), 10% fetal bovine serum (FBS), 2 mM L-glutamine, 100 U/mL penicillin, and 100 μg/mL streptomycin.
2. Trypsin (0.25%).
3. Incubator 5% CO_2, 37 °C.
4. PBS, pH 7.5 (Invitrogen).
5. $CaCl_2$: 2 M. Keep it frozen at −20 °C in 5 mL aliquots and thaw immediately before transfection.
6. 2× HBS: 50 mM HEPES pH 7.12, 10 mM KCl, 280 mM NaCl, 1.5 mM Na_2PO_4.
7. Tris–EDTA (TE) buffer, pH 8.0.
8. Virus concentration solution: Dissolve 40 g PEG8000 in PBS to make a final volume of 100 mL.

2.1.3 Virus Titration

1. HT1080 cells cultured in DMEM high glucose, 10% FBS, 2 mM L-glutamine, 100 U/mL penicillin, and 100 μg/mL streptomycin.
2. Polybrene (4 mg/mL in sterile water and filter-sterilized).
3. Puromycin (2 mg/mL in PBS).
4. Trypsin (0.25%).
5. Fixing/Staining solution: 0.5 g Crystal Violet (0.05% w/v), 27 mL 37% Formaldehyde, 100 mL 10× PBS, 10 mL methanol, and add 863 mL deionized H_2O to 1 L.

2.2 Off-Target Screening Library Preparation

2.2.1 Gene Editing

1. Lipofectamine 2000.
2. Opti-MEM.
3. Single-guide RNA (sgRNA) expression plasmid.
4. Cas9 expression plasmid.
5. Qiagen DNA extraction kit.

2.2.2 nrLAM-PCR

1. Magnetic beads: Dynabeads M-280 Streptavidin, and magnetic stand.
2. T4 RNA ligase 1.
3. Thermal cycler for PCR.
4. HotStar Taq DNA polymerase.
5. dNTPs.
6. Bovine serum albumin (BSA), 0.1% (wt/vol).
7. 3 M LiCl solution: Dissolve 6.36 g of LiCl in 0.5 mL 1 M Tris–HCl (pH 7.5) and 0.1 mL 0.5 M EDTA (pH 8.0) and adjust the volume with water to 50 mL. Sterilize the solution with a 0.45-μm filter.
8. 6 M LiCl solution: Dissolve 12.72 g of LiCl in 0.5 mL 1 M Tris–HCl (pH 7.5) and 0.1 mL 0.5 M EDTA (pH 8.0) and adjust the volume with water to 50 mL. Sterilize the solution with a 0.45-μm filter.
9. Tris–HCl, pH 7.5.
10. PEG8000, 40% (wt/vol).
11. Hexamine cobalt (III) chloride, 10 mM.
12. NaOH, 0.1 N.
13. Agarose LE.
14. TBE buffer: 89 mM Tris (pH 7.6), 89 mM boric acid, and 2 mM EDTA.
15. GelRed nucleic acid gel stain.
16. 100 bp DNA ladder
17. MagnaSep magnetic stand for 1.5-mL tubes.
18. Mircrocon YM-50.
19. Shaker.

2.3 Deep Sequencing

1. Illumina HiSeq2500.
2. Qubit dsDNA high sensitivity assay.
3. Agilent bioanalyzer.
4. Ampure XP beads.

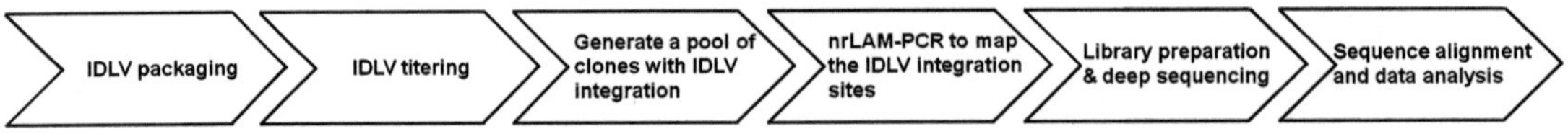

Fig. 1 Flowchart of detecting IDLV-mediated off-target sites generated by CRISPR/Cas9 in mammalian cells. This flowchart shows the main steps for off-target detection. The timeline for each step is indicated. The whole process takes about 9 weeks

3 Methods (Fig. 1)

3.1 IDLV Packaging

3.1.1 Day 1, Seed HEK293T Cells for Virus Production

1. Precaution for virus handling (*see* **Notes 1–4**). HEK293T cells at 80–95% confluence are preferred for virus production.
2. Aspirate HEK293T culture medium (10-cm dish). Rinse with 10 mL PBS and aspirate.
3. Overlay 2 mL 0.25% trypsin prewarmed in 37 °C water bath, and rock the dish to evenly distribute trypsin.
4. Incubate cells in CO_2 incubator for 1 min, or until cells are completely detached from the dish.
5. Perform cell count with a hemocytometer and seed 4×10^6 cells in 10 mL medium per 10 cm dish.

3.1.2 Day 2, Transfect HEK293T Cells with Transfer and Packaging Plasmids

1. Making DNA–$CaCl_2$ mixture.

   Solution A: Total 500 μL made in a 1.5 mL microcentrifuge tube. DNA is mixed as follows:

pHIV7/PGK-Puro	12 μg
pcHELP	12 μg
pCMV-G	4 μg
TE buffer	Make up to 437 μL
$CaCl_2$	63 μL
Total	500 μL

2. Add 500 μL of 2× HBS into a 15-mL tube.
3. Add the DNA–$CaCl_2$ mixture (solution A) dropwise while gently vortexing the tube.
4. Incubate the mixture at room temperature for 30 min to form DNA calcium phosphate coprecipitates.
5. Add 1.0 mL of DNA–calcium phosphate coprecipitates to each 10 cm dish with HEK293T cells.
6. Gently mix the DNA–calcium phosphate coprecipitates with the cells. Do not swirl.
7. Place dishes in a 37 °C, 5% CO_2 incubator.

8. After 6 h, remove the medium containing the DNA–calcium coprecipitates by aspiration, and add 6 mL fresh medium/10 cm dish.

3.1.3 Virus Harvest

1. Collect culture medium containing virus at 48 h after transfection and store the virus at −80 °C.
2. Add 6 mL fresh medium to the dish.
3. Harvest virus again at 72 h after transfection and combine with the previously harvested virus supernatant.

3.1.4 Virus Concentration

1. Spin down cell debris in the medium containing the virus at 800 × *g* for 10 min at room temperature. Transfer the supernatant to a new 50 mL tube.
2. Add stock 40% PEG8000 solution into the virus-containing medium to make a final concentration of 10% PEG8000.
3. Mix well and store at 4 °C for at least 4 h. The virus-containing solution can be stored at 4 °C in PEG for up to 5 days without significantly losing infectivity.
4. Spin the virus-PEG precipitate down at 1600 × *g* for 30 min at room temperature.
5. Carefully remove supernatant without disturbing the pellet.
6. Thoroughly resuspend the viral-PEG pellet into PBS or desired medium without serum and antibiotics with 1/10~1/20 of the original volume by gently pipetting the virus pellet up and down.
7. Aliquot and store the concentrated virus at −80 °C until use.

3.2 Virus Titering

IDLV is integration-defective. Compared with the wild-type lentivirus, the integration activity of IDLV drops by at least three orders of magnitude. The linear genome of lentivirus normally integrates into the host genome via virus-encoded integrase. In the absence of a functional integrase, the IDLV genome is incorporated preferentially into DSB via host cell-mediated nonhomologous end joining (NHEJ). The vector integration sites can then be mapped via nrLAM-PCR. This strategy makes it possible to directly measure CRISPR/Cas9-generated DSBs in the context of different chromatin status and DNA methylation.

3.2.1 Seed HT1080 Cells for Virus Titering

1. Day 1, plate HT1080 cells in two 6-well dishes for each virus (0.5×10^5 cells/well for antibiotic selection).

3.2.2 IDLV Transduction

1. Day 2, count the cell number in one well before transduction.
2. Change to fresh medium with polybrene (1:1000 dilution).

3. Add 5 μL virus at different dilutions: 1:3, 1:9, 1:27, 1: 81, 1:243, 1:729, 1: 2187, 1: 6561, 1: 19683, and mock control.
4. Change to fresh medium 6–24 h after transduction.

3.2.3 Puromycin Selection

1. Day 4, change to fresh medium with puromycin at a concentration of 2 μg/mL.
2. Medium is changed every 3 days with fresh puromycin.
3. Fourteen days after the initiation of selection, puromycin-resistant colonies will become visible.

3.2.4 Colony Staining and Counting

1. Remove medium from the well and wash the well with PBS once.
2. Add sufficient fixing/staining solution to cover the colonies.
3. Stain for 20 min at room temperature.
4. Remove fixing/staining solution.
5. Wash dishes three times with PBS.
6. Air-dry.
7. Count the colonies in each well. Only those wells containing well separated colonies (20–100) will be counted and used to calculate the vector titer.

3.2.5 Titer Calculation

1. Formula for calculation

$$\text{Titer in transduction unit (TU)/mL} = (\text{colony number}) \times (\text{dilution factor/vol of virus solution in mL}).$$

3.3 Generate a Pool of Puromycin-Resistant Clones for the Detection of Gene Editing-Generated Off-Target Effect

This protocol is generally applicable with various gene editing methods, including CRISPR/Cas9, TALEN, and ZFN with the cell line of interest. Here CRISPR/Cas9 mediated gene editing in HEK293T cells is demonstrated as an example:

3.3.1 Gene Editing

1. Day 1, prepare transfection mix: (one 10 cm dish).

A		B	
Lipofectamine 2000	28 μL	Cas9 expression plasmid	15 μg
Opti-MEM	0.5 mL	sgRNA expression plasmid	5 μg
		Opti-MEM	0.5 mL

2. Mix A and B, and leave the DNA–lipid mixture at room temperature for 20 min.

3. Add the DNA–lipid complex into trypsinized HEK293T cell suspension (1×10^7 cells in 10 mL medium) and plate in a 10 cm dish.

3.3.2 IDLV Transduction on Gene Edited Cells

1. Eight hours after transfection, remove concentrated, frozen lentivirus stock from −80 °C freezer and thaw at 37 °C rapidly.
2. Add Polybrene (1:1000 dilution) to a final concentration of 4 μg/mL.
3. Once the virus is fully thawed, transfer 6000 TU IDLV into each dish from Subheading 3.3.1.
4. Day 2: change to fresh medium and carry out the transduction again as **steps 2** and **3**.
5. Day 3, start selection with a final concentration of 1.5 μg/mL puromycin.
6. Day 6, change medium with fresh puromycin.
7. Day 8, remove the medium and collect cells from five dishes using a cell scraper.
8. Gently mix the pooled puromycin-resistant cells and pellet 10% of the cells for genomic DNA extraction using a Qiagen DNA extraction kit.

3.4 nrLAM-PCR to Map the IDLV Integration Sites

See Table 1 and Fig. 2.

3.4.1 Linear PCR to Amplify the Vector LTR and Adjacent Genomic Sequences

1. For each genomic DNA sample, mix the following components for linear PCR initiated from 5-biotinylated vector-specific primer.

Components	Volume (μL)	Final
Water	37	
10× buffer (QIAGEN)	5	1×
dNTPs (0.5 mM)	1	0.01 mM
HotStar Taq (5 U/μL)	0.5	0.05 U
DNA template	5	1000 ng
Primer LTRI (0.17 pmol/μL)	0.5	0.0017 pmol
Total	50	

2. Run linear PCR following the PCR program below.

Table 1
Primer sequences used in this protocol

Primers	Sequences
LTRI	5′-BioAGTAGTGTGTGCCCGTCTGT-3′
LTRII	5′-Bio/GTGTGACTCTGGTAACTAGAG-3′
LTRIII	5′-GATCCCTCAGACCCTTTTAGTC-3′
LCI	5′-GATCTGAATTCAGTGGCACAG-3′
LCII	5′-AGTGGCACAGCAGTTAGG-3′
Linker	5′-P-CCTAACTGCTGTGCCACTGAATTCAGATCTCCCGGGTddC-3′
ILLUF1	5′-CTACACGACGCTCTTCCGATCTCTAGAGATCCCTCAGACCCTTT-3′
ILLUR1	5′-GTGACTGGAGTTCAGACGTGTGCTCTTCCGATCTAGTGGCACAGCAGTTAGG-3′
ILLUF2	5′-AATGATACGGCGACCACCGAGATCTACACTCTTTCCCTACACGACGCTCTTCCGATCT-3′
INDEXR2	5′-CAAGCAGAAGACGGCATACGAGATNNNNNGTGACTGGAGTTC-3′
P7	5′-CAAGCAGAAGACGGCATACG-3′
P5	5′-AATGATACGGCGACACCGA-3′

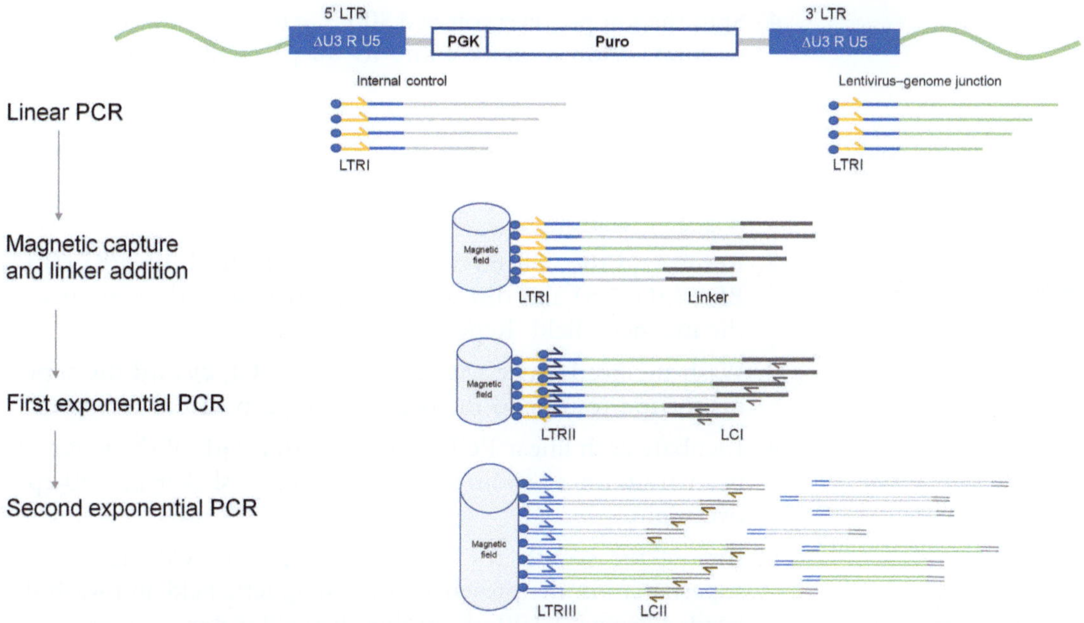

Fig. 2 Schematic flow of nrLAM-PCR to amplify the virus–genomic junction. Linear amplification is conducted with a biotinylated primer. After magnetic capture, a single-stranded linker is ligated to the 3′ end of the linear PCR product. Subsequently, the virus–genome junction sequences are amplified by two rounds of exponential PCR. The first round PCR forward primer is biotinylated. Thus, the first round PCR products can bind to the beads whereas the second round PCR products are unbound

Step 1	95 °C	15 min	
Step 2	94 °C	45 s	
Step 3	58 °C	45 s	
Step 4	72 °C	30 s	Repeat **step 2–4** for 50 cycles
Step 5	72 °C	10 min	
Step 6	4 °C		

3. After the first 50 cycles of linear PCR, add 0.5 μL Taq polymerase and carry out the second round of 50-cycle PCR as described in the above table.

3.4.2 Purify the PCR Product with Mircrocon YM-50 to Remove PCR Amplicons Smaller Than 125 Nucleotides

1. Dilute 50 μL PCR product with 150 μL water, load onto an YM-50 centrifugal filter.
2. Spin the centrifugal filter device at 14,000 × *g* rpm for 15 min in a bench-top centrifuge.
3. Invert the cuvette and put into a new 2 mL tube.

4. Spin the sample reservoir at 1300 × *g* for 3 min to recover the retentate. Adjust the volume to 50 μL with double-distilled water.

3.4.3 PCR Purification of the Linear PCR Product with Magnetic M-280 Streptavidin Beads to Remove Genomic DNA

1. Expose 20 μL of magnetic beads (10 μg/μL) to a magnetic field for 60 s and discard the supernatant in the presence of magnetic stand.
2. Resuspend the collected magnetic beads in 40 μL PBS/0.1% BSA (pH 7.5) and discard the supernatant in the presence of the magnetic field. Repeat this step once.
3. Wash the beads once with 20 μL 3 M LiCl, discard the supernatant and resuspend the beads in 50 μL 6 M LiCl.
4. Incubate each linear PCR product with 50 μL of the magnetic bead solution at room temperature on a shaker at 300 rpm overnight.
5. Expose the sample for 60 s to a magnetic field, discard the supernatant in the presence of the magnetic field and wash the beads once with 100 μL double-distilled water.

3.4.4 Add Linker to the 3′ End of Single Stranded Linear PCR Product

1. Prepare the following ligation mixture to covalently add linker to the linear PCR product.

Components	Volume (μL)	Final
Linker (10 μM)	1	1 μM
T4 RNA ligase 1(20 U/μL)	1	2 U
10× Ligase buffer (New England Biolabs)	1	1×
ATP (10 mM)	1	1 mM
PEG8000 (50%, wt/vol)	5	25%
Hexamine cobalt chloride (10 mM)	1	1 mM
Total	10	

2. Expose the DNA–streptavidin beads complex (from Subheading 3.4.3) to a magnetic field, discard the supernatant, and resuspend the beads in 10 μL of ligation mixture.
3. Incubate the mixture overnight at room temperature on a horizontal shaker set at 300 rpm.
4. Add 90 μL double-distilled water and expose the sample to a magnetic stand for 60 s. Discard the supernatant in the presence of the magnetic stand. Wash the beads with 100 μL double-distilled water once and resuspend them in 10 μL double-distilled water.

3.4.5 First Exponential PCR to Amplify the Linear PCR Product

1. Prepare a 50 μL reaction mixture as shown in the table below to amplify the linear PCR product.

Components	Volume (μL)	Final
Water	37.5	
10× buffer	5	1×
dNTPs (10 mM)	1	0.2 mM
HotStar Taq (5 U/μL)	0.5	0.05 U
DNA template	5	From previous step
LTRII (16.7 μM)	0.5	0.167 μM
LCI (16.7 μM)	0.5	0.167 μM
Total	50	

2. Run PCR following the program below.

Step 1	95 °C	15 min	
Step 2	94 °C	45 s	
Step 3	58 °C	45 s	
Step 4	72 °C	30 s	Repeat **steps 2–4** for 20 cycles
Step 5	72 °C	10 min	
Step 6	4 °C		

3.4.6 Magnetic Capture of the First PCR Product Using Magnetic M-280 Streptavidin Beads

1. Expose 20 μL of magnetic M-280 Streptavidin beads (10 μg/μL) to a magnetic field for 60 s and discard the supernatant in the presence of the magnetic field.
2. Resuspend the beads in 40 μL PBS/0.1% BSA (pH 7.4) and discard the supernatant in the presence of the magnetic field. Repeat this step once.
3. Wash the beads once in 20 μL 3 M LiCl in the presence of the magnetic field, and resuspend them in 20 μL 6 M LiCl.
4. Mix 20 μL of the magnetic bead solution with 20 μL of the PCR product from Subheading 3.4.5.
5. Incubate at room temperature for 1 h on a horizontal shaker set at 300 rpm.
6. Add 60 μL double-distilled water and expose the sample for 60 s to a magnetic field. Discard the supernatant in the presence of the magnetic field and wash the beads once with 100 μL double-distilled water. Discard the supernatant in a magnetic field and repeat the wash once.

7. Expose the sample for 60 s to a magnetic field, discard the supernatant and resuspend the beads in 10 μL of 0.1 N NaOH. Incubate at room temperature for 10 min on a horizontal shaker set at 300 rpm.
8. Expose the sample for 60 s in a magnetic field and transfer the supernatant to a fresh tube.

3.4.7 Second Exponential PCR to Amplify the LTR–Genomic Junction

1. Set up a 50 μL reaction for second round of amplification by a PCR using the linker and 5′-biotinylated vector specific primers as shown in the table below.

Components	Volume (μL)	Final
Double-distilled water	37.5	
10× buffer	5	1×
dNTPs (10 mM)	1	0.2 mM
HotStar Taq (5 U/μL)	0.5	0.05 U
Template	5	From previous step
LTRIII (16.7 μM)	0.5	0.167 μM
LCII (16.7 μM)	0.5	0.167 μM
Total	50	

2. Run PCR following the program below.

Step 1	95 °C	15 min	
Step 2	94 °C	45 s	
Step 3	58 °C	45 s	
Step 4	72 °C	30 s	Repeat **steps 2–4** for 30 cycles
Step 5	72 °C	10 min	
Step 6	4 °C		

3. Conduct electrophoresis on a 2% agarose gel. A successful reaction is indicated by a smear on the gel (Fig. 3).

3.4.8 Purify the PCR Products by Removing Products Smaller Than 150 bp

1. Shake the AMPure XP bottle to resuspend the magnetic particles and add 54 μL AMPure XP paramagnetic beads per 45 μL of PCR product.
2. Mix the PCR product and magnetic beads by pipetting ten times. Let the mixed samples incubate for 5 min at room temperature.

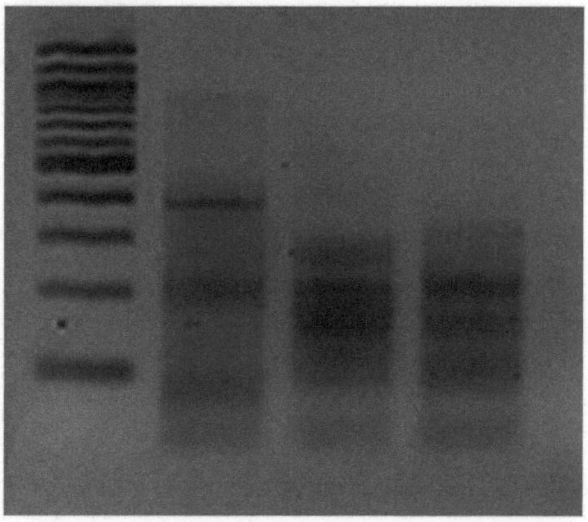

Fig. 3 Agarose gel image of the nrLAM-PCR products which are characterized by a smear. Marker: 100 bp ladder. Lanes 1 and 2 are two CRISPR/Cas9 gene editing samples targeting two genetic loci (*TAT* and *WAS*). Lane 3 is a control created by the same IDLV transduction procedure without gene editing

3. Separate the bound PCR product from unbound contaminants by carefully removing the cleared solution while the tube is situated on a magnetic stand.
4. Wash bound PCR product twice with 70% Ethanol to remove contaminants.
5. Elute the bound PCR product from beads with EB buffer.
6. Transfer the supernatant containing the PCR product to a new tube.
7. Use 2 μL of the PCR product for quantification using Qubit dsDNA high sensitivity assay.

3.5 Library Preparation for Deep Sequencing by Two-Step PCR

3.5.1 Perform First Round PCR to Get Amplicons Tailed with Overhang Adaptor Sequences

1. Set up a 50 μL reaction to add the Illumina adaptors sequence as shown in the table below.

Components	Volume (μL)	Final
Doubled-distilled water	18	
2× Phusion High-Fidelity PCR master mix	25	1×
Template (from previous step)	5	10 ng
ILLUF1 (10 μM)	1	0.2 μM
ILLUR1 (10 μM)	1	0.2 μM
Total	50	

2. Run PCR following the program below.

Step 1	98 °C	30 s	
Step 2	98 °C	10 s	
Step 3	60 °C	30 s	
Step 4	72 °C	30 s	Repeat **steps 2–4** for 3 cycles
Step 5	72 °C	5 min	
Step 6	4 °C		

3.5.2 Purify the PCR Product with Ampure Beads

1. Shake the AMPure XP bottle to resuspend the magnetic particles and add 60 μL AMPure XP paramagnetic beads per 50 μL of PCR product.
2. Mix the PCR product and magnetic beads by pipetting ten times. Let the mixed samples incubate for 5 min at room temperature.
3. Separate the bound PCR product from unbound contaminants by carefully removing the cleared solution while the tube is situated on a magnetic stand.
4. Wash the bound PCR product twice with 70% Ethanol to remove contaminants.
5. Elute purified PCR product from beads with 23 μL EB buffer.
6. Transfer the supernatant containing the PCR product to a new tube.
7. Use 2 μL of the PCR product for quantification with Qubit dsDNA high sensitivity assay.

3.5.3 Perform Second Round PCR to Add Illumina Adaptor and Multiplexing Index Sequences

1. Set up a 50 μL reaction to attach the Illumina adaptors sequence as shown in the table below.

Components	Volume (μL)	Final
Double-distilled water	18	
2× Phusion High-Fidelity PCR master mix	25	1×
Template (from previous step)	5	10 ng
ILLUF2 (10 μM)	1	0.2 μM
INDEXR2 (10 μM)	1	0.2 μM
Total	50	

2. Run PCR following the program below.

Step 1	98 °C	30 s	
Step 2	98 °C	10 s	
Step 3	62 °C	30 s	
Step 4	72 °C	30 s	Repeat **steps 2–4** for 3 cycles
Step 5	72 °C	5 min	
Step 6	4 °C		

3.5.4 Purify the Amplified PCR Product with Ampure Beads

1. Shake the AMPure XP bottle to resuspend the magnetic particles and add 50 μL AMPure XP paramagnetic beads per 50 μL of PCR product.
2. Mix the PCR product and magnetic beads by pipetting ten times. Let the mixed samples incubate for 5 min at room temperature.
3. Separate the bound PCR product from unbound contaminants by carefully removing the cleared solution while the tube is situated on a magnetic stand.
4. Wash the bound PCR product twice with 70% Ethanol to remove the contaminants.
5. Elute the purified PCR product with 25 μL EB.
6. Transfer the supernatant containing the PCR product to a new tube.
7. Validate the PCR product with Agilent bioanalyzer by running 1 μL on a DNA high sensitivity Chip (Fig. 4). The product is characterized as a smear with DNA size enriched in 200–500 bp range.

3.5.5 Quantification of the Library by SYBR Green qPCR Assay Using a LightCycler

1. Set up a 10 μL reaction as shown in the table below for qPCR quantification. Measurements are conducted in triplicate.

Components	Volume (μL)	Final
Double-distilled water	3.6	
2× QuantiFast SYBR green master mix	5	1×
Template (from previous step)	1	10 ng
P7 (10 μM)	0.2	0.2 μM
P5 (10 μM)	0.2	0.2 μM
Total	10	

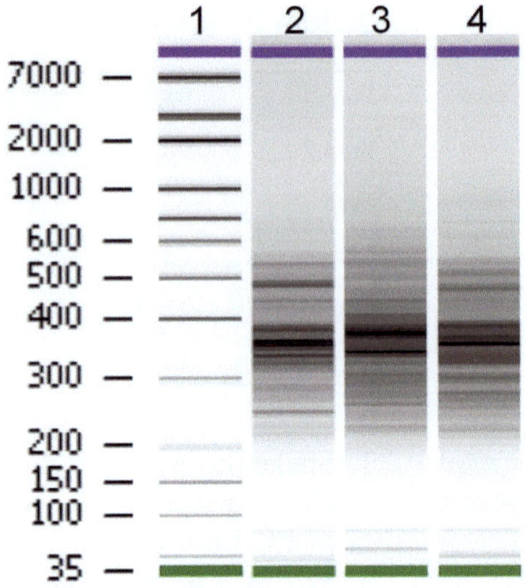

Fig. 4 Electrophoresis image of the nrLAM-PCR product analyzed by Agilent high sensitivity DNA assay. Lane 2 is a sample edited by TALEN targeting the *WAS* gene and lanes 3 and 4 are two CRISPR/Cas9 edited samples targeting two loci (*TAT* and *WAS*). The product is characterized as a smear with DNA size ranging from 200 to 500 bp

2. Run PCR following the program below.

Step 1	95 °C	5 min	
Step 2	95 °C	10 s	
Step 3	60 °C	30 s	Repeat **steps 2–3** for 40 cycles. Fluorescence is measured at the end of elongation
Step 6	4 °C		

3. Calculate the DNA concentration by comparing the measurements to the standard curve using the following formula:

$$\text{Library concentration [molecules per mL]} = \text{inferred concentration [molecules per mL]} \times \text{length of standard/average fragment length in the library (by Agilent bioanalyzer).}$$

3.6 Deep Sequencing and Data Analysis

3.6.1 Deep Sequencing of the Vector–Genome Junction

1. Add NaOH to a final concentration of 0.05 M to denature the PCR product.
2. Dilute the PCR product in the hybridization buffer to a final DNA concentration of 10 pM.
3. Load the sample for paired-end sequencing. 100 cycles of read1, 7 cycles of index read and 100 cycles of read2 sequencing are performed using the Illumina HiSeq2500.

3.6.2 Data Analysis to Map the IDLV Integration Sites

1. Sort the sequence by sample-specific barcodes.
2. Trim the paired-end reads using 5′-adapter sequence (5-′-ctaggatccctcagacccttttagtcagtgtggaaaatctctagcag-3′) and 3-′-adapter sequence (5′-agtggcacagcagttagg-3′). Only reads with these adapters are kept.
3. Identify the vector sequences and remove those readings with 5′-LTR into the vector sequence (Fig. 2).
4. Extract the genomic sequence adjacent to the 3′LTR by removing the barcode, the linker and the lentiviral vector sequences.
5. Remove redundancies by combining all sequence reads that are identical.
6. Align to hg19 genome using Novoalign v2.07.
7. The aligned locations of these reads are then summarized and matched to NCBI's RefSeq database using Bioconductor package "ChIPpeakAnno."
8. The IDLV integration sites with more than 20 reads are counted.
9. Record the chromosome, the integration site, the DNA strand, the length of the sequence, the protein coding gene, targeted intron or exon and the distance to the transcription start site.
10. Combine all the sequences that result in the same integration site.
11. We defined a clustered IDLV integration sites (CLIS) as two or more integration sites separated by no more than 500 bp. Such clusters are considered to be due to the cleavage by the site-specific nuclease or the presence of fragile sites in the genome. Sum up the number of the reads.
12. Sequences with more than 50% similarity to the target sequence are searched in the ±500 bp genomic region as potential off-target sites. Usually the off-target sites have at least 80% sequence similarity to the on-target site.

4 Notes

1. Wear double gloves and lab coat at all times when conducting the experiments with lentivirus.
2. Always work with lentiviral particles in a Class II laminar flow hood.
3. All wastes are decontaminated before disposal by an approved decontamination method such as autoclaving.
4. Liquid waste should be treated with 10% bleach for 30 min before transporting out of the hood.

Acknowledgments

This work is supported by a grant RB3-02161 from California Institute of Regenerative Medicine (J.K.Y.) and a grant from Nesvig Foundation.

References

1. Cong L et al (2013) Multiplex genome engineering using CRISPR/Cas systems. Science 339(6121):819–823
2. Hu JH et al (2018) Evolved Cas9 variants with broad PAM compatibility and high DNA specificity. Nature 556(7699):57–63
3. Cornu TI, Mussolino C, Cathomen T (2017) Refining strategies to translate genome editing to the clinic. Nat Med 23(4):415–423
4. Sheridan C (2018) Sangamo's landmark genome editing trial gets mixed reception. Nat Biotechnol 36(10):907–908
5. Hsu PD et al (2013) DNA targeting specificity of RNA-guided Cas9 nucleases. Nat Biotechnol 31(9):827–832
6. Fu Y et al (2013) High-frequency off-target mutagenesis induced by CRISPR-Cas nucleases in human cells. Nat Biotechnol 31 (9):822–826
7. Akcakaya P et al (2018) In vivo CRISPR editing with no detectable genome-wide off-target mutations. Nature 561(7723):416–419
8. Pattanayak V et al (2011) Revealing off-target cleavage specificities of zinc-finger nucleases by in vitro selection. Nat Methods 8(9):765–770
9. Pattanayak V et al (2013) High-throughput profiling of off-target DNA cleavage reveals RNA-programmed Cas9 nuclease specificity. Nat Biotechnol 31(9):839–843
10. Gabriel R et al (2011) An unbiased genome-wide analysis of zinc-finger nuclease specificity. Nat Biotechnol 29(9):816–823
11. Paruzynski A et al (2010) Genome-wide high-throughput integrome analyses by nrLAM-PCR and next-generation sequencing. Nat Protoc 5(8):1379–1395
12. Wang X et al (2015) Unbiased detection of off-target cleavage by CRISPR-Cas9 and TALENs using integrase-defective lentiviral vectors. Nat Biotechnol 33(2):175–178
13. Mochizuki H et al (1998) High-titer human immunodeficiency virus type 1-based vector systems for gene delivery into nondividing cells. J Virol 72(11):8873–8883
14. Yee JK et al (1994) A general method for the generation of high-titer, pantropic retroviral vectors: highly efficient infection of primary hepatocytes. Proc Natl Acad Sci U S A 91 (20):9564–9568

Chapter 15

Genome-Wide CRISPR Off-Target DNA Break Detection by the BLISS Method

Roberto Ballarino, Britta A. M. Bouwman, and Nicola Crosetto

Abstract

Clustered regularly interspaced palindromic repeat (CRISPR) systems are revolutionizing many areas of biology and medicine, where they are increasingly utilized as therapeutic tools for correcting disease-causing mutations. From a clinical perspective, unintended off-target (OT) DNA double-strand break (DSB) induction by CRISPR nucleases represents a major concern. Therefore, in recent years considerable effort has been dedicated to developing methods for assessing the OT activity of CRISPR nucleases, which in turn can be used to guide engineering of nucleases with minimal OT activity. Here we describe a detailed protocol for quantifying OT DSBs genome-wide in cultured cells transfected with CRISPR enzymes, based on the breaks labeling in situ and sequencing (BLISS) method that we have previously developed. CRISPR-BLISS is versatile and scalable, and allows assessment of multiple guide RNAs in different cell types and time points following cell transfection or transduction.

Key words CRISPR, Off-targets, DNA double-strand breaks, BLISS

1 Introduction

Precision genome editing based on clustered regularly interspaced palindromic repeat (CRISPR) has revolutionized many areas of biology and medicine. CRISPR systems are now routinely used in a variety of applications ranging from gene editing [1–6], transcriptional activation or repression [7–10], localized chromatin modification [11, 12], and chromosome visualization [7, 13]. Furthermore, the clinical application of CRISPR systems as therapeutic tool is highlighted by multiple ongoing clinical trials assessing the efficacy of CRISPR nucleases to correct disease-causing mutations [14]. In this context, the specificity of CRISPR nucleases is of utmost importance, as unwanted DSBs introduced by the nucleases outside of their intended target site might lead, for example, to the formation of potentially oncogenic genomic rearrangements. Although several factors influencing CRISPR specificity have been characterized, predicting off-target (OT) activity of

Tudor A. Fulga et al. (eds.), *CRISPR Guide RNA Design: Methods and Protocols*, Methods in Molecular Biology, vol. 2162, https://doi.org/10.1007/978-1-0716-0687-2_15,

CRISPR nucleases still primarily relies on sequence homology between the single-guide RNA (sgRNA) and its target sequence [2]. Mismatches, gaps, and bulges between sgRNAs and genomic DNA at OT sites have been shown to predict unspecific cleavage of DNA [2, 15–17]. However, binding of Cas9 to certain chromatin regions, as determined by chromatin immunoprecipitation and sequencing (ChIP-seq), cannot be entirely explained by mismatches between the sgRNA and OT sequences, and does not fully predict OT activity [2, 18, 19]. Another indicator of OT activity of CRISPR nucleases is the observed frequency of small insertions or deletions (indels) at OT sites in the genome [15, 20]. In fact, indel frequencies are commonly used to assess both the strength of OT activity of Cas9 and other commonly used CRISPR enzymes [21, 22]. However, the frequency of indels at a given locus is influenced by the local chromatin environment (DNA accessibility, histone modifications) as well as by the type of DSB repair that is dominant in the cell, which in turn depends on the cell cycle phase and other cell-intrinsic properties [22–24]. Thus, indel frequencies do not provide a direct measurement of OT activity. Similarly, methods such as high-throughput genome-wide translocation sequencing (HTGTS) [25, 26], transduction-based integrase-defective lentiviral vector (IDLV) capture [27] and genome-wide unbiased identification of DSBs enabled by sequencing (GUIDE-seq) [16, 22]—all of which have been applied to study the specificity of CRISPR nucleases—provide an indirect measurement of OT activity, by assessing the conversion of CRISPR-induced DSBs to either translocations (HTGTS) or localized insertions (IDLVs and GUIDE-seq) through the action of nonhomologous end-joining DNA repair.

In order to obtain more direct measurements of OT activity, several methods have been developed to detect CRISPR cutting events genome-wide using next-generation sequencing (NGS) (Table 1). We developed breaks labeling in situ, enrichment on streptavidin and sequencing (BLESS) [28] and later breaks labeling in situ and sequencing (BLISS) [39, 45], both of which detect DSBs in their native chromatin context by ligating DSB ends to specialized adapters in cross-linked nuclei. Both BLESS and BLISS have been applied to detect OTs of Cas9 [2, 46] and Cpf1 [39, 40] enzymes, however BLISS is more quantitative and versatile compared to BLESS. In addition, other methods have been developed to detect OT DSBs, although not in their natural chromatin environment. In Digenome-seq [33], for example, purified genomic DNA is digested in vitro with a CRISPR nuclease and OTs are identified through whole-genome sequencing. Similarly, in the more recently developed SITE-seq [22] and CIRCLE-seq [41] methods, purified genomic DNA is digested in vitro. The DNA fragments containing the nuclease cleavage sites are then selected by affinity purification (SITE-seq) or by circularization

Table 1
Methods for CRISPR off-target identification

Method[a]	Main features	Reported CRISPR applications	Ref.
BLESS	Blunting and ligation of DSBs with biotinylated adapters in fixed cells, capture on streptavidin of labeled DSBs	Cas9 OT profiling for sgRNAs against *EMX1*, *VEGFA*, *DNMT1*, *TPCN2*, *NRG2*, *CACNA2D4*, and *ADAMTSL1* genes in HEK293 cells	[2, 28, 29]
ChIP-seq	Immunoprecipitation of Cas9 bound to chromatin, sequencing of captured chromatin	Cas9 and dCas9 OT profiling for 12 sgRNAs targeting different genes in HEK293 cells	[30, 31] [18, 19]
IDLV capture	Integration of viral DNA cassette at DSB sites in living cells	Cas9 OT profiling for sgRNAs against *WAS* and *TAT* genes	[27]
HTGTS and LAM-HTGTS	Translocation between a CRISPR nuclease-induced DSB and endogenous DSBs in living cells, capture on streptavidin of translocation products	Cas9 OT profiling for sgRNAs against *EMX1*, *VEGFA*, and *RAG1B* genes	[25, 26, 32]
GUIDE-seq	Integration of dsDNA oligos at DSB sites in living cells	Cas9 and Cpf1 OT profiling for sgRNAs against *VEGFA*, *EMX1*, *FANCF*, *RUNX1*, *ZNF629*, *DMT1*, *HEK4*, *VEG1* and other genes in multiple cell lines	[16]
Digenome-seq	In vitro digestion of purified genomic DNA, whole-genome sequencing	Cas9 OT profiling for sgRNAs against *HBB*, *VEGFA*, *ABL1*, *EPHB2*, *ERBB3*, *FGFR2*, and *FGFR4* genes in HAP1 and K562 cells	[6, 33–35]
DSBCapture	Blunting, A-tailing and ligation of DSBs with biotinylated adapters in fixed cells, capture on streptavidin of labeled DSBs	No reported CRISPR application	[36]
End-seq	Blunting, A-tailing, and ligation of DSBs with biotinylated adapters in live cells lysed in agarose plugs, capture on streptavidin of labeled DSBs	No reported CRISPR application	[37, 38]
BLISS	Blunting and ligation of DSBs with T7 promoter-containing adapters in fixed cells, selective amplification of labeled DSBs by IVT	Cas9 and Cpf1 OT profiling for sgRNAs against *DNMT1*, *EMX1*, *GRIN2b*, and *VEGFA* genes in HEK293 cells	[39, 40]

(continued)

Table 1
(continued)

Method[a]	Main features	Reported CRISPR applications	Ref.
SITE-seq	In vitro cleavage of purified genomic DNA, ligation of DSBs with biotinylated adapters, whole-genome sequencing	Cas9 OT profiling for sgRNAs against *VEGFA*, *FANCF*, *CD34*, *XRCC5*, *PAPSS2*, *PTPRC* and *CD151* genes in HEK293 cells	[22]
CIRCLE-seq	In vitro cleavage of circularized purified genomic DNA, sequencing of linearized DNA circles	Cas9 OT profiling for sgRNAs against *EMX1*, *RNF1*, *FANCF*, *VEGFA* genes Cas9 and Cas12a OT profiling for sgRNAs against gl2 in maize Recently adopted for VIVO-strategy, discovering *Pcsk9* OT-mutations in mouse liver in vivo	[6, 41–43]
DIG-seq	In vitro cleavage of purified chromatin, whole-genome sequencing	Cas9 OT profiling for 12 pairs of endogenous target sites with the same DNA sequence comparing open and closed chromatin in both HeLa and HEK 293T cells	[24]
TEG-seq	Target-enriched GUIDE-seq through 5′-phosphate primers mediated amplicon selection	Cas9 OT profiling for sgRNAs against *HEK4*, *VEG1*, *SMCHD*, *HBB-2*, *INS*, and *PAH* genes	[44]

[a]The methods are listed in chronological order of their publication

(CIRCLE-seq) before sequencing, which considerably reduces the costs compared to Digenome-seq. (Details on Digenome-Seq, SITE-seq, and IDVL are included in this issue of Methods in Molecular Biology). While genome-wide patterns of in vitro cutting shed light on the biophysical principles that guide CRISPR nuclease cleavage, they might not necessarily reflect the actual OT activity that occurs in vivo, in a cell-type specific manner. On the other hand, methods like BLESS and BLISS, which rely on ligating DSBs inside cross-linked chromatin, are prone to accessibility effects and might miss OT events occurring, for example, in heterochromatin. Ultimately, all of the available methods for OT detection can be considered complementary to some extent, and the choice of which method to use should be guided by the specific question, as well as by type of sample in which OTs need to be identified. Here, we present a protocol for high-throughput OT detection in cultured cells using an adaptation of the BLISS protocol that we previously described [39, 45]. This CRISPR-BLISS assay allows for examination of the OT activity of multiple guide RNAs in parallel, in different cell types and at different time points following cellular transfection or transduction.

2 Materials

2.1 Reagents and Consumables for Cell Culture and CRISPR Reactions

1. Ethanol 100%, molecular biology grade.
2. D10 Medium: 500 ml DMEM, 50 ml heat-inactivated FBS, 5.5 ml HEPES 1 M pH 7.3.
3. TrypLE™ Express Enzyme.
4. Rapidly growing 293FT cells, preferentially low passage (<p20).
5. 24-well Tissue culture plates (tc-treated) for transfection.
6. T75 and T225s Tissue culture plates for maintenance culturing.
7. Falcon™ tubes, 50 and 15 ml.
8. Poly-L-lysine (PLL) solution (Sigma-Aldrich).
9. Manual hemocytometer (preferred: Neubauer improved cell counting chamber) or automated cell counter (e.g., Countess II FL Automated Cell Counter, Thermo Fisher Scientific).
10. Dulbecco's phosphate buffered saline (DPBS).
11. Purified CRISPR RNA (crRNA) and trans-activating crRNA (tracrRNA) as PCR products: 30 ng per well of a 24-well plate.
12. CRISPR effector plasmid, endotoxin-free miniprep: 400 ng per well of a 24-well plate.
13. pMAX-GFP plasmid for transfection control: 10 ng per well.

14. Opti-MEM Reduced Serum Medium (Gibco™).
15. Lipofectamine® 2000 Transfection Reagent.

2.2 Reagents for BLISS

1. T4 Polynucleotide Kinase (New England Biolabs).
2. Methanol-free paraformaldehyde (PFA) 16%.
3. Glycine.
4. Nuclease-free water.
5. Nuclease-free phosphate-buffered saline (10×), pH 7.4.
6. Lysis buffer 1 (LB1): Tris–HCl 10 mM, NaCl 10 mM, EDTA 1 mM, Triton X-100 0.2%, pH 8 at 4 °C.
7. Lysis buffer 2 (LB2): Tris–HCl 10 mM, NaCl 150 mM, EDTA 1 mM, SDS 0.3%, pH 8 at 25 °C.
8. 1× CutSmart® buffer freshly supplemented with 0.1% Triton X-100.
9. Quick Blunting™ Kit (New England Biolabs).
10. BSA (50 mg/ml).
11. ATP Solution (100 mM).
12. T4 DNA Ligase, 5 U/μl (Thermo Fisher Scientific).
13. High-salt wash buffer (HSW): Tris–HCl 10 mM, NaCl 2 M, EDTA 2 mM, Triton X-100 0.5%, pH 8 at 25 °C.
14. TAIL buffer: Tris–HCl 10 mM, NaCl 100 mM, EDTA 50 mM, SDS 1%, pH 7.5 at 25 °C.
15. Proteinase K, molecular biology grade, 800 units / ml.
16. Nuclease-free TE buffer.
17. NEBNext® Ultra II Q5 Master Mix (New England Biolabs).
18. MEGAscript® T7 Transcription Kit (Thermo Fisher Scientific).
19. DNase I, RNase-free (Thermo Fisher Scientific).
20. RiboSafe RNase Inhibitor (Bioline).
21. Agencourt AMPure XP (Beckman Coulter).
22. Agencourt RNAClean XP (Beckman Coulter).
23. T4 RNA Ligase 2, truncated (New England Biolabs).
24. RNaseOUT™ Recombinant Ribonuclease Inhibitor (Thermo Fisher Scientific).
25. SuperScript® IV Reverse Transcriptase (Thermo Fisher Scientific).
26. DTT (100 mM).
27. TruSeq Small RNA Sample Preparation Kit (Illumina). Alternative: adapters and primers with the corresponding sequence obtained from another company (Table 2).

Table 2
Sequence of adapters and primers

Adapter name		Sequence (5′–3′)
BLISS adapter[a]	Upper oligo	GGCGTGATGGCG_NNNNNNNNNNNN_GATCGTCGGACTGTAGAACTCTGAAC_CCCTATAGTGAGTCGTATTACCGGCCTCAATCG_AA
	Bottom oligo	GG_CGATTGAGGCCGGTAATACGACTCACTATAGGG_GTTCAGAGTTCTACAGTCCGACGATC_NNNNNNNNNNNN_CGCCATCACGCC*T
RA3 adapter[b]		TGGAATTCTCGGGTGCCAAGG
RTP primer[b]		GCCTTGGCACCCGAGAATTCCA
Universal RP1 primer[b]		AATGATACGGCGACCACCGAGATCTACACGTTCAGAGTTCTACAGTCCGA
Indexing RPI1 primer[b]		CAAGCAGAAGACGGCATACGAGATCGTGATGTGACTGGAGTTCCTTGGCACCCGAGAATTCCA

[a]As explained in **Note 3**, BLISS adapters are prepared by annealing two partially complementary oligos (upper and bottom). The upper oligo contains the following parts (from 5′ to 3′, separated by "_" in the table above): a sample barcode sequence; a unique molecular identifier (UMI); the Illumina RA5 adapter sequence; the T7 promoter sequence; a dinucleotide overhang to prevent the formation of head-to-tail concatemers during in situ ligation. The bottom oligo contains: a dinucleotide overhang to prevent the formation of head-to-tail concatemers during in situ ligation; the reverse complement of the T7 promoter sequence; the reverse complement of the RA5 adapter sequence; a UMI; the reverse complement of the sample barcode; a T overhang that binds to A-tailed DSBs. We have successfully used 8 and 12 nt UMIs, as well as multiple barcode sequences [39]. Note that the UMIs in the upper and bottom oligos are not complementary. As an example, we show one BLISS adapter which we previously used [39]

[b]Same as in the Illumina's TruSeq Small RNA Sample Preparation Kit. When multiple libraries need to be sequenced in the same run, a different RPI primer should be used for each library. As an example, index sequence #1 is shown as underscored

2.3 Consumables and Equipment for BLISS

1. Microcentrifuge tubes 0.5 ml (preferred: Eppendorf® LoBind).
2. Microcentrifuge tubes 1.5 ml (preferred: Eppendorf® LoBind).
3. Low Retention Filter Tips.
4. Qubit® 2.0 Fluorometer (Thermo Fisher Scientific).
5. Qubit® dsDNA HS, Broad Range, and ssRNA Assay Kit (Thermo Fisher Scientific).
6. Bioanalyzer 2100 (Agilent Technologies).
7. High Sensitivity DNA Kit for Bioanalyzer.
8. RNA 6000 Pico Kit for Bioanalyzer.
9. Cooling incubator (e.g., Binder incubator KB 53).
10. Tabletop centrifuge.
11. Thermoshaker (e.g., Eppendorf® Thermomixer Compact).

12. PCR Thermocycler (e.g., Biometra TRIO combi, Analytikjena).
13. Sonication device (e.g., Bioruptor® Plus, Diagenode).
14. DynaMag™-2 Magnet (Thermo Fisher Scientific).

3 Methods

3.1 Preparation of Cells to Be Transfected (See Note 1)

DAY 1.

1. Prepare the destination 24-well plate by covering each well with 250 μl of PLL solution and gently shake the plate for 15 min.
2. Aspirate the PLL and wash the wells once with 1× DPBS and once with ethanol 70%. Keep the ethanol in the plate until ready to dispense the cell solution (**step 8**). Then, thoroughly remove the ethanol and let the plate air-dry before dispensing the cells.
3. Wash the 293FT cells once with DPBS prewarmed at 37 °C and then trypsinize by covering the entire surface with 3 ml (T75) or 10 ml (T225) of TrypLE™.
4. Aspirate the TrypLE™ solution and place the flask(s) back into the incubator until the sheets of cells move when the flask(s) is tilted.
5. Add 5 ml (T75) or 15 ml (T225) of D10 medium to the flask and pipette ten times up and down to dissociate the cells, then transfer the suspension to a 50 ml Falcon™ tube.
6. Count the cells using a Countess™ Automated Cell counter according to the manufacturer's instructions, or using a hemocytometer.
7. Dilute the cell suspension to reach 250,000 cells/ml.
8. Plate 500 μl of the diluted cell suspension in each well of the coated 24-well plate prepared above. Make sure that the cell suspension is mixed well before pipetting. Gently move the plate 3–4 times along an eight-shaped path to properly distribute the cells inside each well.

DAY 2.

3.2 Transfection of 293FT Cells (See Note 2)

1. Prepare the following transfection mix in tubes or in a PCR plate (volumes for one well, adjust proportionally to make enough for all wells):
   (a) In one tube mix:

Lipofectamine™ 2000	2.0 μl
Opti-MEM® Medium	48 μl

(b) In a second tube mix:

RNA–DNA mix[a]	2.0 μl (max)
Opti-MEM® Medium	48 μl

[a]A mix of crRNA and/or tracrRNA (30 ng), CRISPR effector plasmid (400 ng), and pMAX-GFP (10 ng)

2. Combine the contents of the two tubes and incubate for at least 5 min at room temperature (rt). A longer incubation (up to 30 min) may be used, depending on the plasmid.
3. Gently add the mix to each well, dispensing droplets in a circular fashion on top of the medium. Prevent pipetting out in a single stream.
4. Incubate for 24 h.
5. Check the transfection efficiency by visualizing GFP expression.

3.3 Preparing dsDNA BLISS Adapters (See Note 3)

1. For each adapter, prepare the following phosphorylation mix in a thermocycler-compatible tube or strip:

(a) Nuclease-free water.	58 μl
(b) Forward oligo (100 μM)	10 μl
(c) 10× T4 PNK buffer	10 μl
(d) ATP 10 mM	10 μl
(e) T4 Polynucleotide kinase 10 U/μl	2 μl

2. Incubate the phosphorylation reaction for 1 h at 37 °C in a PCR thermocycler.
3. Add 10 μl of the corresponding reverse oligo (100 μM) and mix well.
4. In a PCR thermocycler, incubate the samples for 5 min at 95 °C for denaturation and then let the oligos anneal by ramping down to 25 °C over a period of 45 min (cooling down approximately 1.55 °C/min).
5. Store the adapters at −20 °C.

DAY 3.

3.4 Cell Fixation

1. After checking the transfection efficiency, gently remove the culture medium and wash the wells twice with DPBS prewarmed at 37 °C.
2. Slowly add 500 μl per well of 4% PFA freshly prepared by diluting 16% methanol-free PFA in 1× DPBS with calcium and magnesium.
3. Incubate for exactly 10 min while gently rocking. We recommend to perform this step inside a laminar flow cabinet.
4. Remove the 4% PFA solution.
5. Gently wash the cells with 125 mM glycine in 1× DPBS.
6. Incubate the cells with 125 mM glycine in 1× DPBS for 5 min on ice.
7. Remove the glycine solution and wash the cells with 1× DPBS.

3.5 Cell Permeabilization

1. Remove the DPBS and add 300 μl of LB1 per well.
2. Incubate for 1 h at 4 °C. In the meantime, prewarm the LB2 solution to 37 °C.
3. Bring the samples to rt. and remove the LB1 solution.
4. Wash each well quickly with 1× CutSmart® buffer with 0.1% Triton X-100 at rt.
5. Add 300 μl per well of LB2 solution.
6. Incubate for 1 h at 37 °C.
7. Gently aspirate the LB2 solution, being careful not to make bubbles.
8. Gently wash the cells twice with 400 μl per well of 1× CutSmart® buffer with 0.1% Triton X-100 at rt.

3.6 In Situ Blunting

1. Incubate the cells with 400 μl per well of 1× CutSmart® with 0.1% Triton X-100 at rt. until the blunting mix is ready.
2. Prepare the blunting mix as follows (volumes for one well, adjust proportionally to make enough for all wells):

(a) Nuclease-free water	187.5 μl
(b) 10× Blunting Buffer	25 μl
(c) Deoxynucleotide Solution Mix (1 mM)	25 μl
(d) BSA (10 mg/ml)	2.5 μl
(e) Blunting Enzyme Mix	10 μl

3. Aspirate the CutSmart® buffer and add 250 μl per well of blunting mix.
4. Incubate for 1 h at rt.

3.7 In Situ A-Tailing

1. Wash the cells twice with 400 μl per well of 1× CutSmart® with 0.1% Triton X-100 for 5 min at rt.
2. Prepare the A-tailing mix (volumes for one well, adjust proportionally to make enough for all wells):

(a) Nuclease-free water	210 μl
(b) NEBNext dA-tailing reaction buffer 10×	25 μl
(c) Klenow fragment (3′ → 5′ exo$^-$)	15 μl

3. Aspirate the CutSmart® buffer and add 250 μl per well of A-tailing mix.
4. Incubate for 30 min at 37 °C.

3.8 In Situ Ligation

1. Wash the cells twice with 400 μl per well of 1× CutSmart® with 0.1% Triton X-100 for 5 min at rt.
2. Incubate the cells with 400 μl per well of 1× T4 ligase buffer at rt. until the ligation mix is ready.
3. Prepare the following ligation mix (volumes for one well, adjust proportionally to make enough for all wells):

(a) Nuclease-free water	171 μl
(a) 10× T4 ligase buffer	25 μl
(a) ATP 10 mM	30 μl
(a) BSA 50 mg/ml	7.5 μl
(a) T4 ligase highly conc. 5 U/μl	12.5 μl

4. Remove the T4 buffer and gently add the mix to each well.
5. Add 4 μl of the desired BLISS adapter to each well.
6. Incubate for 16–18 h at 16 °C.

   DAY 4.

3.9 Washing Out Excess Adapters

1. Bring the plate back to rt. and remove the ligase mix.
2. Wash the cells three times, each of 1 h, with 300 μl HSW per well at 37 °C, gently rocking.
3. Quickly rinse the samples twice with 400 μl per well of 1× CutSmart® with 0.1% Triton X-100 at rt.

3.10 Extraction and Purification of Genomic DNA

1. Rinse the cells once with 400 μl per well of nuclease-free water.
2. Add 200 μl per well of TAIL buffer supplemented with 10 μl Proteinase K (800 U/μl).
3. Incubate the plate for 1 h at 37 °C in a mildly shaking plate incubator.

Table 3
Sonication options for DNA shearing in BLESS and BLISS

Instrument	Sample	Target size (bp)	Settings	Volume, buffer
Bioruptor® Plus	Cells	350	30 s on/90 s off, high-mode, 20 cycles	100 μl, TE
Covaris S-series	Cells	350	Duty 10%, intensity 4, time 30 s, 4 cycle/ burst 200	50 μl, TE

4. Collect the content of each well into a 1.5 ml tube.
5. Add 10 μl of Proteinase K solution to each tube and incubate for 16–18 h at 55 °C in a thermoshaker set to 800 rpm.

DAY 5

6. Purify genomic DNA with a 0.7× ratio of AMPure XP beads or using silica-based columns. Elute the DNA off the beads or column using a volume of TE that is appropriate for the sonication step. We typically use 105 μl TE to be able to continue with exactly 100 μl for Bioruptor sonication (Table 3).

3.11 DNA Fragmentation (See Note 4)

1. Prepare samples for sonication according to the manufacturer's instructions for the system used. Equilibrate exactly 100 μl per sample on ice and sonicate at 4 °C (Table 3). Sonication settings may need to be optimized depending on sample origin and device maintenance (*see* **Note 4**). Aim to achieve fragment sizes between 300 and 800 bp.
2. After sonication, concentrate the sonicated DNA using a 0.8× ratio of AMPure XP beads. Elute the DNA in 10 μl nuclease-free water.
   - Checkpoint: check the size of DNA fragments by running 1 μl of each sample on a High Sensitivity DNA chip on an Agilent Bioanalyzer. Ensure that DNA concentration does not exceed 10 ng/μl by Qubit® measurement and, when necessary, predilute a fraction of the sample before loading the chip.
   - If Bioanalyzer profiles do not meet the expectations, additional cycles of sonication are needed.

3.12 In Vitro Transcription (IVT) (See Note 5)

1. Pool 50 ng of each sample (up to 24 samples together) in a 1.5 ml tube.
2. Fill up with nuclease-free water to reach a final volume of 30 μl. If the total volume exceeds 30 μl, concentrate the solution using a Speedvac or scale up the volumes below accordingly and split the final reaction into multiple tubes, each containing max 80 μl.

3. Prepare the following IVT reaction using the MEGAscript® T7 Transcription Kit:

(a) Purified sonicated DNA	30 μl
(a) rNTPs (premixed in equal volumes)	32 μl
(a) 10 Reaction Buffer	8 μl
(a) T7 Enzyme Mix	8 μl
(a) RiboSafe RNase inhibitor 40 U/μl	2 μl

4. Mix gently and incubate for 14 h at 37 °C.

3.13 Removal of Genomic DNA and Purification of Amplified RNA (aRNA)

1. Bring Agencourt RNAClean XP beads to rt.
2. Add 4 μl of DNaseI, RNase-free (1 U/μl) to each tube and mix well.
3. Incubate for 15 min at 37 °C.
4. Purify the RNA with 1.0× prewarmed Agencourt® RNAClean XP beads according to the supplied protocol.
5. Elute the RNA off the beads with 12 μl nuclease-free water.
6. Transfer 10 μl to a 0.5 ml LoBind tube and proceed directly with library preparation.
   - Recommended checkpoint: use 1 μl of the 2 μl that are left in the tube on the magnet to check the size of the RNA on a Bioanalyzer (we use the RNA 6000 Pico chip) or to measure the RNA concentration on Qubit®.

3.14 Library Preparation (See Note 6)

1. Ligation of the 3′ Illumina sequencing adapter.
   (a) On ice, add 2 μl of the RA3 adapter to 10 μl of purified RNA. Note that when preparing libraries of smaller input samples (from a single well), the RA3 adapter should be prediluted 1:5 in nuclease-free water on ice.
   (b) In a preheated PCR thermocycler, incubate the sample for 2 min at 70 °C, then immediately place it back on ice for 1 min.
   (c) Add 8 μl of the following mix to each sample:

10 T4 RNA Ligase Reaction Buffer	2 μl
RNaseOUT™ 40 U/μl	2 μl
T4 RNA ligase, truncated (T4 Rnl2tr)	2 μl
Nuclease-free water	2 μl

   (d) In a PCR thermocycler, incubate the sample for 2 h at 25 °C. Make sure that the lid is not heated.

(e) Place the tube on ice.

2. First strand synthesis.
   (a) Add 4 μl of reverse transcription primer (10 μM) to the RNA.
   (b) In a preheated PCR thermocycler, incubate the sample for 2 min at 70 °C and then immediately bring the sample back on ice for at least 1 min.
   (c) Add 26 μl of the following mix:

5× SSIV RT buffer	10 μl
dNTPs 12.5 mM (freshly diluted)	2 μl
DTT 0.1 M	4 μl
RNaseOUT™ 40 U/μl	4 μl
SuperScript® IV RT (200 U/μl)	4 μl
Nuclease-free water	2 μl

   (d) Gently mix the sample (the total volume is now 50 μl) and incubate in a PCR thermocycler for 50 min at 50 °C with the lid at 50 °C, followed by 10 min at 80 °C to inactivate SuperScript®. Place the samples back on ice.

3. Library indexing and amplification, part I:
   (a) Transfer the entire cDNA sample (50 μl) into a PCR tube on ice.
   (b) Add 7.5 μl of the desired RPI indexing primer (Table 2), and then 70 μl of the following mix:

Nuclease-free water	10 μl
RP1 primer (common for all samples)	7.5 μl
NEBNext® Ultra™ II Q5® Master Mix	75 μl

   (c) Mix the entire volume thoroughly, and then split it into two PCR tubes. Store one of them at −20 °C for backup for in case anything goes wrong during library preparation. Use the other tube directly for amplification.
   (d) In a fast-cycling PCR thermocycler perform the following steps:

I.	98 °C, 30 s
II.	98 °C, 10 s
III.	60 °C, 30 s
IV.	65 °C, 45 s

(continued)

V.	65 °C, 10 min
VI.	12 °C, hold

(e) Repeat steps II–IV for 6–8 times depending on the initial sample's size.

4. Library purification I (*see* **Note** 7).
   (a) Transfer the PCR product into a 1.5 ml LoBind tube.
   (b) Add 0.8× AMPure XP bead suspension (prewarmed at rt. for 30 min), and mix thoroughly by pipetting 5–6 times up and down.
   (c) Incubate the sample for 10 min at rt.
   (d) Place the sample onto a magnetic stand and wait 5 min or until all the beads have attached to the magnet.
   (e) Aspirate the supernatant with a regular pipet.
   (f) Gently wash the beads with freshly prepared 80% ethanol, while the tube remains on the magnetic stand. Ethanol aspiration can be done with a regular pipet, but preferably with a portable aspiration system such as the Integra VACUSIP.
   (g) Aspirate the ethanol and repeat the wash once.
   (h) Remove as much ethanol as possible, then air-dry the beads for 2–10 min at rt. Avoid overdrying the beads.
   (i) Remove the sample from the magnetic stand and resuspend the beads in 22 μl of nuclease-free water.
   (j) Incubate for 5–15 min at rt.
   (k) Place the tube on the magnetic stand, and let the solution clear for 5 min.
   (l) Transfer 20 μl of the cleared solution into a new PCR tube and continue with the second step of library amplification. If desired, the beads can be eluted with more water and part of this eluate can be used for analyses or comparison to a double-amplified sample (*see* **Note** 7).

5. Library amplification II.
   (a) To the 20 μl of purified PCR product add the following mix:

RPI indexing primer (identical to PCR1)	2.5 μl
RP1 common primer	2.5 μl
NEBNext® Ultra™ II Q5® Master Mix	25 μl

   (b) Perform 4–6 cycles of the same PCR program as indicated in **step 3** of this section.

6. Final library purification (two-sided AMPure purification, *see* **Note** 7).
   (a) Transfer the sample to a 1.5 ml LoBind tube and add 50 μl of nuclease-free water to increase the volume to prevent working with very low volumes in the final steps. Measure with a pipette to ensure that the volume is exactly 100 μl.
   (b) Add 0.5× AMPure XP bead suspension (50 μl) and mix thoroughly by pipetting up and down 5–6 times.
   (c) Incubate the mixture for 15 min at rt.
   (d) Place the tube on a magnetic stand until it clears.
   (e) Transfer the clear supernatant (~150 μl in total) to a new LoBind tube and discard the old tube and beads, which have bound fragments exceeding the expected size of the library.
   (f) To the new tube, add 0.25× AMPure XP bead suspension (*see* **Note** 7). The ratio relates to the initial volume of 100 μl; thus, in the case of a 0.25 ratio, 25 μl should be added here.
   (g) Resuspend thoroughly by pipetting up and down for at least ten times, each time aspirating and dispensing at least 150 μl to make sure that the beads are mixed thoroughly.
   (h) Incubate the mixture for at least 15 min at rt.
   (i) Place the sample on the magnetic stand until it clears.
   (j) Wash the beads twice with freshly prepared 80% ethanol.
   (k) Air-dry the beads and very carefully remove the tube from the magnetic rack. Caution: the beads may jump to the upper part of the tube.
   (l) Resuspend the beads in 10–15 μl nuclease-free water, depending on the expected yield.
   (m) Store the library at −20 °C.
      - Checkpoint: measure library concentration by Qubit® using the high-sensitivity dsDNA reagents. If necessary, dilute 1 μl of library sufficiently to reach a concentration <10 ng/μl. Use 1 μl of undiluted or diluted library on a Bioanalyzer High Sensitivity DNA chip to assess the size distribution of the generated libraries.
      - For sequencing options, *see* Subheading 5.

4 Notes

1. *Cells and format.* CRISPR-BLISS is a multistep procedure in which DSBs induced by Cas9 or another CRISPR enzyme are in situ blunted and ligated to dsDNA adapters that allow their identification through next-generation sequencing. While in classical BLISS cells are attached onto a microscope slide or cover glass, CRISPR-BLISS is done directly in multiwell plates in order to enable transfection or transduction of CRISPR plasmids, and to scale up the number of cells and/or sgRNAs used. Here, we present a protocol adjusted for 293FT cells grown and transfected in 24-well plates, but other cells and formats can be used by adjusting the volumes accordingly.
2. *Transfection of CRISPR plasmids.* We recommend transfecting 293FT cells at 70–80% confluence and fixing them 24 h after transfection. The time it takes to reach this confluence is cell type-dependent, but in the case of 293FT cells we suggest plating approximately 125,000 cells per well, 16–18 h before transfection. We recommend that the cells are evenly distributed throughout each well to ensure optimal transfection. Transfection can be extended to 48 h and possibly up to 72 h to detect late-forming OT DSBs.
3. *Design and preparation of BLISS adapters.* BLISS adapters enable selective amplification of DSB ends using in vitro transcription by the T7 RNA polymerase/promoter system. In addition, the adapters contain (a) a sample barcode that allows multiplexing of the library prep reaction, (b) a sequence of random nucleotides that is used as a unique molecular identifier (UMI) during processing of sequencing reads, and (c) the RA5 adapter sequence consistent with the Illumina TruSeq Small RNA library preparation kit, to make single-end sequencing sufficient for the identification of the ligation site. BLISS dsDNA adapters are produced by phosphorylating a forward oligo using T4 PNK kinase, followed by annealing a reverse complementary oligo with the exception of the UMI sequence. Both oligos are ordered in HPLC-purified state. We advise to use a distinct sample barcode with at least three nucleotides variation in the barcode sequence for each well, allowing for library multiplexing.
4. *DNA sonication.* To prevent biases that may arise from long and differently sized fragments during IVT and downstream library preparation, the purified genomic DNA should be fragmented to reach a more uniform size distribution between 300 and 800 bp, approximately. We have successfully applied the Bioruptor® Plus as well as the Covaris® S2 systems.

5. *Selective linear amplification.* We usually perform IVT for 14 h, regardless of the amount of DNA input. We multiplex library preparation by pooling equal amounts of gDNA from 24 samples that have been in situ labeled with different barcodes. It is however also possible to pool less than 24 samples together, as long as the amount of input DNA is normalized between all the samples and the reaction volumes are scaled proportionally. After IVT, genomic DNA is digested with DNAseI in order to avoid carry-over of DNA into the final library. Purification of RNA generated by IVT can be done using Agencourt RNAClean XP beads or silica-based columns (e.g., RNeasy® MinElute® Cleanup kit, Qiagen).
6. *Library preparation.* The library preparation protocol we describe here is a modification of the protocol for the Illumina TruSeq Small RNA Sample Preparation Kit. The main difference is that, in our protocol, the RA5 adapter sequence is already introduced during the ligation step of BLISS. Purchase of the kit is not necessary as all the required adapters and primers (Table 2) can be purchased separately (for example, we order them through Integrated DNA Technologies Inc.). The optimal number of PCR cycles depends on the initial number of cells and on the expected number of OT DSBs. Typically, for samples of 10^3–10^4 cells we perform a total of 10–14 amplification cycles, divided into two rounds with an AMPure bead purification in between (*see* **Note 7**).
7. *Library purification.* We have experienced that library quality and sequence yield is in general much better when we perform the library amplification in two steps, with a 0.8× AMPure bead purification in between. The protocol specifies the order in which we typically perform library amplification and purification: amplification I (6–8 cycles), 0.8× AMPure XP bead purification, amplification II (four cycles), and a final two-sided AMPure XP bead purification step. Two-sided bead purification removes library fragments on both ends of the fragment size distribution: on the "right side of the curve" it removes molecules larger than we anticipated based on fragment size after sonication plus RA3 adapter length. On the "left side of the curve" it removes small molecules that are likely to disturb sequencing, such as leftovers, dimers, or chimeric products of primers and adapters. Please note that the selected ratios should be adjusted depending on the expected library size, in relation to fragment sizes obtained by sonication. If unsure, we advise to perform a Bioanalyzer assessment of the nonpurified library after amplification. In this protocol, we describe a 0.5 + 0.25× ratio, with the following rationale. First, a 0.5× ratio of beads extracts large fragments from the library ("right-side" cleanup). If required, this ratio can be adjusted to 0.45 or 0.4 to be less stringent, but we advise not

to go below 0.4. Immediately after incubation, we transfer the supernatant, which contains the unbound portion of our amplified library and all the AMPure buffer, to a new tube and discard the beads, which contain the largest fragments and can be discarded. To this new tube with supernatant (100 + 50 μl), new AMPure XP beads are added for the 'left-side' purification that removes small fragments. For the left-side purification, we add 0.25× beads ratio, and it is vital to have this correspond to the starting volume of 100 μl: we add 25 μl, creating a total volume of 175 μl, and then follow regular bead washes and elution. Note that the size-selective binding capacity of AMPure beads is determined by the concentration of AMPure buffer component PEG, and not by the actual number of beads. This 0.5 + 0.25× ratio resembles a regular left-side AMPure XP bead purification with a ratio of 0.75×.

5 Additional Considerations

5.1 Choice of Sequencing Platform and Sequencing Depth

The CRISPR-BLISS protocol described here results in sequencing libraries that can be sequenced on various Illumina platforms, including MiSeq, NextSeq, and NovaSeq. The platform choice depends on the desired sequencing throughput, which ultimately is dictated by the sample size. Based on our experience, for a pool of 24 samples of approx. 250,000 cells, we advise a sequencing depth of 200 million reads per pooled library. For large-scale applications, it may be useful to first perform a round of shallow sequencing in which many pools are sequenced simultaneously, after which libraries with good quality can be sequenced in more depth. Certain applications, such as identifying rare OT events, might require a higher depth and for that purpose libraries with fewer multiplexed samples should be generated.

Acknowledgments

We thank Winston Yan and Feng Zhang from the Broad Institute for support in setting up CRISPR-BLISS.

References

1. Ran FA, Hsu PD, Wright J et al (2013) Genome engineering using the CRISPR-Cas9 system. Nat Protoc 8:2281–2308. https://doi.org/10.1038/nprot.2013.143
2. Ran FA, Cong L, Yan WX et al (2015) In vivo genome editing using *Staphylococcus aureus* Cas9. Nature 520:186–191. https://doi.org/10.1038/nature14299
3. Paquet D, Kwart D, Chen A et al (2016) Efficient introduction of specific homozygous and heterozygous mutations using CRISPR/Cas9. Nature 533:125–129. https://doi.org/10.1038/nature17664

4. Kwart D, Paquet D, Teo S, Tessier-Lavigne M (2017) Precise and efficient scarless genome editing in stem cells using CORRECT. Nat Protoc 12:329–334. https://doi.org/10.1038/nprot.2016.171
5. Kleinstiver BP, Pattanayak V, Prew MS et al (2016) High-fidelity CRISPR-Cas9 nucleases with no detectable genome-wide off-target effects. Nature 529:490–495. https://doi.org/10.1038/nature16526
6. Akcakaya P, Bobbin ML, Guo JA et al (2018) In vivo CRISPR editing with no detectable genome-wide off-target mutations. Nature 561:416–419. https://doi.org/10.1038/s41586-018-0500-9
7. Tanenbaum ME, Gilbert LA, Qi LS et al (2014) A protein-tagging system for signal amplification in gene expression and fluorescence imaging. Cell 159:635–646. https://doi.org/10.1016/j.cell.2014.09.039
8. Cheng AW, Wang H, Yang H et al (2013) Multiplexed activation of endogenous genes by CRISPR-on, an RNA-guided transcriptional activator system. Cell Res 23:1163–1171. https://doi.org/10.1038/cr.2013.122
9. Gilbert LA, Horlbeck MA, Adamson B et al (2014) Genome-scale CRISPR-mediated control of gene repression and activation. Cell 159:647–661. https://doi.org/10.1016/j.cell.2014.09.029
10. Rock JM, Hopkins FF, Chavez A et al (2017) Programmable transcriptional repression in mycobacteria using an orthogonal CRISPR interference platform. Nat Microbiol 2:1–9. https://doi.org/10.1038/nmicrobiol.2016.274
11. Thakore PI, Kabadi AM, Safi A et al (2015) Highly specific epigenome editing by CRISPR-Cas9 repressors for silencing of distal regulatory elements. Nat Methods 12:1143–1149. https://doi.org/10.1038/nmeth.3630
12. Okada M, Kanamori M, Someya K et al (2017) Stabilization of Foxp3 expression by CRISPR-dCas9-based epigenome editing in mouse primary T cells. Epigenet Chrom 10:1–17. https://doi.org/10.1186/s13072-017-0129-1
13. Fu Y, Rocha PP, Luo VM et al (2016) CRISPR-dCas9 and sgRNA scaffolds enable dual-colour live imaging of satellite sequences and repeat-enriched individual loci. Nat Commun 7:1–8. https://doi.org/10.1038/ncomms11707
14. Ghosh D, Venkataramani P, Nandi S, Bhattacharjee S (2019) CRISPR–Cas9 a boon or bane: the bumpy road ahead to cancer therapeutics. Cancer Cell Int 19:1–10. https://doi.org/10.1186/s12935-019-0726-0
15. Lin Y, Cradick TJ, Brown MT et al (2014) CRISPR/Cas9 systems have off-target activity with insertions or deletions between target DNA and guide RNA sequences. Nucleic Acids Res 42:7473–7485. https://doi.org/10.1093/nar/gku402
16. Tsai SQ, Zheng Z, Nguyen NT et al (2015) GUIDE-seq enables genome-wide profiling of off-target cleavage by CRISPR-Cas nucleases. Nat Biotechnol 33:187–198. https://doi.org/10.1038/nbt.3117
17. Newton MD, Taylor BJ, Driessen RPC et al (2019) DNA stretching induces Cas9 off-target activity. Nat Struct Mol Biol. https://doi.org/10.1038/s41594-019-0188-z
18. Kuscu C, Arslan S, Singh R et al (2014) Genome-wide analysis reveals characteristics of off-target sites bound by the Cas9 endonuclease. Nat Biotechnol 32:677–683. https://doi.org/10.1038/nbt.2916
19. O'Geen H, Henry IM, Bhakta MS et al (2015) A genome-wide analysis of Cas9 binding specificity using ChIP-seq and targeted sequence capture. Nucleic Acids Res 43:3389–3404. https://doi.org/10.1093/nar/gkv137
20. Kosicki M, Tomberg K, Bradley A (2018) Repair of double-strand breaks induced by CRISPR–Cas9 leads to large deletions and complex rearrangements. Nat Biotechnol 36. https://doi.org/10.1038/nbt.4192
21. Fu Y, Foden JA, Khayter C et al (2013) High-frequency off-target mutagenesis induced by CRISPR-Cas nucleases in human cells. Nat Biotechnol 31:822–826. https://doi.org/10.1038/nbt.2623
22. Cameron P, Fuller CK, Donohoue PD et al (2017) Mapping the genomic landscape of CRISPR-Cas9 cleavage. Nat Methods 14:600–606. https://doi.org/10.1038/nmeth.4284
23. Knight SC, Xie L, Deng W et al (2015) Dynamics of CRISPR-Cas9 genome interrogation in living cells. Science (80) 350:823–826. https://doi.org/10.1126/science.aac6572
24. Kim D, Kim JS (2018) DIG-seq: a genome-wide CRISPR off-target profiling method using chromatin DNA. Genome Res 28:1882–1893. https://doi.org/10.1101/gr.236620.118
25. Chiarle R, Zhang Y, Frock RL et al (2011) Genome-wide translocation sequencing reveals mechanisms of chromosome breaks and

rearrangements in B cells. Cell 147:107–119. https://doi.org/10.1016/j.cell.2011.07.049

26. Frock RL, Hu J, Meyers RM et al (2015) Genome-wide detection of DNA double-stranded breaks induced by engineered nucleases. Nat Biotechnol 33:179–188. https://doi.org/10.1038/nbt.3101
27. Wang X, Wang Y, Wu X et al (2015) Unbiased detection of off-target cleavage by CRISPR-Cas9 and TALENs using integrase-defective lentiviral vectors. Nat Biotechnol 33:175–179. https://doi.org/10.1038/nbt.3127
28. Crosetto N, Mitra A, Silva MJ et al (2013) Nucleotide-resolution DNA double-strand break mapping by next-generation sequencing. Nat Methods 10:361–365. https://doi.org/10.1038/nmeth.2408
29. Slaymaker IM, Yan WX, Gao L et al (2015) Rationally engineered Cas9 nucleases with improved specificity. Science (80) 351:84–88. https://doi.org/10.1126/science.aad5227
30. Duan J, Lu G, Xie Z et al (2014) Genome-wide identification of CRISPR/Cas9 off-targets in human genome. Cell Res 24:1009–1012. https://doi.org/10.1038/cr.2014.87
31. Wu X, Scott DA, Kriz AJ et al (2014) Genome-wide binding of the CRISPR endonuclease Cas9 in mammalian cells. Nat Biotechnol 32:670–676. https://doi.org/10.1038/nbt.2889
32. Martin F, Sánchez-Hernández S, Gutiérrez-Guerrero A et al (2016) Biased and unbiased methods for the detection of off-target cleavage by CRISPR/Cas9: an overview. Int J Mol Sci 17. https://doi.org/10.3390/ijms17091507
33. Kim D, Bae S, Park J et al (2015) Digenome-seq: genome-wide profiling of CRISPR-Cas9 off-target effects in human cells. Nat Methods 12:237–243. https://doi.org/10.1038/nmeth.3284
34. Kim D, Kim S, Kim S et al (2016) Genome-wide target specificities of CRISPR-Cas9 nucleases revealed by multiplex digenome-seq. Genome Res 26:406–415. https://doi.org/10.1101/gr.199588.115.Freely
35. Kim D, Kim D, Lee G et al (2019) Genome-wide target specificity of CRISPR RNA-guided adenine base editors. Nat Biotechnol:1. https://doi.org/10.1038/s41587-019-0050-1
36. Lensing SV, Marsico G, Hänsel-Hertsch R et al (2016) DSBCapture: in situ capture and sequencing of DNA breaks. Nat Methods 13:855–857. https://doi.org/10.1038/nmeth.3960
37. Canela A, Sridharan S, Sciascia N et al (2016) DNA breaks and end resection measured genome-wide by end sequencing. Mol Cell 63:898–911. https://doi.org/10.1016/j.molcel.2016.06.034
38. Canela A, Maman Y, Jung S et al (2017) Genome organization drives chromosome fragility. Cell 170:507–521.e18. https://doi.org/10.1016/j.cell.2017.06.034
39. Yan WX, Mirzazadeh R, Garnerone S et al (2017) BLISS is a versatile and quantitative method for genome-wide profiling of DNA double-strand breaks. Nat Commun 8:15058. https://doi.org/10.1038/ncomms15058
40. Gao L, Cox DBT, Yan WX et al (2017) Engineered Cpf1 variants with altered PAM specificities. Nat Biotechnol 35:789–792. https://doi.org/10.1038/nbt.3900
41. Tsai SQ, Nguyen NT, Malagon-Lopez J et al (2017) CIRCLE-seq: a highly sensitive in vitro screen for genome-wide CRISPR-Cas9 nuclease off-targets. Nat Methods 14:607–614. https://doi.org/10.1038/nmeth.4278
42. Lazzarotto CR, Nguyen NT, Tang X et al (2018) Defining CRISPR–Cas9 genome-wide nuclease activities with CIRCLE-seq. Nat Protoc 13:2615–2642. https://doi.org/10.1038/s41596-018-0055-0
43. Lee K, Zhang Y, Kleinstiver BP et al (2019) Activities and specificities of CRISPR/Cas9 and Cas12a nucleases for targeted mutagenesis in maize. Plant Biotechnol J 17:362–372. https://doi.org/10.1111/pbi.12982
44. Tang PZ, Ding B, Peng L et al (2018) TEG-seq: an ion torrent-adapted NGS workflow for in cellulo mapping of CRISPR specificity. BioTechniques 65:259–267. https://doi.org/10.2144/btn-2018-0105
45. Mirzazadeh R, Kallas T, Bienko M, Crosetto N (2018) Genome – wide profiling of DNA double – strand breaks by the BLESS and BLISS methods. Methods Mol Biol:167–194
46. Iannelli F, Galbiati A, Capozzo I et al (2017) A damaged genome's transcriptional landscape through multilayered expression profiling around in situ-mapped DNA double-strand breaks. Nat Commun 8:15656. https://doi.org/10.1038/ncomms15656

Index

Tudor A. Fulga et al. (eds.), *CRISPR Guide RNA Design: Methods and Protocols*, Methods in Molecular Biology, vol. 2162, https://doi.org/10.1007/978-1-0716-0687-2, © Springer Science+Business Media, LLC, part of Springer Nature 2021

D

E

F

G

H

I

L

M

N

O

P

Q

R

S

T

U

V

W

Z

MIX
Papier aus verantwortungsvollen Quellen
Paper from responsible sources
FSC® C105338

If you have any concerns about our products,
you can contact us on
ProductSafety@springernature.com

In case Publisher is established outside the EU,
the EU authorized representative is:
Springer Nature Customer Service Center GmbH
Europaplatz 3, 69115 Heidelberg, Germany

Printed by Libri Plureos GmbH
in Hamburg, Germany